AF581651

Housing Environment and Farm Animals' Well-Being

Housing Environment and Farm Animals' Well-Being

Editors

Lilong Chai
Yang Zhao

MDPI • Basel • Beijing • Wuhan • Barcelona • Belgrade • Manchester • Tokyo • Cluj • Tianjin

Editors
Lilong Chai
Poultry Science
University of Georgia
Athens
United States

Yang Zhao
Animal Science
The University of Tennessee
Knoxville
United States

Editorial Office
MDPI
St. Alban-Anlage 66
4052 Basel, Switzerland

This is a reprint of articles from the Special Issue published online in the open access journal *Animals* (ISSN 2076-2615) (available at: www.mdpi.com/journal/animals/special_issues/Housing_Environment_Farm_Animals).

For citation purposes, cite each article independently as indicated on the article page online and as indicated below:

LastName, A.A.; LastName, B.B.; LastName, C.C. Article Title. *Journal Name* **Year**, *Volume Number*, Page Range.

ISBN 978-3-0365-4586-8 (Hbk)
ISBN 978-3-0365-4585-1 (PDF)

Cover image courtesy of Lilong Chai

Contents

About the Editors

Lilong Chai

Dr. Lilong Chai is an Assistant Professor and Engineering Specialist in the Department of Poultry Science at UGA since 2018. He is a member of UGA Phenomics and Plant Robotics Center & Integrative Precision Agriculture team. His primary research/extended interests include animal environmental engineering and precision poultry farming. Prior to joining UGA, Chai was a postdoc research associate in the department of agricultural and biosystems engineering at Iowa State University (ISU). At ISU, Chai led the field study on commercial farms for two USDA-NIFA programs (both as Co-PI) for improving indoor the air quality of cage-free egg production facilities and exploring practical means for preventing/controlling airborne infectious poultry diseases such as avian influenza. Chai's contributions include 110 scientific publications, PI/Co-PI of 20 grants/contracts, and 15 awards/honors including the 2021 Sunkist Young Designer Award from the American Society of Agricultural and Biological Engineers (ASABE). Chai is currently the Coordinator/program Chair of Georgia Precision Poultry Farming Conference & Georgia Layer Conference, and Vice Chair of ASABE-Environmental Air Quality Committee.

Yang Zhao

Dr. Yang Zhao is an Assistant Professor at Animal Science department at the University of Tennessee. Dr. Zhao received his B.S. and M.S. degrees from China Agricultural University, and a PhD degree from Wageningen University in the Netherlands. Before joining The University of Tennessee, he was an assistant professor of Agricultural and Biological Engineering at Mississippi State University for three and half years. Dr. Zhao serves as the chair and member in several academic and award committees of American Society of Agricultural and Biological Engineers (ASABE) and is an associate editor of the Transactions of the ASABE. His research focuses on smart poultry farming that addresses challenges in poultry production regarding sensoring, automation, behavior monitoring, welfare assessment, disease prevention, and environment management. His work has been published in over 120 scientific articles. Dr. Zhao is the awardee of ASABE Sunkist Young Designer Award, AOC Early Career Award, and Gamma Sigma Delta Research Award. Other recognitions of his work include ASABE Superior Paper Awards, Poultry Science Highly Cited Paper of the Year, Poultry Science Article of Editor's Choice, and ASABE Outstanding Reviewer.

Preface to "Housing Environment and Farm Animals' Well-Being"

The global population is expected to reach 9.5 billion in 2050, and the demand for global animal products (e.g., milk, meat, and eggs) will have increased by over 70% by 2050 as compared to 2005. Concentrated animal feeding operations (CAFOs) plays a key role in producing animal protein to meet the increasing world population. The housing environment affects the health and well-being of farm animals raised in CAFOs. For instance, poor air quality (e.g., high levels of indoor ammonia, particulate matter, and airborne microorganisms) may deteriorate animals'health and well-being over time in animal houses. Mitigating air pollutant generation and suppressing levels of housing air pollutants are critical for maintaining the well-being of farm animals. To collect the updated information and research results of poultry and livestock housing, environment management, and wellbeing worldwide, the editors of this book invited researchers from different countries to contribute their latest findings. The primary goal of book is to share information and references for researchers, students, and animal agriculture producers to enhance on-farm environmental management and animal well-being.

Lilong Chai and Yang Zhao
Editors

 *animals*

Article

Visual Sensor Placement Optimization with 3D Animation for Cattle Health Monitoring in a Confined Operation

Abdullah All Sourav and Joshua M. Peschel *

Department of Agricultural and Biosystems Engineering, Iowa State University, Ames, IA 50010, USA; sourav@iastate.edu
* Correspondence: peschel@iastate.edu

Simple Summary: This paper introduces a new method of finding the best locations to place video cameras inside large cattle barns to monitor the behavior and health of the animals. Current approaches to livestock video monitoring rely on mounting cameras in the most convenient places for installation, but those locations might either be impractical for actual barns and/or might not capture the best views. This work showed that there is short list of the best placement options for the cameras to choose from which will provide the best camera views.

Abstract: Computer vision has been extensively used for livestock welfare monitoring in recent years, and data collection with a sensor or camera is the first part of the complete workflow. While current practice in computer vision-based animal welfare monitoring often analyzes data collected from a sensor or camera mounted on the roof or ceiling of a laboratory, such camera placement is not always viable in a commercial confined cattle feeding environment. This study therefore sought to determine the optimal camera placement locations in a confined steer feeding operation. Measurements of cattle pens were used to create a 3D farm model using Blender 3D computer graphic software. In the first part of this study, a method was developed to calculate the camera coverage in a 3D farm environment, and in the next stage, a genetic algorithm-based model was designed for finding optimal placements of a multi-camera and multi-pen setup. The algorithm's objective was to maximize the multi-camera coverage while minimizing budget. Two different optimization methods involving multiple cameras and pen combinations were used. The results demonstrated the applicability of the genetic algorithm in achieving the maximum coverage and thereby enhancing the quality of the livestock visual-sensing data. The algorithm also provided the top 25 solutions for each camera and pen combination with a maximum coverage difference of less than 3.5% between them, offering numerous options for the farm manager.

Keywords: livestock monitoring; camera coverage optimization; sensor placement; genetic algorithm

Citation: Sourav, A.A.; Peschel, J.M. Visual Sensor Placement Optimization with 3D Animation for Cattle Health Monitoring in a Confined Operation. *Animals* **2022**, *12*, 1181. https://doi.org/10.3390/ani12091181

Academic Editors: Lilong Chai and Yang Zhao

Received: 20 March 2022
Accepted: 3 May 2022
Published: 5 May 2022

1. Introduction

1.1. Use of Visual Sensors in Livestock Monitoring

There are two major hardware operations in cattle health monitoring with computer vision: data collection and processing units. Data are often collected through sensors or cameras, followed by processing on a personal computer. From a literature review, the cameras used in previous computer vision in livestock monitoring studies can be divided into the following categories: depth cameras, digital cameras, RGBD (red–green–blue color information with per-pixel depth information) cameras, and Closed-circuit Television (CCTV) or surveillance cameras. The use of digital and CCTV or surveillance cameras has been dominant in livestock health monitoring. These have been used for live weight estimation, lameness detection, individual cattle identification, behavior monitoring, and the tracking of pigs and cattle [1–6].

The camera or sensor installation location also varies based on the purpose of the research and the structure of the livestock housing. Cameras with a wide field of view or 360-degree lens are often installed on the ceiling [1,7–11]. Such installation facilitates capturing the whole cell and does not have occlusion issues while cattle or pigs stand behind one another. The camera installed in the ceiling is mostly practiced in the lab or research environment, as the average commercial feedlot does not have enough ceiling height for such an installation.

In most cases, the digital RGBD and depth camera is installed on the livestock housing ceiling to collect images and videos for further analysis. The captured data has shown promising results in estimating live weight [1,12], individual pig and cattle identification [8,11], aggressive behavior detection [13], mounting behaviors detection [9], standing behaviors detection [14], and so on. In contrast, digital and surveillance cameras have also been installed to collect side-view videos and images. Multiple studies successfully detected lameness, locomotion, and cattle and pig feeding behaviors using such setups [15,16].

1.2. Camera Placement Optimization

Regardless of the type, the camera is a valuable tool in computer vision systems to record and transmit spatiotemporal data in image and video format. The system also provides real-time information on livestock's movement, posture, and behaviors [5,10,11,17]. Livestock behavior data collection with camera and quantification is an important tool for welfare monitoring and related research. Jackson et al. recorded used an optical camera to record piglet pen for a specific length of time [18]. On the other hand, Heiderscheit et al. recorded behavioral data of steer in video format from the Beef Nutrition Farm [19]. In both studies, time spent on drinking, eating, lying down, and displacement were calculated by visually observing the recorded video. Thus, accurate placement of the cameras for such research, as well as livestock monitoring, is crucial.

The purchase and maintenance cost of a surveillance camera system is often expensive [20]. In addition, changing the location of a surveillance system after installation is also inconvenient [21]. Thus, proper camera layout must be determined beforehand to calculate the number of cameras and their locations to be installed and minimize modification costs [22].

Achieving maximum camera coverage by minimizing the number of the cameras with a set of constraints is a complex optimization problem; thus, numerous studies have been conducted in this domain [23]. The camera optimization problems are similar to the Art Gallery Problem (AGP). The AGP is a well-studied computational geometric optimization problem finding the minimum number of guards with their restricted viewpoint required to cover all parts of the gallery interior. It is assumed that the guards/sensors have a 360-degree visual angle and unlimited viewpoint [24]. However, the camera visibility is limited due to its field of view angle and limited visual distance. The multi-camera coverage calculation problem treats each camera coverage differently, and all camera coverage is merged to maximize the total coverage. This multi-camera coverage optimization belongs to the class of non-deterministic polynomial-time hard (NP-hard) combinatorial optimization problem. Thus, computational complexity is expected to solve large instances and deal with multiple objectives [25,26].

The earlier studies in camera placement optimization solely focused on maximizing fixed camera coverage for building and indoor monitoring while considering the region of interest as a 2D plane [23,24,26,27]. However, in application, the camera covers a 3D space, and optimizing camera coverage in such an environment is computationally more complex than optimizing in a 2D plane [27]. The research paradigm has recently been shifted towards maximizing camera coverage in a 3D environment while minimizing the overall project cost and meeting certain constraints [28–31]. The process is computationally expensive but proved to be useful in different domains.

Kim et al. [28] performed a hybrid simulation of camera placement optimization to monitor construction job sites where the primary objectives were to maximize the coverage and minimize the cost. The objectives were also subjected to a certain budget, minimum coverage, and accessibility to power and data transmission constraints. The research work provided three solutions for three camera combinations for different price levels for the stakeholders. However, the work used Microsoft Excel to design the job site, which does not offer full-featured 3D modeling of the objects for precise camera coverage calculation. On the other hand, the job site was modeled using blocks of 1 m in size, which is relatively large and cannot yield very precious camera coverage calculation. Albahri and Hammad [29] proposed a coverage calculation method with the same primary objectives, but constraints were mostly regarding limiting the camera's position to specific locations, pan, and tilt angle. In that simulation-based study, building information modeling (BIM) software played a crucial role in calculating the camera coverage by deriving geometrical constraints (e.g., ceiling, walls, and columns) and instrumental constraints (e.g., vibration caused by heating, ventilation, and air conditioning system). The study was overly dependent on building information and required two different programs, BIM and Unity 3D, to work in harmony for coverage calculation.

Other research has shown promising results in camera placement optimization for maximizing multi-camera coverage in indoor spaces (e.g., residential buildings, metro stations, and hospitals) and outdoor areas (e.g., construction sites, open urban areas, traffic intersections, and open sea) [22,24,28–30]. Despite such a wide variety of studies conducted, to the best of our knowledge, no camera placement study has been conducted to date on the welfare monitoring of livestock in a farm environment. This study focuses on achieving two objectives: (i) simulation-based camera coverage calculation for a confined environment of cattle and (ii) camera placement optimization simulation for achieving optimum camera coverage in a given budge.

This paper is organized as follows. Section 2 discusses the development of a 3D model and camera coverage workflow, followed by a genetic algorithm implementation to achieve maximum camera coverage at a given budget. Results of camera coverage workflow and the implementation of genetic algorithm with different camera and budget combinations are discussed in detail in Section 3. Section 4 presents this study's research findings and limitations, followed by our conclusion in Section 5.

2. Materials and Methods

2.1. A Case Study for Camera Coverage Calculation

2.1.1. 3D Environment Creation

The first objective of this study was to calculate camera coverage for a given position and environment using camera properties. It has been observed that a majority of studies which have attempted to address camera placement optimization considered designing the environment in 2D space, whereas it is a 3D environment in reality. Calculating camera coverage using a 3D model close to an actual farm environment is also necessary for higher accuracy. In addition, some permanent physical obstructions are often overlooked in 2D and a poorly designed 3D scenarios. In our study, the 3D scene of the pen was created using the 3D animation software Blender. Blender is a free and open-source 3D creation cross-platform software that supports the entirety of the 3D pipeline—modeling, rigging, animation, simulation, rendering, compositing and motion tracking, even video editing and game creation [32]. It also supports Python scripting and access to Blender's data, classes, and functions from its own Python modules (e.g., bpy and mathutils). This software was used in this study due to its scripting capabilities and simplicity for creating 3D scenes without expert knowledge. In the early stage of this study, measurements of a steer pen were collected from the Beef Nutrition Farm at Iowa State University (Figure 1). Six steers were usually housed in a 43 ft × 11 ft × 15 ft pen with a drinking trough and a feeding trough. The back of each pen was open to allow steer movement into the nearby open field, and on the front side there was a 9 ft open area for farmworkers and machine movement. A 3D farm scene with physical structures, fences, and dividing walls was created based on these measurements (Figure 2). The single pen was then copied and used to create the multi-pen scenario.

Figure 1. Aerial view of Beef Nutrition Farm at Iowa State University (GoogleMap).

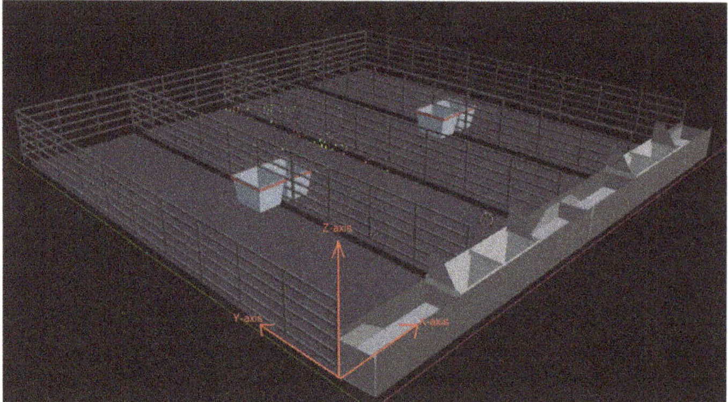

Figure 2. A sample confined farm scenario with four pens.

Blender has a default camera tool to adjust the location, view type, focal length, and field of view parameters. A 3D object with the exact shape and size of the camera was used to calculate the camera coverage. For the camera coverage calculation in this case study, we used the most common parameters of CCTV cameras available in the market. The cameras selected for this study had a field of view (FOV) of 76 degrees and 86 degrees. Once the camera was in place with the appropriate FOV, a camera shape object was created to represent the same physical properties of the Blender camera. A cone-shaped object with four vertices was created, and manually edited to have the exact shape and size of the Blender camera tool. This cone represented the total enclosed space recorded by each camera in a 3D environment. As shown in Figure 3, the yellow lines represent the Blender camera outlines, whereas the blue shape represents the cone created to represent the camera. The Blender objects are hollow, and if any object falls inside this hollow cone, it can be safely assumed that the object is visible on the Blender camera. The camera was enlarged in multiple folds, usually 60 ft in length, to check the visibility of all parts of a pen. As this study focuses on single and multi-camera setups, two cameras with 76 degrees and 86 degrees FOV were created, and their shapes were copied onto respective cones.

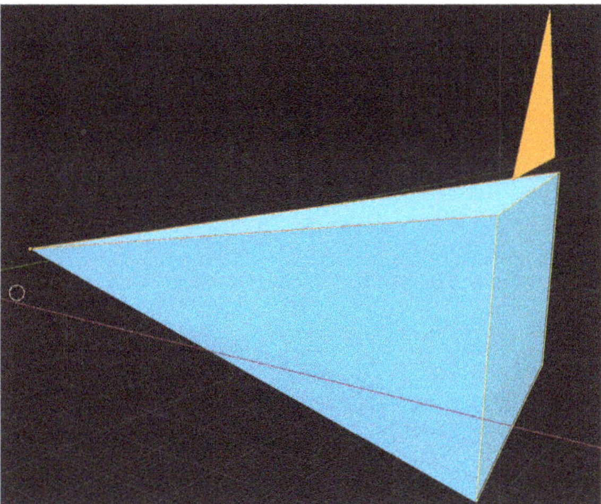

Figure 3. Duplicating the Blender camera properties on a cone shape Blender object with four vertices.

2.1.2. Camera Placement

The camera placement locations were determined by the shape and size of the cattle pen. In our study, each pen had a size of 43 ft x 11 ft x 15 ft. The camera could be installed anywhere within the boundary of the pen. However, the back of each pen at the Beef Nutrition Farm was open for facilitating steer movement to the nearby field. Installing cameras at the back was not feasible, as the camera would be exposed to the rain and snow. In addition, cameras in such a position would not provide significant details of feeding behaviors, as feeding troughs were located at the front. Thus, we decided to use every other location except the backside as a viable camera placement location. The front side of the pen had a 9 ft clearance for instrument movement, and a camera could also be set up at 12 and 15 ft height on the opposite wall of the pathway. Viable camera location could be defined in 3D environment in terms of the X, Y, and Z-axis (Figure 2). Here, X and Y-axis represent the width and length of the pen, respectively. The Z-axis represents the height of the camera from the ground. Although there is an infinite number of XY locations throughout the boundary of a pen, a spacing between two viable camera locations was used to reduce the computational complexity. Initially, each camera location was set to be 3 feet apart on the X-axis and Y-axis. There are two possible Z-axis values: 15 ft and 12 ft. In addition, all the coverage calculations were completed starting at the (0,0) point and extended toward the X and Y-axis, as shown in Figure 2.

Each pose (O) represents camera coordinates (x,y,z), which indicates the camera's exact location. The yaw and pitch angles represent the orientation of the camera (Figure 4). The yaw angle (γ) represents the camera rotation on the horizontal axis, ranging between 0 and 360 degrees. The pitch angle (ϱ) represents camera rotation on the vertical axis, ranging between 0 and 360 degrees, as shown in Figure 4. The camera angles were limited to a specific range to only point it towards the region of interest or the cattle pen to avoid additional camera coverage calculation without significant coverage gain. The pitch angles were limited from 20 degrees and 60 degrees and the yaw angle was limited to the range 40 degrees to -40 degrees.

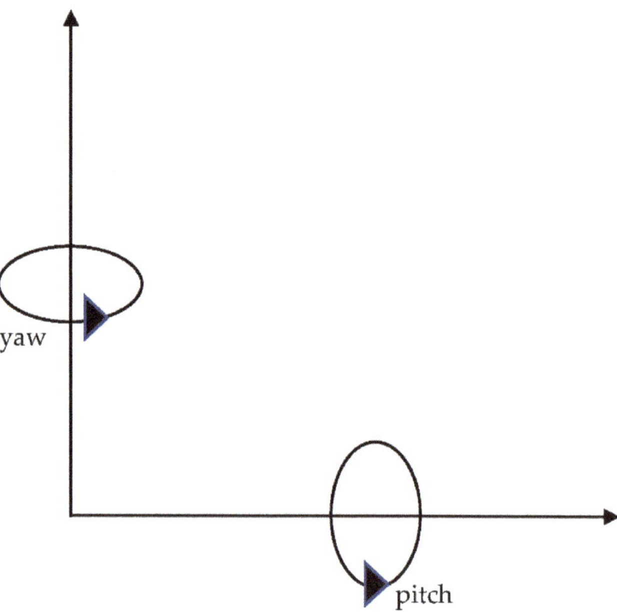

Figure 4. Yaw and pitch angle of the camera.

The area inside the cattle pen was the area of interest which could be assumed to be a large rectangular 3D object comprised of many smaller cubes. The center point of the cell (C_{ijk}) expresses the location of the cell. In the 3D environment, i represents the *X*-axis value of the cell center, j represents the *Y*-axis value of the cell center, and k represents the *Z*-axis value of the cell center. If the center of the cell has a value of (4,9,5) then it is 4 units away from the (0,0) point on the positive x axis, 9 units away from the (0,0) point on the positive *Y*-axis, and 5 units above the floor. Each cell in the cattle pen was evaluated to check its visibility by the camera by checking the cell's center point. In addition, calculating camera coverage for cells up to the roof of the pen is computationally expensive and will not provide any useful information. Thus, we used the average height of the steer to determine the height of the region of interest. In this case study, the height of the region of interest (ROI) was estimated to be 6 ft, whereas the approximate adult steer height was around 5 ft.

2.1.3. Coverage Calculation

Once the camera-shaped cone was positioned at a location ($CM_{ijk\gamma\varrho}$) with specific x, y, and z values and directed to a particular direction with the yaw and pitch angle, each cell's center was examined to determine if it falls entirely inside the cone. The cone created using the camera's dimensions was extended to a 60 ft length, so only cells visible to the camera were inside the cone. As physical structure of the pen such as the drinking trough, feeding trough, fence, pools, and other physical structures can block the view of the camera on the pen (Figure 5), ray cast, a native Blender function, was used to check for any visibility interruption of a particular cell. A ray was cast from the center of each cell (CC_{XcYcZc}) toward the camera's position (CM_{XiYjZk}). The physical structures created using Blender were merged to form one mesh object. A cell was counted toward camera coverage if the ray did not intercept the physical structures of the feedlot. As shown in Figure 5, a cell could be partially visible to the camera but could not be counted toward camera coverage as the center of the cell is only considered for the total coverage calculation.

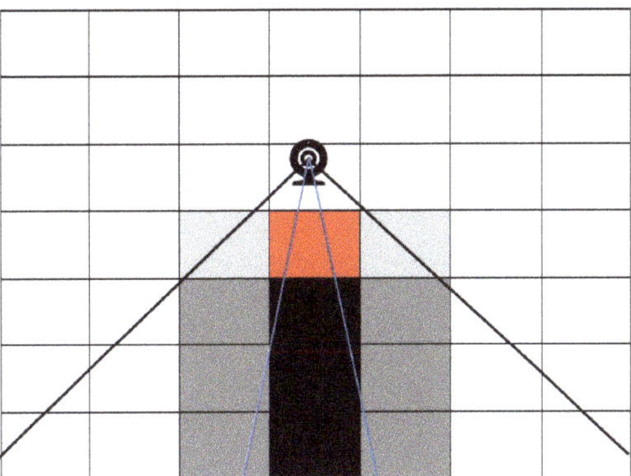

Figure 5. Occlusion due to physical structure; grey cells—partially visible, black cells—completely/mostly invisible, and red cell—physical obstruction.

While camera coverage is expressed in the percentage of the cells visible to the camera, all the ROI cells might not have the same level of importance for specific research work. For example, in the Beef Nutrition Research Center of the Beef Nutrition Farm, the researchers are interested in the steer's feeding and drinking behaviors for animal nutritional studies. Thus, setting the camera to monitor the steer approaching the feeding trough and eating is the primary objective. To facilitate this research, we counted the number of the weighted total cell on the total coverage. The complete working principle of this camera coverage calculation is provided below in Equation (1);

$$C_c = \frac{\sum_{i=1}^{n} w_i \times c_i}{T_c} \times 100 \quad (1)$$

Here, C_c = percentage of the total weighted camera coverage on a specific location, W_i = weight of a pixel in a particular group, C_i = number of pixels on the camera coverage belongs to a specific group, and T_c = number of total pixels inside the ROI.

The C_c is expressed as a percentage, but the final number could be above 100% due to the different weight values assigned to the important pixels.

A cell visible to each camera was counted separately for multiple camera coverage calculations, followed by the union of two sets of camera coverage cells. A cell was counted toward the camera coverage calculation when it was visible to one or more than one camera, followed by a weighted coverage calculation.

2.2. Multi-camera Placement Optimization

Multiple camera placement was optimized using a genetic algorithm which is specifically designed to solve the multi-camera placement solution. In this method, a fixed number of cameras and their positions were evaluated and the highest camera coverage was delineated by considering a series of different camera locations. The camera position evaluation is based on multiple objectives set by the user, thus known as multi-objective genetic algorithms.

Genetic algorithms (GAs) were first proposed by Holland as a computational optimization model based on the principle of natural evaluation [33]. The two main ideas of evaluation those genetic algorithms borrowed are; (a) passing information from one generation to the next generation, also known as inheritance, and (b) competition for survival, or survival of the fittest. The main advantages of using genetic algorithms to solve optimization problems are adaptations and parallelism. Adaption works the best in finding a set of good solutions that might not be the best, and parallel calculation can be achieved without much communication overhead.

It is very challenging to achieve multi-objective goals in a real problem using the Genetic Algorithm, as objectives can conflict and lead to unacceptable results for a particular objective [32]. Konak et al. demonstrated two possible solutions based on previous studies in this domain to achieve acceptable objectives [32]. The first standard method is to move one objective into the set of constraints. The second method is to optimize a weighted sum of the objective functions. In this research, both approaches were followed. In the first case, the cost optimization objective function is moved to the set of constraints. In the second case, the total cost of the camera setup was multiplied by a weight followed by subtraction from the total coverage, which was being maximized. Thus, the higher cost setup was penalized more than lower cost setups for the same camera coverage. The goal is to select the genes with higher coverage but lower cost.

2.2.1. Approach 1: Coverage Optimization with Budget Constraints

In the first case, the installation cost is used as the constraint rather than the objectives of the algorithm. The sole objective, in this case, is to maximize the coverage with a set of constraints

$$\max \sum C_{c1} \cup C_{c2} \cup C_{c3} \cup \ldots\ldots\ldots\ldots \cup C_{cn} + C_{awd} \tag{2}$$

C_C represents the camera coverage by camera C. In each generation, some selected genes with the highest coverage was passed to the next analysis stage. C_{awd} is the coverage awarded to the total coverage based on the secondary coverage and number of cells of the region of interest present on the coverage. The details of the secondary and awarded coverage are discussed in Section 2.2.4. The optimization function has the following constraints along with the common constraints discussed in Section 2.2.3.

$$CIM_1 + CIM_2 + CIM_3 \ldots\ldots\ldots + CIM_n \leq B_c \tag{3}$$

Here, *CIM* represents the installation and maintenance of the camera and B_c represents the total budget of the farm manager.

2.2.2. Approach 2: Weighted Sum of Coverage and Budget Optimization

Camera coverage maximization and cost minimization could be conflicting objectives. Thus, both objectives were unified into one with a fixed weight. Determining the most appropriate weight selection is challenging because the solutions can be changed.

$$\max \sum (C_{c1} \cup C_{c2} \cup C_{c3} \cup \ldots\ldots \cup C_{cn}) * W_T + C_{awd} \tag{4}$$

Here, W_T is the weight associated with the budget $CIM_1 + CIM_2 + CIM_3 \ldots\ldots\ldots + CIM_n$ and W_T is the weight for the associated cost, with a value between 0 and a specific percentage. If it represents the maximum possible cost, the penalty is a particular percent, 20% or 30% in this study. For zero-cost, the weight is 0; everything in between is interpolated based on the two extreme values.

2.2.3. Common Constraints for Both Approaches

The abovementioned functions have the following constraints

$$C_{c1} \cup C_{c2} \cup C_{c3} \cup \ldots\ldots\ldots\ldots \cup C_{cn} > CC_{min} \tag{5}$$

Here, C_C represents the camera coverage by camera C and CC_{min} represents the minimum required total weighted camera coverage for each camera position combination to be considered acceptable. This function is added to eliminate the combination with very low camera coverage.

Each camera must have a minimum coverage above a certain threshold defined by the user based on the number of pens and cameras the manager is planning to use.

$$C_{cn} \leq threshold \tag{6}$$

The other constraints involved were

$$1 \leq i \leq n \tag{7}$$

where *i* is the number of cameras; the user can define the maximum number of cameras.

$$0 \leq yaw \leq 360 \tag{8}$$

$$\frac{fov}{2} \leq pitch \leq 90 - \frac{fov}{2} \tag{9}$$

The camera should focus on the region of interest. When the camera is positioned at 0 degrees, one side will focus on the ground. On the other hand, when the camera is positioned at 180 degrees, the upper side will focus parallel to the ground. An angle more than 90 degrees will point the camera upward. Thus, the pitch angle is fixed between a certain value to position the camera toward the region of interest.

Each camera placement combination cannot have more than one camera installed at any given location.

$$x_1y_1z_1 \neq x_2y_2z_2 \neq \ldots\ldots\ldots\ldots\ldots \neq x_ny_nz_n \tag{10}$$

2.2.4. Camera Coverage Award

As shown in Equation (4), the final adjusted weighted camera coverage is considered for genetic algorithm optimization. The adjustments were made based on the number of cameras placed, capturing data from high-priority surveillance areas, and the percentage of overlapping camera coverage areas. A similar approach was followed by Altahir et al. to optimize a multiview surveillance system [33]. Multiple sensors can capture the same location in visual sensor placement optimization studies. The common area or cells on the coverage is known as common coverage or secondary coverage.

As the coverage of each cell is in binary format, the total coverage is calculated using a binary OR operator, and the common coverage or secondary coverage can be attained by using an AND operator. The cell number under the common coverage was used to award a certain percentage of coverage. In this study, the awarded coverage is a linear interpolation between 0 and total number of cells on the *ROI* divided by the number of cameras. If all cells on the *ROI* fall onto the secondary coverage, the award was the total number of cells divided by the number of cameras used in the optimization.

$$0 \leq C_s \leq \frac{ROI_{cc}}{C_n} \tag{11}$$

As mentioned earlier, often the camera or sensor placement has some specific objectives, for example monitoring the feeding behaviors. In such a case, the closer the camera to the food container, the better the visual data. The camera coverage is subject to a certain award if the regions of interest can be visualized closely. For this study, if the camera covers the region of interest from the maximum allowable distance (pen length), then the number of cells covered will be counted only once. If it covers from the lowest possible distance, the cells inside the region of interest will be covered a specific number of times based on the user input. In this study, the maximum award was five times. Everything else in between is interpolated based on the two extreme numbers.

$$C_{awd} = C_{roi} + C_s \tag{12}$$

C_{roi} is the number award for a specific region of interest camera coverage. C_s is the secondary camera coverage award.

2.2.5. Genetic Algorithm Implementation

The objective function of this problem is to maximize coverage per expenditure. The genetic algorithm developed in this study has the following steps.

1. Generate random camera location: Each camera position has six parameters; camera, x, y, and z location, yaw angle, and pitch angle. Each random gene created had six parameters. For a single pen, single-camera scenario, a camera can be placed in an infinite number of locations. To reduce the resources and time required, each camera parameter was subject to some limitations. In the *X* and *Y* directions, feasible locations for camera placement were 3 ft apart. In addition, the *Z*-axis height could be either 12 ft or 15 ft based on the physical structure of the cattle pen. On the other hand, the yaw and pitch angle also had limitations, as shown in Equations (8) and (9).

Each of the parameters was selected randomly to create n number of genes. Each gene had the following formats:

$$Camera\ name + X + Y + Z + Yaw + Pitch$$

2. Check gene fitness: Survival of the fittest is the main motto of the genetic algorithm. Each of the N-genes generated randomly in the first step was checked to see if it had the minimum percentage of the required coverage. The user provides the threshold value, in this case, 100 cells. In a given location, no more than one camera can exist; multiple cameras at one location was also checked. Each camera parameter, or chromosome of the gene, was used to position a camera to a specific location and calculate the coverage and mean distance of the cells located inside the food container from the camera. Total adjusted camera coverage was calculated based on the total camera coverage and award and penalty coverage.

3. Offspring generation: Two parent genes were randomly selected from the pool of eligible genes to create offspring. Chromosomes/parameters were randomly selected from parents and merged.

4. Merging: A specific number of properties of the randomly selected genes were changed to prevent the algorithm from getting stuck at local minima or maxima. The default number of chromosomes to be changed in a gene was chosen as 4.

5. The highest scoring genes proceed to the next steps. The camera coverage for each gene was calculated. All genes were ranked based on the coverage optimization function. Only a specific number of genes with the highest adjusted camera coverage value were passed into the next steps.

6. Steps 2 to 5 were repeated for i iterations, which is defined by the user.

The above steps were implemented in Python 3.6. The program requires Blender's native python libraries, mathutils, bpy, and bmesh, for camera placement in a specific location and coverage calculation.

The genetic algorithm was used to maximize the coverage while minimizing the cost for two different camera setups at the Beef Nutrition Farm at Iowa State University. The cameras were selected based on their price point and field of view. It was also assumed that camera A had FOV of 86 degrees and an installation cost of 200 USD. On the other hand, camera B had a 76 degrees FOV and an installation cost of 125 USD. The cameras were designed in Blender using the abovementioned properties on camera properties, followed by creating a cone-shaped camera. A comparison between the three cones shaped camera view are shown in Figure 6.

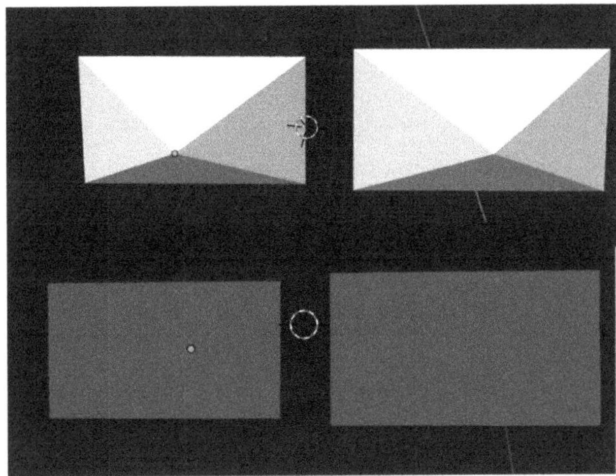

Figure 6. Camera view constructed in Blender Camera B (**left**) and Camera A (**right**) in back view (**top**) and front view (**bottom**).

3. Results

3.1. A Case Study of Camera Coverage for Single Camera

A single pen at the Beef Nutrition Farm at Iowa State University equipped with a single camera was used as a case study for evaluation and validation of the working principle of this study. The camera properties used in this case study were determined by observing the most common properties found in the surveillance camera system available at Amazon.com, Inc. USA and priced below 500 USD. It is assumed that the camera had a resolution of 2560 × 1440 pixels and a field of view either 86 degrees (Camera A) or 76 degrees (Camera B). The following parameters were designed to create a camera object, as shown in Figure 6. The camera coverage calculations algorithm was executed, and the algorithm provided the top 20 results with the highest camera coverage. The best results for each cell size and each camera are provided in Table 1.

Table 1. Calculated camera coverage of two different camera.

Pen	Camera	Cell Size	Camera Position	Optimal Camera Angles		Weighted Coverage (%)	Time Required
				Pitch	Yaw		
Single	A	0.5	7, −9, 15	60	0	99.69	613
		1	7, −9, 15	60	10	95.96	117
		2	3, −9, 15	60	−30	102.2	56.29
	B	0.5	11, −9, 15	60	30	97.05	1765
		1	0, −9, 15	60	−30	94.44	332
		2	11, −9, 15	60	30	99.05	128
Double	A	0.5	12, −9, 15	60	0	98.28	4236
		1	12, −9, 15	60	0	95.18	688
		2	0, −9, 15	60	−30	100	196
	B	0.5	12, −9, 15	60	10	91.43	1325
		1	12, −9, 15	60	10	90.82	218
		2	0, −9, 15	60	−30	97.4	190

As shown in Table 1, the cell size of the region of interest was changed from 0.5 ft × 0.5 ft to 2 ft × 2 ft. The smaller cell size required significantly more time to calculate coverage. In addition, the camera with a larger FOV, camera A, provided significantly higher camera coverage than camera B with a smaller FOV. The result demonstrated a methodology for selecting the optimal location for a single camera with given parameters.

3.2. *Multi-Camera Coverage Optimization with Genetic Algorithm*

3.2.1. Coverage Optimization with Budget as a Constraint

Camera A and Camera B were used to find the best possible placement combination for optimal camera coverage. Using a genetic algorithm, two sets of camera and pen combinations were evaluated to find the best possible camera location. The first case used two cameras installed in the eight-pen setup as shown in Figure 7. In this setup, Camera A had a 86 degrees FOV and Camera B had a 76 degrees FOV, and both cameras had a resolution of 2560 × 1440 pixels. The genetic algorithm was used to find the optimal location for maximum camera coverage at different pen setups and two different budgets. In this scenario, a cell size of 1 was used.

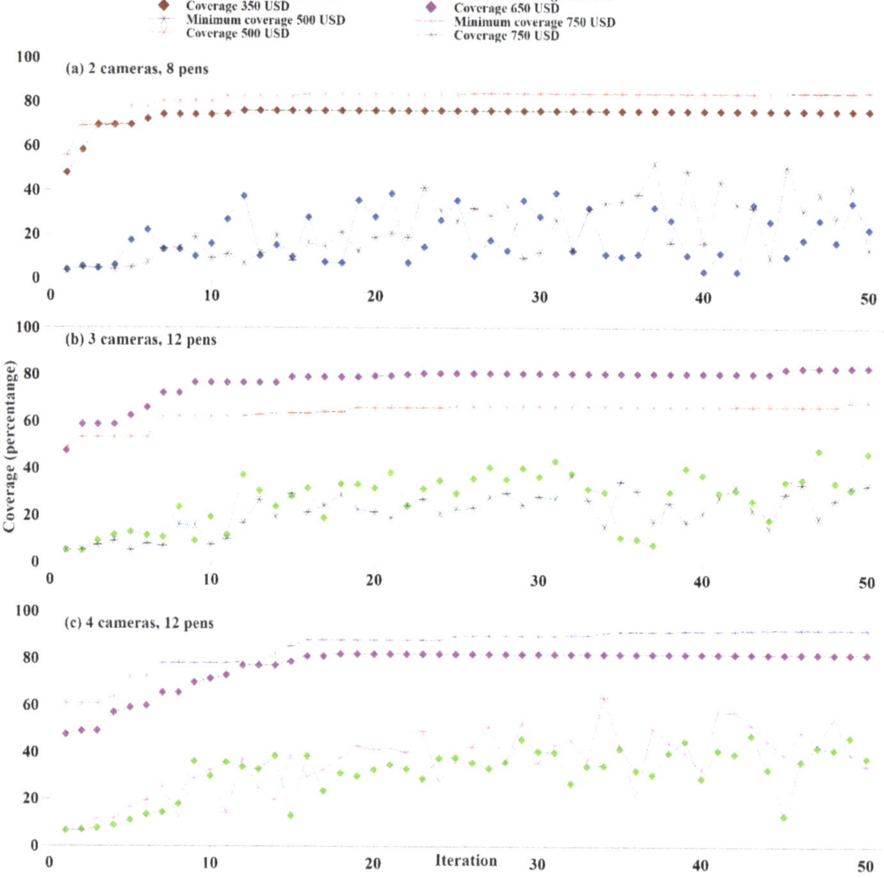

Figure 7. Camera coverage optimization for cameras at different budgets on eight pens setup. Subfigures (**a**–**c**) illustrate 2 cameras across 8 pens, 3 cameras across 12 pens, and 4 cameras across 12 pens, respectively.

The camera placement budgets were 350 USD and 500 USD, respectively. The optimal camera coverage was achieved after 34 and 45 iterations, respectively, as shown in Figure 7. Combined camera coverage was initially increased significantly, mostly for the first ten iterations. However, the improvement slowed down thereafter. There was a significant fluctuation in the lowest camera coverage among the different generations. The camera coverage and time required to run the algorithm were not always the same for each run, as genes were randomly created and edited in the different parts of the algorithm. The maximum camera coverage for 350 USD and 500 USD budgets were 76.1% and 84.3%, respectively. The camera coverage was increased for the 500 USD budget as the higher budget allowed use of camera A with a higher FOV.

The genetic algorithms were also used to maximize the coverage for four different budgets, 350 USD, 500 USD, 650 USD, and 750 USD, and two pen environments, eight pens and twelve pens. In the eight pens environment, the budgets and setups were 350 USD for two cameras and 500 USD for three cameras. In the twelve pens environment, the budgets and setups were 500 USD for three cameras, 650 USD for three cameras, 650 USD for four cameras, and 750 USD for four cameras. As shown in Figure 7, for all combinations, the camera coverage increased drastically over the first ten iterations, followed by a slight gain in camera coverage. For three cameras, the difference between maximum camera coverage of 500 USD for a two camera setup and 650 USD for a two camera setup was significant at 14.53%. This was due to the higher budgets allowed for selecting Camera A for all three allowed cameras. The four camera setups showed a drastic difference in maximum coverage with a higher budget. A budget increase of 100 USD allowed a gain of 10.5% of camera coverage. However, the four cameras with a 650 USD budget had 1.32% less maximum coverage than the three-camera coverage with the same budget, as the higher number of cameras forced the algorithm to select cameras with lower FOV. This difference shows that increasing the number of cameras does not always guarantee coverage gain after a certain point. The farm manager can efficiently decide on using a specific budget and number of cameras to achieve the desired coverage within their budget.

The time required to complete each iteration of camera coverage calculation was drastically reduced as the iteration progressed, as shown in Figure 8. The main reason for this was that our genetic algorithm saves the camera coverage for each camera combination in the Random Access Memory (RAM). The algorithm searches in the previous record of a given camera combination before calculating camera coverage. If it previously calculated the coverage, then recalculation of the same camera coverage was avoided to save time and computational resources. Thus, for two camera setups, the first five iterations took almost 20 s, the three and four cameras setups required 30 to 100 s, and then it came down from five seconds to a fraction of a second as the iteration progressed.

Camera coverage solutions were ranked from high to low based on the percentage of area covered by the camera. The difference in the percentage of camera coverage between 1st and 25th on the ranked list is plotted in Figure 9. The figure illustrates that the difference between ranked 1st and 25th solutions was lowest at 0.7% for four cameras with a 650 USD budget for twelve pens and highest at 3.3% for three cameras with a 650 USD budget for twelve pens. Such a low difference shows that there were numerous camera placement options without significant percentage of coverage difference.

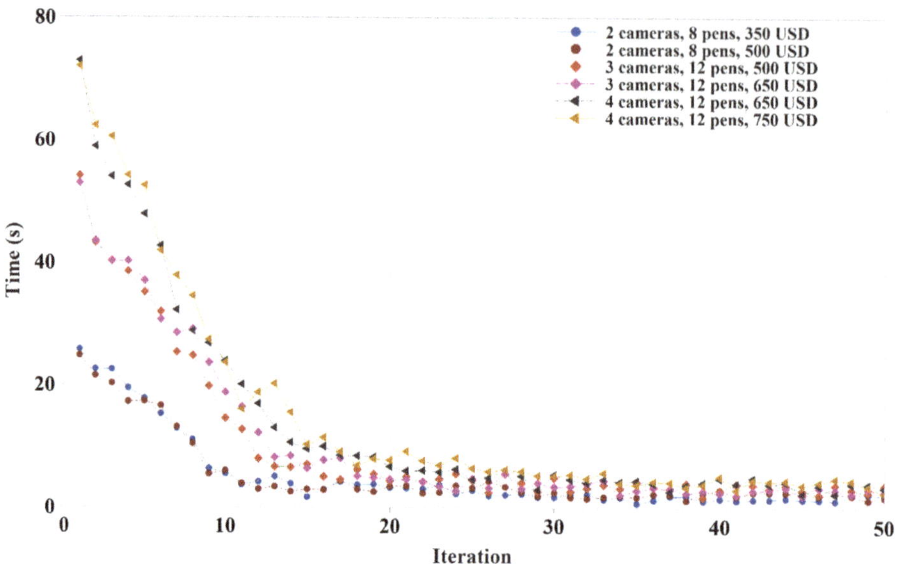

Figure 8. Time required to complete each iteration of different camera combinations in a given budget.

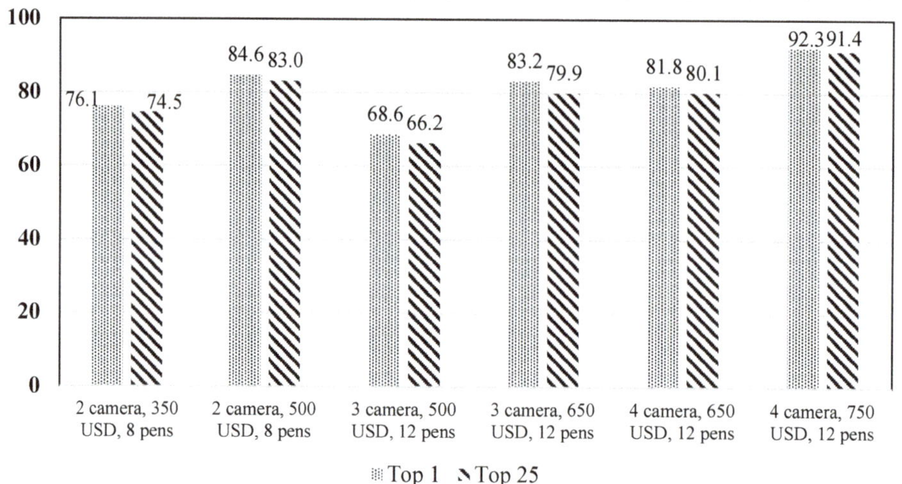

Figure 9. Difference in camera coverage between the top 25 camera placement solutions with budget as constraints.

3.2.2. Coverage Optimization with Budget Integrated into the Optimization Function

The second approach to camera coverage optimization includes budget constraints in the optimization function. As described earlier in Section 2, the adjusted camera coverage was penalized based on the specific camera setup budget. The penalty was ranging from 0 to a specific percentage of the coverage. If the cost is zero, the penalty is zero; if cost is the maximum possible cost for a particular number of cameras, the penalty is a selected percentage. In this study, the maximum penalty amounts were selected as 20% or 30%, as shown in Figure 10. Users can define this percentage.

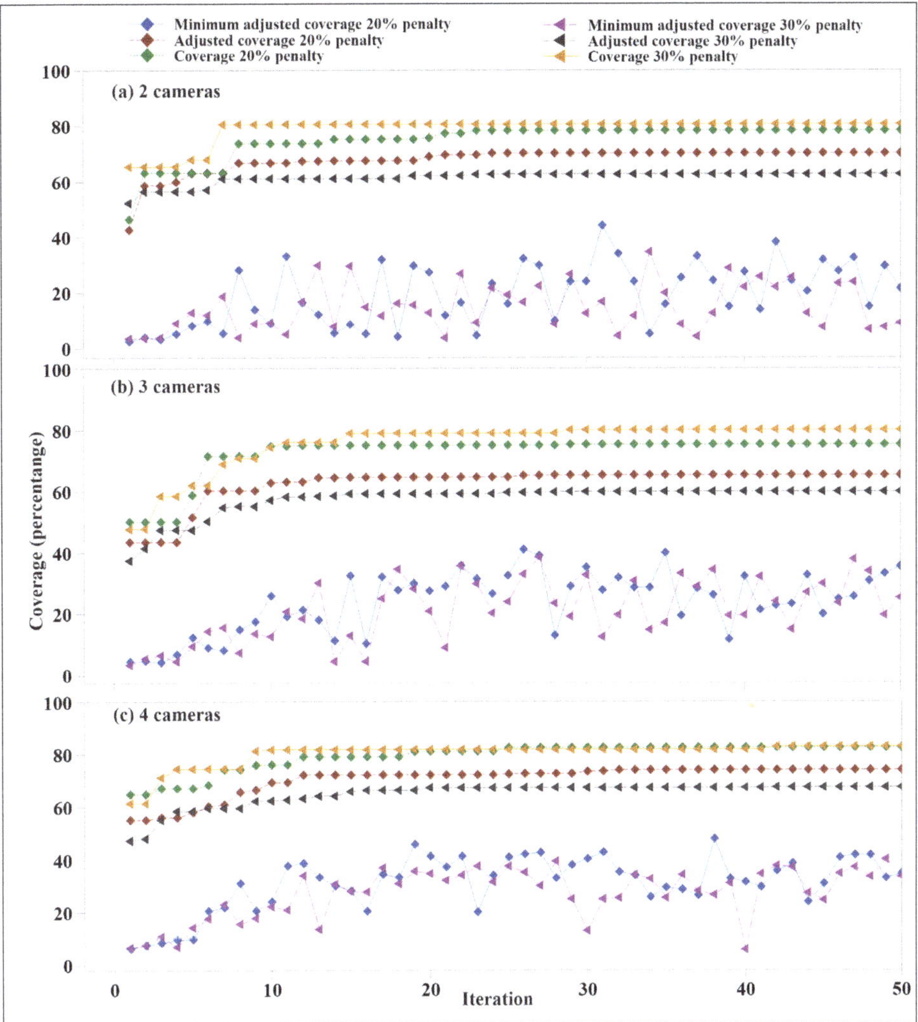

Figure 10. Camera coverage optimization for (**a**) eight pens–two camera coverage, (**b**) twelve pens–three cameras setup, and (**c**) twelve pens–four cameras setup.

As shown in Figure 10, the eight pens–two cameras coverage, twelve pens–three cameras coverage, and twelve pens–four cameras coverage showed that the differences in total coverage in 20% cost penalty and 30% cost penalty were very low, less than 5%. This difference exists because the optimization algorithm was trying to maximize only the adjusted coverage. However, the adjusted coverage for the 20% penalty rate was significantly higher than for the 30% penalty rate coverage. It was also observed that the maximum adjusted and actual coverage did not reach their peaks simultaneously, because the algorithm focused on maximizing the adjusted coverage by changing the location and camera parameters. Like coverage optimization with a given budget, the coverage and adjusted coverage increased drastically at the beginning, mostly during the first ten iterations (Figure 11). The rate of additional coverage gains was very low afterward. Similar trends were observed for the three cameras–twelve pens and four cameras–twelve pens setups. The difference between maximum coverage was either small or zero for 20% penalty and 30% penalty. However, the difference between adjusted coverage was very significant in both cases.

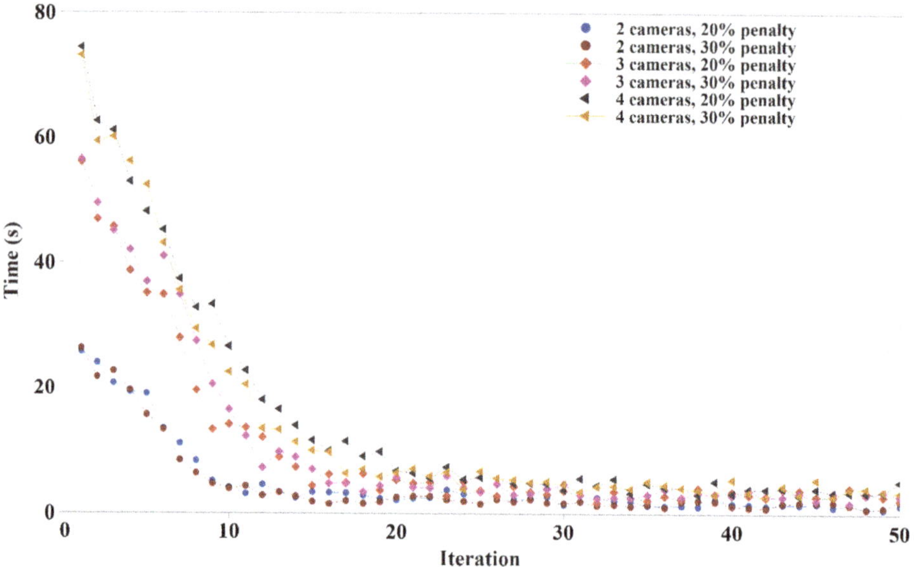

Figure 11. Time required to complete each iteration of different camera combination.

In Figure 11, the time required to complete each iteration in the optimization follows a similar pattern to that shown in Figure 8. Size of the pen and the number of unique camera combinations or genes of the genetic algorithm to process dictated the required time. Initially, there was a relatively higher number of new genes or camera combinations. As optimization proceeded, the number of new genes or camera combinations not presented in previous iterations was low and were skipped to avoid duplicate processing. In addition, twelve pen combinations had 50% more cells than eight pen combinations, resulting in a higher processing time.

Figure 12 shows the difference between the 1st and 25th solutions raked from high to low based on the percentage of camera coverage provided by the genetic algorithm with budget integrated on the optimization algorithm. Figure 12 exhibited a trend very similar to that for the budget as constraint-based optimization illustrated in Figure 9. In this case, the lowest difference was only 0.5% for four cameras–twelve pens with 20% penalty, and the highest was 1.9% for four cameras–twelve pens with 30% penalty. Such a low difference in percentage can also offer numerous feasible solutions for camera placement.

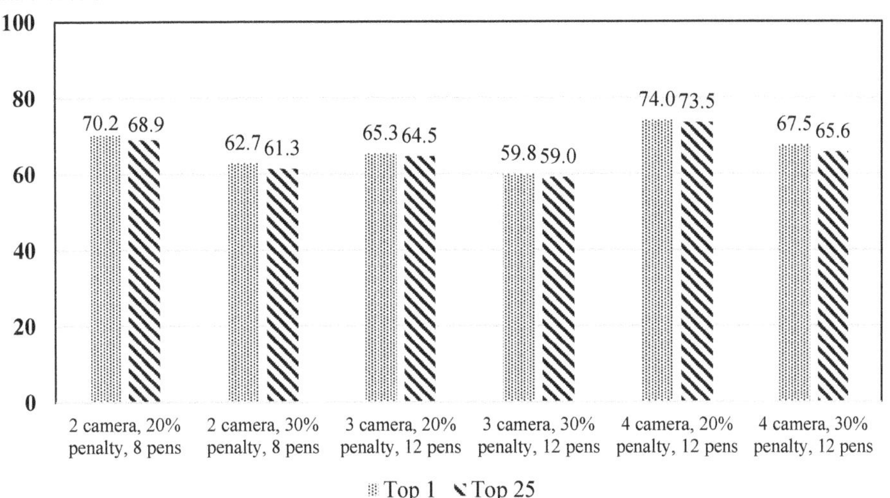

Figure 12. Difference in camera coverage between the top 25 camera placement solutions with budget integrated into the optimization function.

4. Discussions

4.1. Findings

This study's novel camera placement optimization methodology showed the efficacy of 3D animation software combined with an optimization algorithm to find the optimal solution with specific constraints in a large space. The results demonstrated that the optimal placement location could be derived for both a single camera and muti-camera setup in real-farm environment that takes the occlusion due to physical structure into account. The study pursued two different avenues of multi-objective genetic algorithms; coverage optimization with a given budget as a constraint and integrated budget function within the coverage optimization function. The results showed that the coverage difference in the 25 possible solutions, sorted based on percentage of camera coverage, was minimal, offering the user various options to choose from without significantly sacrificing total camera coverage. The study also addressed two major shortcomings of earlier studies in camera coverage optimization; taking the real 3D scenario into account and considering occlusion due to physical structures.

4.2. Limitations and Recommendations

The time required to complete this algorithm is its main limitation. The genetic algorithm was run with a cell size of 1 square ft, which is small compared to 1 square meter of some of the earlier studies. However, while a smaller cell size would yield higher precious coverage calculation, the time required for each iteration calculation increases significantly when the cell size is decreased. Occlusion was calculated based on the center of each cell, not the overall cell itself, which poses the risk of omitting a complete cell from the coverage for fractional occlusion due to the location of the cell center. While this study focused on optimizing camera placement in a relatively simple environment—a pen extending only on one side—this study can be adapted for a more complex environments with multistorey buildings and pens extending in different directions.

5. Conclusions

Surveillance data quality plays a pivotal role in cattle welfare monitoring with computer vision. In this study, a confined cattle farm environment was designed to determine the optimized camera location for the data collection. The multi-camera combination was solved by employing the genetic algorithm. Two approaches were followed to find optimal placement for maximizing camera placement; one with installation budget as a constraint and one with budget integrated into the optimization algorithm. The genetic algorithm showed that the optimal camera location could be determined within the first few iterations of camera coverage. In addition, the difference between the 1st and 25th result, when ranked from high to low based on percentage of camera coverage, also proved that the genetic algorithm could provide several optical camera locations to choose from. It was also observed that the FOV of the camera played the most crucial role in total coverage. The methodology also demonstrated that the approach can be adapted for camera coverage calculation in other domains with the versatile genetic algorithm and powerful Blender 3D software.

Author Contributions: J.M.P. conceived and designed the study; A.A.S. performed the simulations; A.A.S. and J.M.P. analyzed the data; A.A.S. and J.M.P. wrote the paper. All authors have read and agreed to the published version of the manuscript.

Funding: This research was funded by the Iowa Beef Industry Council. The APC was funded by Iowa State University.

Institutional Review Board Statement: Not applicable.

Data Availability Statement: The data presented in this study are available on request from the corresponding author.

Conflicts of Interest: The authors declare no conflict of interest.

References

1. Wongsriworaphon, A.; Arnonkijpanich, B.; Pathumnakul, S. An approach based on digital image analysis to estimate the live weights of pigs in farm environments. *Comput. Electron. Agric.* **2015**, *115*, 26–33. [CrossRef]
2. Poursaberi, A.; Bahr, C.; Pluk, A.; Van Nuffel, A.; Berckmans, D. Real-time automatic lameness detection based on back posture extraction in dairy cattle: Shape analysis of cow with image processing techniques. *Comput. Electron. Agric.* **2010**, *74*, 110–119. [CrossRef]
3. Andrew, W.; Greatwood, C.; Burghardt, T. Aerial Animal Biometrics: Individual Friesian Cattle Recovery and Visual Identification via an Autonomous UAV with Onboard Deep Inference. In Proceedings of the IEEE International Conference on Intelligent Robots and Systems, Macau, China, 3–8 November 2019.
4. Kashiha, M.A.; Bahr, C.; Ott, S.; Moons, C.P.H.; Niewold, T.A.; Tuyttens, F.; Berckmans, D. Automatic monitoring of pig locomotion using image analysis. *Livest. Sci.* **2014**, *159*, 141–148. [CrossRef]
5. Nilsson, M.; Herlin, A.H.; Ardö, H.; Guzhva, O.; Aström, K.; Bergsten, C. Development of automatic surveillance of animal behaviour and welfare using image analysis and machine learned segmentation technique. *Animal* **2015**, *9*, 1859–1865. [CrossRef] [PubMed]
6. Viazzi, S.; Bahr, C.; Schlageter-Tello, A.; Van Hertem, T.; Romanini, C.E.B.; Pluk, A.; Halachmi, I.; Lokhorst, C.; Berckmans, D. Analysis of individual classification of lameness using automatic measurement of back posture in dairy cattle. *J. Dairy Sci.* **2013**, *96*, 257–266. [CrossRef]
7. Matthews, S.G.; Miller, A.L.; Clapp, J.; Plötz, T.; Kyriazakis, I. Early detection of health and welfare compromises through automated detection of behavioural changes in pigs. *Vet. J.* **2016**, *217*, 43–51. [CrossRef]
8. Lu, M.; Xiong, Y.; Li, K.; Liu, L.; Yan, L.; Ding, Y.; Lin, X.; Yang, X.; Shen, M. An automatic splitting method for the adhesive piglets' gray scale image based on the ellipse shape feature. *Comput. Electron. Agric.* **2016**, *120*, 53–62. [CrossRef]
9. Nasirahmadi, A.; Hensel, O.; Edwards, S.A.; Sturm, B. Automatic detection of mounting behaviours among pigs using image analysis. *Comput. Electron. Agric.* **2016**, *124*, 295–302. [CrossRef]
10. Kashiha, M.A.; Bahr, C.; Ott, S.; Moons, C.P.H.; Niewold, T.A.; Ödberg, F.O.; Berckmans, D. Automatic identification of marked pigs in a pen using image pattern recognition. *Comput. Electron. Agric.* **2013**, *93*, 111–120. [CrossRef]
11. Andrew, W.; Hannuna, S.; Campbell, N.; Burghardt, T. Automatic individual holstein friesian cattle identification via selective local coat pattern matching in RGB-D imagery. In Proceedings of the International Conference on Image Processing, Phoenix, AZ, USA, 25–28 September 2016.
12. Tasdemir, S.; Urkmez, A.; Inal, S. Determination of body measurements on the Holstein cows using digital image analysis and estimation of live weight with regression analysis. *Comput. Electron. Agric.* **2011**, *76*, 189–197. [CrossRef]
13. Viazzi, S.; Ismayilova, G.; Oczak, M.; Sonoda, L.T.; Fels, M.; Guarino, M.; Vranken, E.; Hartung, J.; Bahr, C.; Berckmans, D. Image feature extraction for classification of aggressive interactions among pigs. *Comput. Electron. Agric.* **2014**, *104*, 57–62. [CrossRef]

14. Kim, J.; Chung, Y.; Choi, Y.; Sa, J.; Kim, H.; Chung, Y.; Park, D.; Kim, H. Depth-based detection of standing-pigs in moving noise environments. *Sensors* **2017**, *17*, 2757. [CrossRef] [PubMed]
15. Brown-Brandl, T.M.; Eigenberg, R.A. Development of a livestock feeding behavior monitoring system. *Trans. ASABE* **2011**, *54*, 1913–1920. [CrossRef]
16. Pluk, A.; Bahr, C.; Leroy, T.; Poursaberi, A.; Song, X.; Vranken, E.; Maertens, W.; Van Nuffel, A.; Berckmans, D. Evaluation of Step Overlap as an Automatic Measure in Dairy Cow Locomotion. *Trans. ASABE* **2010**, *53*, 1305–1312. [CrossRef]
17. Martinez-Ortiz, C.; Everson, R.; Mottram, T. Video tracking of dairy cows for assessing mobility scores. In Proceedings of the Precision Livestock Farming 2013—Papers Presented at the 6th European Conference on Precision Livestock Farming, ECPLF, Leuven, Belgium, 10–12 September 2013.
18. Kelly, H. After Boston: The Pros and Cons of Surveillance Cameras. Available online: https://www.cnn.com/2013/04/26/tech/innovation/security-cameras-boston-bombings/index.html (accessed on 4 May 2022).
19. Chen, H.T.; Wu, S.W.; Hsieh, S.H. Visualization of CCTV coverage in public building space using BIM technology. *Vis. Eng.* **2013**, *1*, 5. [CrossRef]
20. Zhang, Y.; Luo, H.; Skitmore, M.; Li, Q.; Zhong, B. Optimal Camera Placement for Monitoring Safety in Metro Station Construction Work. *J. Constr. Eng. Manag.* **2019**, *145*, 04018118. [CrossRef]
21. Erdem, U.M.; Sclaroff, S. Automated camera layout to satisfy task-specific and floor plan-specific coverage requirements. *Comput. Vis. Image Underst.* **2006**, *103*, 156–169. [CrossRef]
22. Yabuta, K.; Kitazawa, H. Optimum camera placement considering camera specification for security monitoring. In Proceedings of the IEEE International Symposium on Circuits and Systems, Seattle, WA, USA, 18–21 May 2008.
23. Cheng, B.; Cui, L.; Jia, W.; Zhao, W.; Gerhard, P.H. Multiple region of interest coverage in camera sensor networks for tele-intensive care units. *IEEE Trans. Ind. Inform.* **2016**, *12*, 2331–2341. [CrossRef]
24. Kim, K.; Murray, A.T.; Xiao, N. A multiobjective evolutionary algorithm for surveillance sensor placement. *Environ. Plan. B Plan. Des.* **2008**, *35*, 935–948. [CrossRef]
25. Indu, S.; Bhattacharyya, A.; Mittal, N.R.; Chaudhury, S. Optimal sensor placement for surveillance of large spaces. In Proceedings of the 2009 3rd ACM/IEEE International Conference on Distributed Smart Cameras, Como, Italy, 30 August–2 September 2009.
26. Kim, J.; Ham, Y.; Chung, Y.; Chi, S. Systematic Camera Placement Framework for Operation-Level Visual Monitoring on Construction Jobsites. *J. Constr. Eng. Manag.* **2019**, *145*, 04019019. [CrossRef]
27. Albahri, A.H.; Hammad, A. Simulation-Based Optimization of Surveillance Camera Types, Number, and Placement in Buildings Using BIM. *J. Comput. Civ. Eng.* **2017**, *31*, 04017055. [CrossRef]
28. Yang, X.; Li, H.; Huang, T.; Zhai, X.; Wang, F.; Wang, C. Computer-Aided Optimization of Surveillance Cameras Placement on Construction Sites. *Comput. Civ. Infrastruct. Eng.* **2018**, *33*, 1110–1126. [CrossRef]
29. Rebai, M.; Le Berre, M.; Hnaien, F.; Snoussi, H. Exact Biobjective Optimization Methods for Camera Coverage Problem in Three-Dimensional Areas. *IEEE Sens. J.* **2016**, *16*, 3323–3331. [CrossRef]
30. The Blender Foundation Blender. Available online: https://www.blender.org/ (accessed on 9 February 2021).
31. Holland, J.H. Genetic algorithms. *Sci. Am.* **1992**, *267*, 66–73. [CrossRef]
32. Konak, A.; Coit, D.W.; Smith, A.E. Multiobjective optimization using genetic algorithms: A tutorial. *Reliab. Eng. Syst. Saf.* **2006**, *91*, 992–1007. [CrossRef]
33. Altahir, A.A.; Asirvadam, V.S.; Hamid, N.H.B.; Sebastian, P.; Saad, N.B.; Ibrahim, R.B.; Dass, S.C. Optimizing visual sensor coverage overlaps for multiview surveillance systems. *IEEE Sens. J.* **2018**, *18*, 4544–4552. [CrossRef]

MDPI

Review

Effects of Key Farm Management Practices on Pullets Welfare—A Review

Xiaohui Du [1], Pingwu Qin [1], Yanting Liu [1], Felix Kwame Amevor [1], Gang Shu [2], Diyan Li [1] and Xiaoling Zhao [1,3,*]

1 College of Animal Science and Technology, Sichuan Agricultural University, Chengdu 611130, China; 10400@sicau.edu.cn (X.D.); qinpingwu1998@163.com (P.Q.); lyanting0212@163.com (Y.L.); amevorfelix@gmail.com (F.K.A.); diyanli@sicau.edu.cn (D.L.)
2 Department of Basic Veterinary Medicine, Sichuan Agricultural University, Chengdu 611130, China; dyysg2005@sicau.edu.cn
3 Farm Animal Genetic Resources Exploration and Innovation Key Laboratory of Sichuan Province, Sichuan Agricultural University, Chengdu 611130, China
* Correspondence: zhaoxiaoling@sicau.edu.cn; Tel.: +86-028-8629101

Simple Summary: Studies on animal behavior and welfare have reported that improving the management practices of pullets can enhance their growth, as well as their physical and mental condition, thus benefiting the productivity of laying hens. Therefore, in this review, we elaborated on the key effective farm management measures, including housing type and matching, flock status, and environmental management and enrichment, to provide the necessary information to incorporate welfare into chicken rearing and its importance in production, with the aim of improving the quantity and quality of chicken products.

Abstract: Studies on animal behavior and welfare have reported that improving the management practices of pullets can enhance their growth, as well as their physical and mental condition, thus benefiting the productivity of laying hens. There is growing confidence in the international community to abandon the conventional practices of "cage-rearing and beak-trimming" to improve the welfare of chickens. Therefore, in this review, we summarized some of the effective poultry management practices that have provided welfare benefits for pullets. The results are as follows: 1. Maintaining similar housing conditions at different periods alleviates fear and discomfort among pullets; 2. Pullets reared under cage-free systems have better physical conditions and temperaments than those reared in cage systems, and they are more suitable to be transferred to similar housing to lay eggs; 3. Improving flock uniformity in appearance and body size has reduced the risk of pecking and injury; 4. Maintaining an appropriate population (40–500 birds) has reduced flock aggressiveness; 5. A combination of 8–10 h of darkness and 5–30 lux of light-intensity exposure via natural or warm white LED light has achieved a welfare–performance balance in pullets. (This varies by age, strain, and activities.); 6. Dark brooders (mimicking mother hens) have alleviated fear and pecking behaviors in pullets; 7. The air quality of the chicken house has been effectively improved by optimizing feed formulation and ventilation, and by reducing fecal accumulation and fermentation; 8. Complex environments (with litter, perches, straw bales, slopes, platforms, outdoor access, etc.) have stimulated the activities of chickens and have produced good welfare effects. In conclusion, the application of comprehensive management strategies has improved the physical and mental health of pullets, which has, in turn, improved the quantity and quality of poultry products.

Keywords: welfare; pullet status; housing type; environmental management

Citation: Du, X.; Qin, P.; Liu, Y.; Amevor, F.K.; Shu, G.; Li, D.; Zhao, X. Effects of Key Farm Management Practices on Pullets Welfare—A Review. *Animals* **2022**, *12*, 729. https://doi.org/10.3390/ani12060729

Academic Editors: Lilong Chai and Yang Zhao

Received: 10 February 2022
Accepted: 11 March 2022
Published: 14 March 2022

1. Introduction

In recent years, animal welfare campaigns and studies have increased worldwide. A series of "group-rearing and untrimmed" strategies for poultry farming, combined with the selection of good-tempered pullets, the integrated application of various environmental enrichments, and the adoption of appropriate management measures, can effectively promote the welfare of chickens, especially at early ages [1,2]. Pullets with well-developed musculoskeletal and nervous systems easily adjust to various complex environmental changes, thereby significantly reducing the incidence of fear and pecking during laying (even if kept in traditional laying cages after rearing) [3,4], and they exhibit better health and production performance, thereby producing quality products [5–7]. Therefore, the aim of this review is to elaborate on key, effective farm management measures, focusing on three main aspects, including housing type and matching, flock status, and environmental management, to provide information needed to incorporate welfare into chicken rearing and to discuss its importance in production.

2. Housing Type and Matching

Currently, in poultry production, pullets are rearing in multilayer colony cages or cage-free houses, and they are later transferred to laying systems consisting of four options (cages, barns, aviaries, and outdoor ranges). Such a shift requires the flock to adapt to the new environment, which can be more or less stressful depending on the matching of different housing types. Major shortfalls and solutions to current laying housing systems and management practices that require immediate attention are discussed in these two rearing systems, as shown in Table 1.

Table 1. The matching effects of different housing types on transfers from rearing to laying.

Rearing Type	Housing Conversion	Matching Effects	References
Cage rearing	To aviaries	Higher risk of feed waste, dehydration, and ground eggs	Tauson 2005 [8]
	To aviaries	Prone to flight accidents, keel fractures, and vent pecking	Gunnarsson et al., 1999 [9]
	To perches	chicks exposed to perches earlier behaved better at moving between the layers later.	Gunnarsson et al., 2000 [10]
	To floor barns with perch	Delaying access to perches for at least 10 weeks	Mitchell et al., 2015 [11]
	To enriched colony cages *	Reduces discomfort, enhances the development of bone mass parameters better than those of the traditional cage layers	Regmi et al., 2016 [12]
Cage-free rearing	From aviaries or cages to the same housing type or enriched cages	Total medullary and pneumatic bone weight and ash content scores from high to low were A-A, C-E, A-C and C-C hens, respectively	Neijat et al., 2019 [13]
	From aviaries to furnished cages at 16 weeks	Mortality (20–76 wk) is higher (5.52% vs. 2.48%) than cage-reared birds	Tahamtani et al., 2014 [14]
	From aviaries to cages	Early transfer (16 weeks or earlier) could reduce mortality and increase nest eggs	Janczak et al., 2015 [15]
	From aviaries to enriched cages *	Fewer acceleration events and collisions during daytime at 21 and 35 weeks of age, and more high-perching compared to conventional cages	Pulin et al., 2020 [16]
	From aviaries to enriched cages at 16 weeks *	Lower levels of fearfulness indicated by spending less time away from the novel object at 19 and 21 weeks compared to conventional cages	Brantsæter et al., 2016 [17]
	From aviaries to * aviaries	More eggs in the nest compared to barn-reared hens	Colson et al., 2008 [18]
	To outdoor *	The high outdoor hens showed the highest spleen and empty gizzard weights	Md Saiful et al., 2020 [19]
	To modified cages (with 2 nests each) *	Expressed a full repertoire of pre-laying activities; displacement behaviors and pacing were less frequent; more eggs in the nest than conventional cages without nests	Shervin et al., 1993 [20]

* Indicates the best matching effects.

2.1. Cage Rearing and Transfer Effects

Cage rearing system is the most predominant system of rearing chickens worldwide, in which birds are fed and watered in prefabricated cages with specific dimensions for efficient management [8–10].

The cage rearing system of keeping chickens is faced with key problems that affect the welfare of the birds. First, due to the lack of life experience in three dimensions, cage-raised pullets that lay eggs in aviaries are at a higher risk for feed waste, dehydration, and ground eggs [8], as basic needs (such as feed, water, perches, and nest boxes) are usually placed at different tiers in the aviary. Flocks without aviary experience are also prone to flight accidents after transition, resulting in keel fractures and a high rate of vent pecking [9]. In addition, chicks exposed to perches from hatching behaved better at moving between the layers of the coop at 16 weeks than those first exposed to perches at 8 weeks [10]. Delaying access to perches and laying nests until 25 weeks of age usually affects movement, use of vertical space, and the nocturnal falls of pullets for at least the next 10 weeks. Moreover, dust particles, management level, and total costs of aviaries are much higher than those of colony cages (including perches, dust bath mats, and laying boxes during the laying period) and traditional floor barns [11]. Therefore, it is important to avoid transferring cage-raised pullets to aviaries for laying in the production system. However, if necessary, the transition should be completed before 17 weeks of age with a supply of perches and laying nests [21,22].

Nevertheless, transferring cage-reared pullets to enriched colony cages for laying greatly reduces the discomfort, as well as the density and hardness of the tibial cortex of hens in the later laying period, over that of traditional cage-layers, indicating that the enriched cages provide opportunities for the hens to exercise, which enhance the development of bone-mass parameters [12].

2.2. Cage-Free Rearing and Transfer Effects

Compared with cage-reared pullets, cage-free-reared pullets show better adaptability and bone-bearing capacity [13], and they are usually transferred to one of the four current laying systems: battery cages, conventional barns, aviaries, or outdoor ranges.

Previous studies have shown that, in the first few days after the transition of pullets reared in an aviary to a laying system consisting of one of the four options (cages, barns, aviaries, and outdoor ranges), transfer results in crouching; however, there is no difference after 15 days of transition [23,24], which means hens habituate to new spaces in a relatively short period. Meanwhile, compared to the cage-reared birds, mortality in furnished cages caused by frustration and feather pecking is higher among aviary-reared pullets, suggesting that their later welfare may be compromised [14,25,26]; however, early transfer to a laying aviary (16 weeks or earlier) could reduce this effect [15]. In addition, pullets transferred from aviaries to enriched cages exhibit less fear, more dust bath activity, and use higher perches compared to cage-reared pullets [16,17]. Similarly, a study showed that birds reared in an aviary have higher laying rates in the nest, as compared to those reared in a barn-housed group [18].

Compared with the pullets kept indoors, outdoor-range-reared pullets showed improved growth and development because they were exposed to available natural light. The reason for this is that those pullets that are kept outdoors are free to exercise, which readies them to utilize the available outdoor area as adults and thus improve their health [19,27]. A study has achieved similar effects by adopting aviary or other enrichment methods during rearing to optimize the growth and development of pullets, and the follow-up effect of outdoor stocking. This could help pullets to adapt to the laying environment, thereby encouraging hens to lay eggs in their nests [20].

In summary, cage-free-reared pullets are raised with abundant environmental stimulation and provided with enough space for exercise, improving skeletal characteristics, reducing fear, and enhancing adaptability, compared to the cage-reared ones. Worldwide, we can find producers and technicians who need to be convinced about these systems. It is therefore recommended that pullets raised in cages should not be transferred to aviaries to lay eggs, but rather should preferably be transferred to enriched cages. Again, it is recommended that if the goal is to lay eggs in an aviary, it is better to rear pullets in an aviary. If aiming for free-range laying, it is better to provide pullets with the necessary outdoors experience as early as possible.

3. Flock Status

Good feather and physical condition of pullets are parameters that are used to predict high egg production and low mortality rates, and the welfare of pullets can be improved by ensuring flock uniformity, appropriate flock size, distribution, and stocking density [1,28].

3.1. Flock Uniformity

Flock uniformity includes consistency in body weight, feather color, and state. Studies have shown that preventing underweight conditions in pullets, increasing flock uniformity, and preventing pain and lameness among chickens could reduce the risk of pecking and cannibalism [15]. Furthermore, a study by Janczak et al. (2015), reported that Lohmann hens with a uniformity above 90% at 15 weeks old have a lower mortality rate during the laying period than flocks with uniformity rates between 85 and 90% or below 85%, whereas the average body weight at 15 weeks old had no effect on mortality [15]. In addition, a survey of 122 Canadian egg farms found that isolating chickens with abnormal conditions (such as being underweight, having messy or prominent plumage, or suffering from surface lesions or lameness) could reduce the risk of severe feather pecking and cannibalism (chickens pluck feathers from their companions for fiber rather than simply swallowing naturally shed feathers), as chickens with different appearances or behaviors are vulnerable to attack [29]. Therefore, the level of harmful pecking behavior can be significantly reduced through a combination of lighting, feeding, and flock adjustment [30], and a "no trimming" policy is expected to be incorporated into the sustainable egg-production system [31]. However, these management tasks are highly dependent on the conscientiousness and competence of the farm staff.

3.2. Flock Size and Distribution

Since social animals usually behave synchronously, resources should be allocated on the basis of flock size, stocking density, and behavior, especially in non-cage production systems. Several studies on the effects of feather pecking and aggression on the social behavior of pullets and laying hens have shown that aggressive pecking is most common in smaller flocks, possibly because these birds usually form social hierarchies, whereas birds in larger groups often have trouble distinguishing between familiar and unfamiliar companions and then develop tolerance strategies, so they tend to be less aggressive [32–34]. Furthermore, with increasing group size, chickens became less vigilant, with more hens resting on the lower perch [35], fewer birds roosting on higher perches, and more birds engaging in preening on the floor [36]. In cage-free systems, smaller flocks tend to have a higher proportion of outdoor activities [37], which can reduce pecking injuries. Therefore, the flock size should be controlled at fewer than 500 [38,39] and more than 40 individuals [32–34].

A study on the behavioral synchronization and spatial clustering of commercial pullets reported that the relative synchronization degree of ingestion, drinking water, resting, and preening declined exponentially as flock size (5–120 individuals per group) increased although the absolute number of pullets increased. Among these behaviors, preening was the most synchronous behavior (more than twice that of the least synchronized behavior), and feeding was the most clustered behavior in space (three times more clustered than the other behaviors) [40]. This demonstrates the importance of providing sufficient space (especially feeding space) for individual birds, which is more important in small flocks (more synchronous) than in large or cage-free flocks.

In general, ensuring that all pullets are occupied during the day, as well as avoiding their pecking at each other, is very important to improve the welfare of pullets, and this can be achieved by adopting proper housing systems and using the correct stocking density.

3.3. Stocking Density

Stocking density can influence the health condition and adaptability of pullets during the rearing stage. Usually, decreased stocking density and the provision of proper environmental enrichment can reduce plumage damage. The maximum suitable density of 16-week-old laying pullets is 11–14 per m^2 without environmental enrichment, and it is important to adjust the density required by different breeds of chickens through behavioral observation [41]. In the presence of enrichment (pecking stones, pecking blocks, and alfalfa bales), the stocking density can be properly increased, and the increasing effect can be better than simply decreasing the stocking density (from 22 to 17 pullets per m^2) [42,43]. Crowded pullets exhibit more anxious behaviors and elevated corticosterone levels in their plasma and feathers, which could impair the adaptability of the pullets and cause long-term, adverse consequences [44].

In addition, studies have shown that different hens occupy different spaces when expressing different behaviors. On average, compared with a white hen, a brown hen needs 89.6 cm^2 more space when standing, 81.5 cm^2 more space when lying, 572 cm^2 more space when flapping, 170.3 cm^2 more space when dust bathing, and 3.38 cm more length while perching. Hens of all strains were wider when roosting than the recommended 15 cm per hen. Therefore, various factors (including breed type, body size, flock size, stocking density, environmental management and enrichment, facility layout, and synchronicity) should be comprehensively considered in the development of industry guidelines and regulations [45,46].

4. Environmental Management

Chickens are naturally disposed to fearfulness and sensitivity, whereas young red jungle fowls exhibit fear responses and flight behavior for a few days post-hatch [47]. The peak for fear responses exhibited by chicks is within 10 days of hatching, and this is due to their visual development and novelty evaluation [48,49]. Stress reduces feed intake and growth and impairs immune response and function, resulting in high disease susceptibility [50]. Therefore, appropriate environmental management in production systems is beneficial for alleviating fear and stress among chickens.

4.1. Lighting Management

Light is an important environmental factor which could influence the behavior, growth, productivity, and welfare of poultry via three characteristics based on the natural photoperiod: duration, intensity, and color/wavelength, with each consisting of multiple practices. This makes the use of an artificial lighting regimen a popular and complex option for manipulating laying hen production; however, few studies have been published on the important early stages of laying hens [51–53]. With the emergence of multiple housing types and the current accumulation of knowledge of chicken behavior, early light management of pullets is worth attempting, through adjustments of the three aspects with the concern of the chicken's welfare in mind.

4.1.1. Light Duration

Light duration is important for the growth, reproduction, and welfare of chickens. In poultry production, chickens are confronted with long-light phases. However, extremely long-light cycles are associated with reduced performance. In the European Union, an uninterrupted darkness of 8 h for laying hens is mandatory to maintain normal circadian rhythms and promote maximum rest because intermittent darkness may affect the rest and feeding of the chickens, resulting in a variety of metabolic and immune disorders [15,54]. Hence, keeping chickens under short-light conditions could lead to stronger responsiveness against bacterial infections and better responses to vaccinations [53], and decrease the risk of vent pecking [55]. Given the freedom to choose different light intensities (<1 lux–100 lux), W-36 laying hens (23–30 weeks) spent an accumulation of 10.0 h in darkness (<1 lux) per day, and dark hours were distributed intermittently throughout the day, which differed from the typical commercial practice of providing continuous dark periods for certain parts of the day (e.g., 8 h at night) [56].

On the contrary, pullets reared on long durations (14 h to 17 h) mature faster than birds reared on constant 10 h [57]. Introducing 2 h midnight lighting (ML, 2 h + 12 h) late in the growing period (12–18 wk) also induced early maturity and had the least egg production (302) from 18 to 70 wk, whereas providing 2 h ML from 0–18 wk resulted in the greatest number of eggs (317), and ML given only from 0–12 wk of age had the effect of delaying maturity and produced a middling number of eggs (310–312) [58]. However, the midnight lighting treatments had quite minor effects on the growth and feed intake of pullets according to another study by the same authors [59].

4.1.2. Light Intensity

Light intensity may affect laying hen behaviors and production performance. For caged, laying hen production, the dominant light intensity (LI) regime is 20 lux in the early stage, 5–10 lux in the growing stage, and 10–15 lux in the laying period [52].With the concerns of behavior and welfare, a study revealed that pullets spent more time preening at 50 lux than at 10 lux, spent more time wall pecking at 10 lux than at 50 lux, and had higher jumping frequency at 30 lux than at 10 lux [60]. When exposed to extremely high LI (500 lux [61], 121.8 lux [62]), layers produced smaller eggs in size and total egg mass, which indicated inadequate feed intake under high LI conditions. When exposed to low LI (1, 5, 11.9), a reduced rate of egg production resulted [61,62]. Given the freedom to choose LI, W-36 laying hens (23–30 weeks) generally spent more time in lower LI per day (an accumulation of 10.0 h at <1 lux, 6.4 h at 5 lux, 3.0 h at 15 lux, 3.1 h at 30 lux, and 1.5 h at 100 lux), and ingested the highest amount of feed at 5 lux (28.4 g/hen, 32.5% daily total) and the lowest amount at 100 lux (5.8 g/hen, 6.7%) [56]. The above studies revealed that pullets generally preferred a different intensity of light for different activities; and it suggested providing light intensities varying between 5 and 30 lux at different locations to achieve welfare–performance balance.

4.1.3. Light Color/Wavelength

Birds can perceive colored light (400–700 nm) as well as the ultraviolet (UV) portion of the spectrum (100–400 nm) due to the presence of external retinal cones in their eyes [63]. Light colors may affect poultry behaviors, well-being, and performance and produce different results. Current mainstream light-emitting diode (LED) lights are better than incandescent lamps and fluorescent light in energy savings, longevity, and other aspects [64]. With LED lamps widely applied in poultry housing systems, specific light colors (white, red, blue, green, or combinations) have been investigated as an additional management tool to achieve better performance and welfare of chickens [53].

Studies have confirmed that pullets exposed to longer wavelengths of light (LWL, red/yellow/orange) have increased egg production later in life compared to shorter wavelengths of light (SWL, blue/green) although responses may vary depending on bird strain and the intensity of the light used [63,65]. In addition, it was monitored that follicle-stimulating hormone (FSH) concentration, ovarian weight, and follicle number increased in hens raised under LWL [66]. The reason is that LWL contains more energy and can penetrate deeper tissues more easily, thus, releasing more gonadotropin-releasing hormone (GnRH) and FSH, increasing egg production. Some researchers have speculated that SWL plus higher intensities may produce the same effect as red light, which has been demonstrated to stimulate an increase of luteinizing hormone (LH) in quails [67].

It was reported that increasing the intensity at shorter wavelengths can compensate for differences in light-color effects. Moreover, SWL could reduce the activities and fear responses of chickens, with younger chickens being more sensitive to green light, whereas older chickens are more sensitive to blue light [68]. Moreover, providing layer pullets (54–82 d) freedom to stay under 4 LED color lights, they preferred blue light the most and red light the least [69], and most pullets preferred to drink under the blue and white lights [70]. This is understandable because chicks in their natural state often stay under hens' wings and rarely encounter red lights. (See the following section on the dark brooder). Based on the above studies, Wei et al. (2020) exposed chickens to combinations of LED lights (white light (400–700 nm, WL), blue/green (435–565 nm, BG), and yellow/orange (565–630 nm, YO), in the following patterns: BG at 1 D–13 wk + YO at 14–20 wk (BG–YO)) on pullets from 1 D to 20 wk, and revealed that, compared with the other treatments, the YO treatment significantly increased the bone-mineral density of the layer ($p < 0.05$) and reached 50% egg production age first; BG–YO treatment promoted the development of the sexual organs (oviducts and ovaries) of the laying hens at the age of 20 wk ($p < 0.05$); the BG–YO and BG treatment had higher serum Ig concentrations at 13 wk of age ($p < 0.05$) [65]. The results are consistent with the above studies, suggesting and demonstrating that appropriately using LED lights during brooding and rearing periods could have positive effects on the immune performance, bone development, and later production performance of pullets.

Currently, transforming the color of the light to red or dimming the light has been regarded as an effective method to alleviate the fear responses and reduce the risk of feather pecking in layers and mortality from cannibalism because it reduces the birds' ability to recognize blood and bare skin [71]. However, the red light may interfere with other wavelengths penetrating the eyes [72], or dim light may not provide enough stimulation to develop the reproduction system of the pullets. Therefore, red light is best used as a short-term, curative measure, and not as a long-term, preventive measure, while green or blue light could have some comforting effects based on the age of birds [68,73]. However, further studies are needed to confirm the impact of color LEDs as a production practice on pullets.

Chickens can sense UV light [74], and natural-light-reared pullets showed a preference for natural light [19] and natural-like light (white LEDs or red + green + blue LEDs to match the forest understory spectra, 4500 K, with UV) were associated with more active behaviors and better plumage in laying hens, compared with artificial commercial lighting (warm white LED, 3000 K, no UV) [75]. The benefits of UV light are mediated through its ability to activate cholecalciferol from 7-dehydroxycholesterol in the skin, resulting in improvements in eggshell quality [63]. However, providing sunlight in addition to standard lighting [76,77] or increasing UV illuminance during the laying period could increase the risk of pecking because of excessive light intensity [78]. However, lack of UV light can negatively affect basal corticosterone levels and exploration [79,80], encouraging feather damage and cannibalism [81].

In conclusion, light duration and intensity are the most important factors that impact the development, activity, production, and welfare of chickens. This is due to the fact that light intensity can compensate for short-wave lights (blue/green) to achieve similar effects of long-wave lights (red/yellow/orange). Pullets can achieve welfare–performance balance with a guaranteed 8–10 h of darkness, 5–30 lux light intensity exposed by natural-like or warm white LED (which varies by age, strain, and activities). Red or dim light is a short-term measure to prevent pecking addiction, whereas short-wave lights have some comforting benefits on pullets, but this requires further study.

4.2. Dark Brooder

In poultry production, chicks and hens are usually separated, leading to high levels of stress, fear, crowding, suffocation, injury, and pecking [82]; especially, young free-range flocks are prone to panic [83].

The presence of hens during the rearing period has an important influence on the behavioral development of chicks, such as increasing their foraging activity, reducing fear [84], and avoiding danger [85,86]. During the first week, the brooding hens pecked the ground four times as much as their chicks, presumably stimulating pecking behavior. Such pecking did not result in a higher feather pecking frequency because the chicks were guided by the hens to peck at materials such as feed and bedding materials [87]. Pullets with hen-care during their first 53 days of life are less fearful and more socially motivated between 14 and 29 weeks of age than non-brooded pullets [88]. In addition, brooding hens guide their chicks to roosting earlier (3.5 days earlier), which reduces the risk of pecking due to ground congestion [87].

Chicks are more comfortable when resting under the dark wings of hens, but the development of hen-raising systems that do not endanger animal health or efficiency is a challenge for research and industry. Hence, the creation of dark brooders (DB, mimicking hens) with heating and shading effects was introduced. Studies have shown that chicks of laying hens raised in DB were similar to the effect of brood hens [89,90], with significant synchronous activities (longer active period and resting period) [91], better feathers and skin conditions at 23 of weeks age (greatly reducing feather pecking frequency), and lower mortality [87,92]. Such effects have been reported in commercial chicken production, from 1 to 35 weeks old, and the pecking rate of DB chickens was significantly lower than that of the control group, while the feather condition was improved with no adverse effects on the growth rate, weight uniformity, or production performance [73,93].

In conclusion, dark brooders can better simulate the brooding effect of hens, reduce fear and pecking among chicks, have no adverse effects on production performance, and have good commercial applications.

4.3. Manure Management

Harmful gases inevitably accumulate in intensive chicken farms and interact with other environmental components, especially at the stage of sensitive chicks with reduced ventilation during the cold season [53]. Continuous exposure to high concentrations of harmful gases can impair the immune system of chickens (resulting in widespread and secondary infections) [94], behavioral capacity [77], production characteristics, and the death of chickens [95,96]. Ultimately, this could cause acidification and eutrophication of ecosystems, which could harm human beings [97].

Usually, the air pollution of chicken houses is mainly caused by NH_3 emission produced by urease degradation of uric acid in feces [98,99]. Therefore, manure management (especially ammonia control) is a major concern in the poultry industry. The European Union established Directive 2010/75/EU, and some developed countries have conducted inventories and measurements of odorants emitted from animal farms [100,101]. In addition, NH3 emissions usually increase under warm and wet conditions and worsen due to global warming [97].

Studies have shown that NH3 concentrations are generally high and exceed 20 ppm (the maximum allowable concentration in European countries) in litter-based housing types, including floor housing (66–120 ppm) and aviary systems (21–42 ppm), and low concentrations in furnished cages (3–12 ppm) [95,102]. Moreover, a study showed that human exposure to ammonia concentration associated with significant pulmonary function decrements was 12 ppm [103]. Therefore, under different circumstances, it is particularly important to maintain an ammonia concentration below 10–20 ppm in a chicken house [95,101]. Studies have revealed three main series of methods that can effectively reduce ammonia emissions in chicken farms: optimizing feed formula and additives to reduce uric acid excretion; proper drying, collection, or removal of poultry droppings from the chicken house in time to avoid uric acid decomposition; and centrally decomposing poultry droppings with manure additives.

Since uric acid excreted by birds is highly related to undigested proteins, it is essential to reduce the amount of uric acid in feces by avoiding overfeeding chickens with protein and improving their digestive capacity. Therefore, addition of essential amino acids (such as lysine and methionine) to substitute part of the protein, is a practical approach for reducing the protein level of diets. Furthermore, the addition of feed additives (probiotics) [104], wheat bran [105], and enzymes [106] (phytase, xylanase, and proteases) can promote feed digestibility.

To avoid accumulation and decomposition of feces in the chicken houses, a variety of methods for timely management of manure, such as the timely removal of feces by belts, scrapers, catchers, or other technological equipment [98]; replacing old litter with dry (dust-free), clean, fluffy, and mildew-free material; improving barn ventilation while maintaining an appropriate ambient temperature, or pump in heated air before ventilation in winter [97], and avoiding decomposition by keeping stacked manure dry in chicken houses (using fans, no leakage on waterline) [95].

Chicken manure is usually centrally collected and processed in special processing rooms, with continuous inoculation of probiotics to decompose feces at an appropriate temperature and humidity [107]. Moreover, spraying on or mixing additives (bentonite, sugarcane bagasse, and saline additives) with litter can reduce air pollution in chicken farms [108,109].

Therefore, manure management is an important factor in promoting quality air in the poultry industry, and an optimal solution can be found through comprehensive measures, such as optimizing feed formulation, manure ventilation, collection and removal facilities, and manure fermentation technology.

4.4. Complex Environment (CE)

In addition to the above environmental factors, there are other combinations of facilities that pullets have a high incentive to enjoy, collectively termed 'complex environments' (CE, with perches, litter, dark brooders, straw bales, slopes, platforms, outdoor access, etc.). CE has profound and long-lasting benefits for the welfare and stress adaptation of chickens, especially in the early stage. Studies have revealed that aviary-reared birds have low levels of fearfulness and use elevated areas of the pen more often compared with cage-reared birds [110], and CE (perches and dark brooder) birds had a higher resting behavior and more optimistic response (better resilience), approached ambiguous cues more quickly, and had lower heterophil/lymphocyte ratios after stressful challenges than birds reared in a simple environment [111]. Furthermore, when subjected to a predator test at 42 days of age, chicks (from one day old) reared in CE (perches and litter materials) were characterized by decreased fearfulness, lower plasma corticosterone, improved gut microbial functions, lower relative mRNA expression of GR, and elevated mRNA expressions of stress-related genes CRH, BDNF, and NR2A in the hypothalamus, compared to a litter-materials group or a barren environment group, thus enabling CE birds to comfortably cope with any future challenge [112].

Of all the CE, litter is considered unimportant, because it is prone to accumulating feces, as well as encouraging the spread of diseases, thereby prompting farmers to adopt the cage-rearing system. However, studies have shown that ectoparasite significantly increases preen behavior in caged chickens due to the lack of litter material, resulting in messy plumage, skin lesions, anemia, slow growth, lower egg production, and higher chances of being pecked [42,113,114]. Recently, studies have reported that changes in intestinal flora affect neurological diseases (anxiety, depression, etc.) through neural or hormonal pathways, and environmental enrichment in early life can affect adult behaviors, stress physiology [1,115], musculoskeletal and neurological development, health, egg quality, and other long-term benefits of pullets through the brain–gut axis, which could improve its response capacity in complex environments [116]. Therefore, litter material may have profound effects by establishing certain connections with nerves or hormones through the gut flora.

Chickens are selective about bedding material, depending on their physiological status and the behavior of their companions. They preferred materials such as peat, sand, and wood chips (which easily seeped into their feathers) for dustbathing, and preferred straws (long straws are better than short straws) for foraging [117,118]. Even in cage rearing, the provision of paper to chicks from day 1 led to less feather damage and fear at 30 wks than their counterparts [119,120]. Partially or completely removing chicken paper from the cage without providing other foraging materials caused the foraging behavior of the chicks to decrease, and the frequency of severe pecking to increase [121]. Therefore, the provision of appropriate litter materials and litter quality is of great significance for promoting foraging, dustbathing, and reducing pecking injuries among chickens.

5. Conclusions

In general, the welfare of pullets has been enhanced by several management practices, such as rearing pullets under cage-free systems, ensuring proper bird's status and proper flock size (40–500 birds), providing 8–10 h of darkness and 5–30 lux of different light-intensity exposure via natural-like LED lights, furnishing appropriate complex environments (litter, perches, dark brooders, straw bales, slopes, platforms, etc.), improving the air quality of the chicken farm through good manure management. However, using appropriate management systems to promote the welfare of birds is complex due to the costs of materials, different chicken strains, ages, and diverse activities, as well as different housing conditions and seasons in different geographical locations. Therefore, there is a need for intensive education and the development of affordable strategies that will aid chicken farmers to ensure proper management practices by making use of any available resources to meet the welfare requirements of pullets and improve productivity.

Author Contributions: X.D. conceptualized and drafted the manuscript. P.Q., Y.L., F.K.A., G.S., D.L. and X.Z. contributed to the literature review; X.D., F.K.A., P.Q., Y.L., G.S., D.L. and X.Z. critically reviewed, edited, and approved the final version of the manuscript. All authors have read and agreed to the published version of the manuscript.

Funding: This study was funded by the Key Research & Development Plan of the Department of Science and Technology of Tibet Autonomous Region (XZ202101ZY0002N); the Projects Funded by the Central Government to Guide Local Scientific and Technological Development from Guizhou province (QIANKEZHONGYINDI, 2021, 4003).

Institutional Review Board Statement: Not applicable.

Informed Consent Statement: Our review did not involve human patients.

Data Availability Statement: Not applicable.

Acknowledgments: We thank Tina Widowski and her research team at the University of Guelph who have impressed us with their excellent work in this area.

Conflicts of Interest: The authors declare that they have no competing interests.

References

1. Campderrich, I.; Nazar, F.N.; Wichman, A.; Marin, R.H.; Estevez, I.; Keeling, L.J. Environmental complexity: A buffer against stress in the domestic chick. *PLoS ONE* **2019**, *14*, e0210270. [CrossRef]
2. Rodenburg, T.B.; Van Krimpen, M.M.; De Jong, I.C.; De Haas, E.N.; Kops, M.S.; Riedstra, B.J.; Nicol, C.J. The prevention and control of feather pecking in laying hens: Identifying the underlying principles. *World's Poult. Sci. J.* **2013**, *69*, 361–374. [CrossRef]
3. de Haas, E.N.; Bolhuis, J.E.; Kemp, B.; Groothuis, T.G.; Rodenburg, T.B. Parents and early life environment affect behavioral development of laying hen chickens. *PLoS ONE* **2014**, *9*, e90577. [CrossRef] [PubMed]
4. Mellor, E.; Brilot, B.; Collins, S. Abnormal repetitive behaviours in captive birds: A Tinbergian review. *Appl. Anim. Behav. Sci.* **2018**, *198*, 109–120. [CrossRef]
5. Rubolini, D.; Romano, M.; Boncoraglio, G.; Ferrari, R.P.; Martinelli, R.; Galeotti, P. Effects of elevated egg corticosterone levels on behavior, growth, and immunity of yellow-legged gull (*Larus michahellis*) chicks. *Horm. Behav.* **2005**, *47*, 592–605. [CrossRef] [PubMed]
6. Henriksen, R.; Groothuis, T.G.; Rettenbacher, S. Elevated plasma corticosterone decreases yolk testosterone and progesterone in chickens: Linking maternal stress and hormone-mediated maternal effects. *PLoS ONE* **2011**, *6*, e23824.
7. Henriksen, R.; Rettenbacher, S.; Ton, G.G.G. Maternal corticosterone elevation during egg formation in chickens (*Gallus gallus domesticus*) influences offspring traits, partly via prenatal undernutrition. *Gen. Comp. Endocrinol.* **2013**, *191*, 83–91. [CrossRef]
8. Tauson, R. Management and housing systems for layers—Effects on welfare and production. *World's Poult. Sci. J.* **2005**, *61*, 477–490. [CrossRef]

9. Gunnarsson, S.; Keeling, L.J.; Svedberg, J. Effect of rearing factors on the prevalence of floor eggs, cloacal cannibalism and feather pecking in commercial flocks of loose housed laying hens. *Br. Poult. Sci.* **1999**, *40*, 12–18. [CrossRef]
10. Gunnarsson, S.; Yngvesson, J.; Keeling, L.J.; Forkman, B. Rearing without early access to perches impairs the spatial skills of laying hens. *Appl. Anim. Behav. Sci.* **2000**, *67*, 217–228. [CrossRef]
11. Mitchell, D.; Arteaga, V.; Armitage, T.; Mitloehner, F.; Tancredi, D.; Kenyon, N. Cage Versus Noncage Laying-Hen Housings: Worker Respiratory Health. *J. Agromed.* **2015**, *20*, 256–264. [CrossRef]
12. Regmi, P.; Smith, N.; Nelson, N.; Haut, R.C.; Orth, M.W.; Karcher, D.M. Housing conditions alter properties of the tibia and humerus during the laying phase in Lohmann white Leghorn hens. *Poult. Sci.* **2016**, *95*, 198–206. [CrossRef] [PubMed]
13. Neijat, M.; Casey-Trott, T.M.; Robinson, S.; Widowski, T.M.; Kiarie, E. Effects of rearing and adult laying housing systems on medullary, pneumatic and radius bone attributes in 73-wk old Lohmann LSL lite hens1. *Poult. Sci.* **2019**, *98*, 2840–2845. [CrossRef]
14. Tahamtani, F.M.; Hansen, T.B.; Orritt, R.; Nicol, C.; Moe, R.O.; Janczak, A.M. Does rearing laying hens in aviaries adversely affect long-term welfare following transfer to furnished cages? *PLoS ONE* **2014**, *9*, e107357.
15. Janczak, A.M.; Riber, A.B. Review of rearing-related factors affecting the welfare of laying hens. *Poult. Sci.* **2015**, *94*, 1454–1469. [CrossRef]
16. Pullin, A.N.; Temple, S.M.; Bennett, D.C.; Rufener, C.B.; Blatchford, R.A.; Makagon, M.M. Pullet Rearing Affects Collisions and Perch Use in Enriched Colony Cage Layer Housing. *Animals* **2020**, *10*, 1269. [CrossRef]
17. Brantsæter, M.; Tahamtani, F.M.; Moe, R.O.; Hansen, T.B.; Orritt, R.; Nicol, C. Rearing Laying Hens in Aviaries Reduces Fearfulness following Transfer to Furnished Cages. *Front. Vet. Sci.* **2016**, *3*, 13. [CrossRef]
18. Colson, S.; Arnould, C.; Michel, V. Influence of rearing conditions of pullets on space use and performance of hens placed in aviaries at the beginning of the laying period. *Appl. Anim. Behav. Sci.* **2008**, *111*, 286–300. [CrossRef]
19. Bari, M.S.; Laurenson, Y.C.; Cohen-Barnhouse, A.M.; Walkden-Brown, S.W.; Campbell, D.L. Effects of outdoor ranging on external and internal health parameters for hens from different rearing enrichments. *PeerJ* **2020**, *8*, e8720. [CrossRef] [PubMed]
20. Sherwin, C.M.; Nicol, C.J. A descriptive account of the pre-laying behaviour of hens housed individually in modified cages with nests. *Appl. Anim. Behav. Sci.* **1993**, *38*, 49–60. [CrossRef]
21. Ali, B.A.; Toscano, M.; Siegford, J.M. Later exposure to perches and nests reduces individual hens' occupancy of vertical space in an aviary and increases force of falls at night. *Poult. Sci.* **2019**, *98*, 6251–6262. [CrossRef]
22. MacLachlan, S.S.; Ali, A.B.A.; Toscano, M.J.; Siegford, J.M. Influence of later exposure to perches and nests on flock level distribution of hens in an aviary system during lay. *Poult. Sci.* **2019**, *99*, 30–38. [CrossRef]
23. Faure, J.M. Rearing conditions and needs for space and litter in laying hens. *Appl. Anim. Behav. Sci.* **1991**, *31*, 111–117. [CrossRef]
24. Craig, J.V.; Okpokho, N.A.; Milliken, G.A. Floor- and cage-rearing effects on pullets' initial adaptation to multiple-hen cages. *Appl. Anim. Behav. Sci.* **1988**, *20*, 319–333. [CrossRef]
25. Nicol, C.J.; Caplen, G.; Edgar, J.; Browne, W.J. Associations between welfare indicators and environmental choice in laying hens. *Anim. Behav.* **2009**, *78*, 413. [CrossRef]
26. Nicol, C.J.; Caplen, G.; Edgar, J.; Richards, G.; Browne, W.J. Relationships between multiple welfare indicators measured in individual chickens across different time periods and environments. *Anim. Welf.* **2011**, *20*, 133–143.
27. Grigor, P.N.; Hughes, B.O.; Appleby, M.C. Effects of regular handling and exposure to an outside area on subsequent fearfulness and dispersal in domestic hens. *Appl. Anim. Behav. Sci.* **1995**, *44*, 47–55. [CrossRef]
28. Milisits, G.; Szász, S.; Donkó, T.; Budai, Z.; Almási, A.; Pőcze, O. Comparison of Changes in the Plumage and Body Condition, Egg Production, and Mortality of Different Non-Beak-Trimmed Pure Line Laying Hens during the Egg-Laying Period. *Animals* **2021**, *11*, 500. [CrossRef]
29. Decina, C.; Berke, O.; van Staaveren, N.; Baes, C.F.; Widowski, T.M.; Harlander-Matauschek, A. An Investigation of Associations Between Management and Feather Damage in Canadian Laying Hens Housed in Furnished Cages. *Animals* **2019**, *9*, 135 [CrossRef] [PubMed]
30. Lambton, S.L.; Nicol, C.J.; Friel, M.; Main, D.C.J.; McKinstry, J.L.; Sherwin, C.M. A bespoke management package can reduce levels of injurious pecking in loose-housed laying hen flocks. *Vet. Rec.* **2013**, *172*, 423. [CrossRef]
31. Kaukonen, E.; Valros, A. Feather Pecking and Cannibalism in Non-Beak-Trimmed Laying Hen Flocks—Farmers' Perspectives. *Animals* **2019**, *9*, 43. [CrossRef] [PubMed]
32. Campderrich, I.; Liste, G.; Estevez, I. Group size and phenotypic appearance: Their role on the social dynamics in pullets. *Appl. Anim. Behav. Sci.* **2017**, *189*, 41–48. [CrossRef]
33. Buchwalder, T.; Huber-Eicher, B. Effect of group size on aggressive reactions to an introduced conspecific in groups of domestic turkeys (*Meleagris gallopavo*). *Appl. Anim. Behav. Sci.* **2005**, *93*, 251–258. [CrossRef]
34. Nicol, C.J.; Gregory, N.G.; Knowles, T.G.; Parkman, I.D.; Wilkins, L.J. Differential effects of increased stocking density, mediated by increased flock size, on feather pecking and aggression in laying hens. *Appl. Anim. Behav. Sci.* **1999**, *65*, 137–152. [CrossRef]
35. Bilcík, B.; Keeling, L.J. Relationship between feather pecking and ground pecking in laying hens and the effect of group size. *Appl. Anim. Behav. Sci.* **2000**, *68*, 55–66. [CrossRef]
36. Newberry, R.C.; Estevez, I.; Keeling, L.J. Group size and perching behaviour in young domestic fowl. *Appl. Anim. Behav. Sci.* **2001**, *73*, 117–129. [CrossRef]
37. Gilani, A.M.; Knowles, T.G.; Nicol, C.J. Factors affecting ranging behaviour in young and adult laying hens. *Br. Poult. Sci.* **2014**, *55*, 127–135. [CrossRef]

38. Hegelund, L.; Sørensen, J.T.; Kjaer, J.B.; Kristensen, I.S. Use of the range area in organic egg production systems: Effect of climatic factors, flock size, age and artificial cover. *Br. Poult. Sci.* **2005**, *46*, 1–8. [CrossRef]
39. Bestman, M.W.P.; Wagenaar, J.P. Farm level factors associated with feather pecking in organic laying hens. *Livest. Prod. Sci.* **2003**, *80*, 133–140. [CrossRef]
40. Keeling, L.J.; Newberry, R.C.; Estevez, I. Flock size during rearing affects pullet behavioural synchrony and spatial clustering. *Appl. Anim. Behav. Sci.* **2017**, *194*, 36–41. [CrossRef]
41. Spindler, B.; Clauss, M.; Briese, A.; Hartung, J. Planimetric measurement of floor space covered by pullets. *Berl. Munch. Tierarztl. Wochenschr.* **2013**, *126*, 156–162. [PubMed]
42. Liebers, C.J.; Schwarzer, A.; Erhard, M.; Schmidt, P.; Louton, H. The influence of environmental enrichment and stocking density on the plumage and health conditions of laying hen pullets. *Poult. Sci.* **2019**, *98*, 2474–2488. [CrossRef]
43. Zepp, M.; Schwarzer, A.; Helmer, F.; Louton, H.; Erhard, M.; Schmidt, P. The influence of stocking density and enrichment on the occurrence of feather pecking and aggressive pecking behavior in laying hen chicks. *J. Vet. Behav.* **2018**, *24*, 9–18. [CrossRef]
44. von Eugen, K.; Nordquist, R.E.; Zeinstra, E.; van der Staay, F.J. Stocking Density Affects Stress and Anxious Behavior in the Laying Hen Chick During Rearing. *Animals* **2019**, *9*, 53. [CrossRef]
45. Riddle, E.R.; Ali, A.B.A.; Campbell, D.L.M.; Siegford, J.M. Space use by 4 strains of laying hens to perch, wing flap, dust bathe, stand and lie down. *PLoS ONE* **2018**, *13*, e0190532. [CrossRef] [PubMed]
46. Zimmerman, P.H.; Lindberg, A.; Pope, S.J.; Glen, E.; Bolhuis, J.; Nicol, C.J. The effect of stocking density, flock size and modified management on laying hen behaviour and welfare in a non-cage system. *Appl. Anim. Behav. Sci.* **2006**, *101*, 111–124. [CrossRef]
47. Kruijt, J.P. *Ontogeny of Social Behaviour in Burmese Red Junglefowl (Callus Gallus Spadìceus) Bonnaterre*; University of Groningen: Groningen, The Netherlands, 1964; pp. ix+201.
48. Campbell, D.L.M.; de Haas, E.N.; Lee, C. A review of environmental enrichment for laying hens during rearing in relation to their behavioral and physiological development. *Poult. Sci.* **2019**, *98*, 9–28. [CrossRef]
49. Andrew, R.J.; Brennan, A. The lateralization of fear behaviour in the male domestic chick: A developmental study. *Anim. Behav.* **1983**, *31*, 1166–1176. [CrossRef]
50. Abo-Al-Ela, H.G.; El-Kassas, S.; El-Naggar, K.; Abdo, S.E.; Jahejo, A.R.; Al Wakeel, R.A. Stress and immunity in poultry: Light management and nanotechnology as effective immune enhancers to fight stress. *Cell Stress Chaperones* **2021**, *26*, 457–472. [CrossRef]
51. Ray, A.; Pradhan, R.K. Importance of Light in Poultry Industry. *Res. J. Sci. Technol.* **2012**, *4*, 172–177.
52. Manser, C.E. Effects of lighting on the welfare of domestic poultry: A review. *Anim. Welf.* **1996**, *5*, 341–360.
53. Hofmann, T.; Schmucker, S.S.; Bessei, W.; Grashorn, M.; Stefanski, V. Impact of Housing Environment on the Immune System in Chickens: A Review. *Animals* **2020**, *10*, 1138. [CrossRef] [PubMed]
54. Hieke, A.S.C.; Hubert, S.M.; Athrey, G. Circadian disruption and divergent microbiota acquisition under extended photoperiod regimens in chicken. *PeerJ* **2019**, *7*, e6592. [CrossRef]
55. Poetzsch, C.J.; Lewis, K.; Nicol, C.J.; Green, L.E. A cross-sectional study of the prevalence of vent pecking in laying hens in alternative systems and its associations with feather pecking, management and disease. *Appl. Anim. Behav. Sci.* **2001**, *74*, 259–272. [CrossRef]
56. Ma, H.; Xin, H.; Zhao, Y.; Li, B.; Shepherd, T.A.; Alvarez, I. Assessment of lighting needs by W-36 laying hens via preference test. *Animal* **2016**, *10*, 671–680. [CrossRef]
57. Lewis, P.D.; Morris, T.R.; Perry, G.C. A model for the effect of constant photoperiods on the rate of sexual maturation in pullets. *Br. Poult. Sci.* **1998**, *39*, 147–151. [CrossRef] [PubMed]
58. Leeson, S.; Caston, L.J.; Summers, J.D. Performance of layers given two-hour midnight lighting as growing pullets. *J. Appl. Poult. Res.* **2003**, *12*, 313–320. [CrossRef]
59. Leeson, S.; Caston, L.J.; Summers, J.D. Potential for midnight lighting to influence development of growing leghorn pullets. *J. Appl. Poult. Res.* **2003**, *12*, 306–312. [CrossRef]
60. Chew, J.A.; Widowski, T.; Herwig, E.; Shynkaruk, T.; Schwean-Lardner, K. The Effect of Light Intensity, Strain, and Age on the Behavior, Jumping Frequency and Success, and Welfare of Egg-Strain Pullets Reared in Perchery Systems. *Animals* **2021**, *11*, 3353. [CrossRef]
61. Renema, R.A.; Robinson, F.E.; Feddes, J.J.R.; Fasenko, G.M.; Zuidhoft, M.J. Effects of light intensity from photostimulation in four strains of commercial egg layers: 2. Egg production parameters. *Poult. Sci.* **2001**, *80*, 1121–1131. [CrossRef]
62. Kadir, E.; Sarıca, M.; Moise, N.; Dur, M.; Aslan, R. Effect of light intensity and stocking density on the performance, egg quality, and feather condition of laying hens reared in a battery cage system over the first laying period. *Trop. Anim. Health Prod.* **2021**, *53*, 320.
63. England, A.; Ruhnke, I. The influence of light of different wavelengths on laying hen production and egg quality. *World's Poult. Sci. J.* **2020**, *76*, 443–458. [CrossRef]
64. Liu, K.; Xin, H.; Settar, P. Effects of light-emitting diode light v. fluorescent light on growing performance, activity levels and well-being of non-beak-trimmed W-36 pullets. *Animal* **2018**, *12*, 106–115. [CrossRef] [PubMed]
65. Wei, Y.; Zheng, W.; Li, B.; Tong, Q.; Shi, H. Effects of a two-phase mixed color lighting program using light-emitting diode lights on layer chickens during brooding and rearing periods. *Poult. Sci.* **2020**, *99*, 4695–4703. [CrossRef] [PubMed]

66. Hassan, M.R.; Sultana, S.; Ho Sung, C.; Ryu, K.S. Effect of Monochromatic and Combined Light Colour on Performance, Blood Parameters, Ovarian Morphology and Reproductive Hormones in Laying Hens. *Ital. J. Anim. Sci.* **2013**, *12*, e56. [CrossRef]
67. Foster, R.G.; Follett, B.K. The involvement of a rhodopsin-like photopigment in the photoperiodic response of the Japanese quail. *J. Comp. Physiol. A* **1985**, *157*, 519–528. [CrossRef]
68. Sultana, S.; Hassan, M.R.; Choe, H.S.; Ryu, K.S. The effect of monochromatic and mixed LED light colour on the behaviour and fear responses of broiler chicken. *Avian Biol. Res.* **2013**, *6*, 207. [CrossRef]
69. Li, G.; Li, B.; Zhao, Y.; Shi, Z.; Liu, Y.; Zheng, W. Layer pullet preferences for light colors of light-emitting diodes. *Animal* **2019**, *13*, 1245–1251. [CrossRef]
70. Li, G.; Ji, B.; Li, B.; Shi, Z.; Zhao, Y.; Dou, Y.J. Assessment of layer pullet drinking behaviors under selectable light colors using convolutional neural network. *Comput. Electron. Agric.* **2020**, *172*, 105333. [CrossRef]
71. Shi, H.; Li, B.; Tong, Q.; Zheng, W.; Zeng, D.; Feng, G. Effects of LED Light Color and Intensity on Feather Pecking and Fear Responses of Layer Breeders in Natural Mating Colony Cages. *Animals* **2019**, *9*, 814. [CrossRef]
72. Lind, O.; Kelber, A. Avian colour vision: Effects of variation in receptor sensitivity and noise data on model predictions as compared to behavioural results. *Vis. Res.* **2009**, *49*, 1939–1947. [CrossRef] [PubMed]
73. Nicol, C.J.; Bestman, M.; Gilani, A.M.; De Haas, E.N.; De Jong, I.C.; Lambton, S. The prevention and control of feather pecking: Application to commercial systems. *World's Poult. Sci. J.* **2013**, *69*, 775–788. [CrossRef]
74. Osorio, D.; Vorobyev, M.; Jones, C.D. Colour vision of domestic chicks. *J. Exp. Biol.* **1999**, *202*, 2951–2959. [CrossRef]
75. Wichman, A.; De Groot, R.; Håstad, O.; Wall, H.; Rubene, D. Influence of Different Light Spectrums on Behaviour and Welfare in Laying Hens. *Animals* **2021**, *11*, 924. [CrossRef]
76. Bestman, M.; Koene, P.; Wagenaar, J.P. Influence of farm factors on the occurrence of feather pecking in organic reared hens and their predictability for feather pecking in the laying period. *Appl. Anim. Behav. Sci.* **2009**, *121*, 120–125. [CrossRef]
77. Drake, K.A.; Donnelly, C.A.; Dawkins, M.S. Influence of rearing and lay risk factors on propensity for feather damage in laying hens. *Br. Poult. Sci.* **2010**, *51*, 725–733. [CrossRef]
78. Spindler, B.; Weseloh, T.; Eßer, C.; Freytag, S.K.; Klambeck, L.; Kemper, N. The Effects of UV-A Light Provided in Addition to Standard Lighting on Plumage Condition in Laying Hens. *Animals* **2020**, *10*, 1106. [CrossRef]
79. Shimmura, T.; Suzuki, T.; Hirahara, S.; Eguchi, Y.; Uetake, K.; Tanaka, T. Pecking behaviour of laying hens in single-tiered aviaries with and without outdoor area. *Br. Poult. Sci.* **2008**, *49*, 396–401. [CrossRef]
80. Maddocks, S.A.; Cuthill, I.C.; Goldsmith, A.R.; Sherwin, C.M. Behavioural and physiological effects of absence of ultraviolet wavelengths for domestic chicks. *Anim. Behav.* **2001**, *62*, 1013–1019. [CrossRef]
81. Prescott, N.B.; Wathes, C.M.; Jarvis, J.R. Light, vision and the welfare of poultry. *Anim. Welf.* **2003**, *12*, 269–288.
82. Barrett, J.; Rayner, A.C.; Gill, R.; Willings, T.H.; Bright, A. Smothering in UK free-range flocks. Part 1: Incidence, location, timing and management. *Vet. Rec.* **2014**, *175*, 19. [CrossRef] [PubMed]
83. Richards, G.J.; Brown, S.N.; Booth, F.; Toscano, M.J.; Wilkins, L.J. Panic in free-range laying hens. *Vet. Rec.* **2012**, *170*, 519. [CrossRef] [PubMed]
84. Campo, J.L.; María García, G.; Sara García, D.V. Comparison of the tonic immobility duration, heterophil to lymphocyte ratio, and fluctuating asymmetry of chicks reared with or without a broody hen, and of broody and non-broody hens. *Appl. Anim. Behav. Sci.* **2014**, *151*, 61–66. [CrossRef]
85. Roden, C.; Wechsler, B. A comparison of the behaviour of domestic chicks reared with or without a hen in enriched pens. *Appl. Anim. Behav. Sci.* **1998**, *55*, 317–326. [CrossRef]
86. Riber, A.B.; Nielsen, B.L.; Ritz, C.; Forkman, B. Diurnal activity cycles and synchrony in layer hen chicks (*Gallus gallus domesticus*). *Appl. Anim. Behav. Sci.* **2007**, *108*, 276–287. [CrossRef]
87. Riber, A.B.; Wichman, A.; Braastad, B.O.; Forkman, B. Effects of broody hens on perch use, ground pecking, feather pecking and cannibalism in domestic fowl (*Gallus gallus domesticus*). *Appl. Anim. Behav. Sci.* **2007**, *106*, 39–51. [CrossRef]
88. Perre, Y.; Wauters, A.M.; Richard-Yris, M.A. Influence of mothering on emotional and social reactivity of domestic pullets. *Appl. Anim. Behav. Sci.* **2002**, *75*, 133–146. [CrossRef]
89. Riber, A.B.; Guzman, D.A. Effects of different types of dark brooders on injurious pecking damage and production-related traits at rear and lay in layers. *Poult. Sci.* **2017**, *96*, 3529–3538. [CrossRef]
90. Riber, A.B.; Guzman, D.A. Effects of Dark Brooders on Behavior and Fearfulness in Layers. *Animal* **2016**, *6*, 3. [CrossRef] [PubMed]
91. Rodenburg, T.B.; Komen, H.; Ellen, E.D.; Uitdehaag, K.A.; van Arendonk, J.A.M. Selection method and early-life history affect behavioural development, feather pecking and cannibalism in laying hens: A review. *Appl. Anim. Behav. Sci.* **2008**, *110*, 217–228. [CrossRef]
92. Jensen, A.B.; Palme, R.; Forkman, B. Effect of brooders on feather pecking and cannibalism in domestic fowl (*Gallus gallus domesticus*). *Appl. Anim. Behav. Sci.* **2006**, *99*, 287–300. [CrossRef]
93. Gilani, A.M.; Knowles, T.G.; Nicol, C.J. The effect of dark brooders on feather pecking on commercial farms. *Appl. Anim. Behav. Sci.* **2012**, *142*, 42–50. [CrossRef]
94. Chi, Q.; Chi, X.; Hu, X.; Wang, S.; Zhang, H.; Li, S. The effects of atmospheric hydrogen sulfide on peripheral blood lymphocytes of chickens: Perspectives on inflammation, oxidative stress and energy metabolism. *Environ. Res.* **2018**, *167*, 1–6. [CrossRef] [PubMed]

95. David, B.; Mejdell, C.; Michel, V.; Lund, V.; Moe, R.O. Air Quality in Alternative Housing Systems may have an Impact on Laying Hen Welfare. Part II—Ammonia. *Animals* **2015**, *5*, 886–896. [CrossRef] [PubMed]
96. Roque, K.; Shin, K.-M.; Jo, J.-H.; Kim, H.-A.; Heo, Y. Relationship between chicken cellular immunity and endotoxin levels in dust from chicken housing environments. *J. Vet. Sci.* **2015**, *16*, 173–177. [CrossRef] [PubMed]
97. Jiang, J.; Stevenson, D.S.; Uwizeye, A.; Tempio, G.; Sutton, M.A. A climate-dependent global model of ammonia emissions from chicken farming. *Biogeosciences* **2021**, *18*, 135–158. [CrossRef]
98. Shepherd, T.A.; Xin, H.; Stinn, J.P.; Hayes, M.D.; Zhao, Y.; Li, H. Ammonia and Carbon Dioxide Emissions of Three Laying-Hen Housing Systems as Affected by Manure Accumulation Time. *Trans. ASABE* **2017**, *60*, 229–236.
99. Zhao, Y.; Shepherd, T.A.; Li, H.; Xin, H. Environmental assessment of three egg production systems—Part I: Monitoring system and indoor air quality. *Poult. Sci.* **2015**, *94*, 518–533. [CrossRef]
100. German, G.S.; Karlis, P. *Developing EU Environmental Standards for the Food, Drink and Milk Industries: Key Environmental Issues and Data Collection*; Research Square: Durham, UK, 2020.
101. Matheus, D.O.; Fernanda, C.S.; Jair, O.S.; Arele, A.C.; Ilda Fátima Ferreira, T.; Antônio, P.S.C. Ammonia Emission in Poultry Facilities: A Review for Tropical Climate Areas. *Atmosphere* **2021**, *12*, 1091.
102. Nimmermark, S.; Lund, V.; Gustafsson, G.; Eduard, W. Ammonia, dust and bacteria in welfare-oriented systems for laying hens. *Ann. Agric. Environ. Med.* **2009**, *16*, 103–113.
103. Donham, K.J.; Cumro, D.; Reynolds, S.J.; Merchant, J.A. Dose-response relationships between occupational aerosol exposures and cross-shift declines of lung function in poultry workers: Recommendations for exposure limits. *J. Occup. Environ. Med.* **2000**, *42*, 260–269. [CrossRef] [PubMed]
104. Mahardhika, B.P.; Mutia, R.; Ridla, M. Effort to reduce ammonia gas in the broiler chicken excreta with the addition of probiotic as substitute for antibiotic growth promoter. *IOP Conf. Ser. Earth Environ. Sci.* **2021**, *883*, 012013. [CrossRef]
105. Such, N.; Csitári, G.; Stankovics, P.; Wágner, L.; Koltay, I.A.; Farkas, V. Effects of Probiotics and Wheat Bran Supplementation of Broiler Diets on the Ammonia Emission from Excreta. *Animals* **2021**, *11*, 2703. [CrossRef] [PubMed]
106. Sharma, N.K.; Choct, M.; Wu, S.; Swick, R.A. Nutritional effects on odour emissions in broiler production. *World's Poult. Sci. J.* **2017**, *73*, 257–280. [CrossRef]
107. Zheshi, K.; Jie, C.; Xiangjie, Z.; Xianli, L. Study Progress on Animal Feces Treatment by Microorganism. *Meteorol. Environ. Res.* **2014**, *5*, 56–58.
108. Vela-Aparicio, D.; Forero, D.F.; Hernández, M.A.; Brandão Pedro, F.B.; Cabeza, I.O. Simultaneous biofiltration of H_2S and NH_3 using compost mixtures from lignocellulosic waste and chicken manure as packing material. *Environ. Sci. Pollut. Res. Int.* **2021**, *28*, 24721–24730. [CrossRef]
109. Redding, M.R. Bentonite can decrease ammonia volatilisation losses from poultry litter: Laboratory studies. *Anim. Prod. Sci.* **2013**, *53*, 1115–1118. [CrossRef]
110. Brantsæter, M.; Nordgreen, J.; Rodenburg, T.B.; Tahamtani, F.M.; Popova, A.; Janczak, A.M. Exposure to Increased Environmental Complexity during Rearing Reduces Fearfulness and Increases Use of Three-Dimensional Space in Laying Hens (*Gallus gallus domesticus*). *Front. Vet. Sci.* **2016**, *3*, 14. [CrossRef]
111. Zidar, J.; Campderrich, I.; Jansson, E.; Wichman, A.; Winberg, S.; Keeling, L. Environmental complexity buffers against stress-induced negative judgement bias in female chickens. *Sci. Rep.* **2018**, *8*, 5404. [CrossRef]
112. Yan, C.; Hartcher, K.; Liu, W.; Xiao, J.; Xiang, H.; Wang, J.; Zhao, X. Adaptive response to a future life challenge: Consequences of early-life environmental complexity in dual-purpose chicks. *J. Anim. Sci.* **2020**, *98*, skaa348. [CrossRef]
113. Vestergaard, K.S.; Kruijt, J.P.; Hogan, J.A. Feather pecking and chronic fear in groups of red junglefowl: Their relations to dustbathing, rearing environment and social status. *Anim. Behav.* **1993**, *45*, 1127–1140. [CrossRef]
114. Murillo, A.C.; Abdoli, A.; Blatchford, R.A.; Keogh, E.J.; Gerry, A.C. Parasitic mites alter chicken behaviour and negatively impact animal welfare. *Sci. Rep.* **2020**, *10*, 8236. [CrossRef]
115. Mens, A.J.W.; van Krimpen, M.M.; Kwakkel, R.P. Nutritional approaches to reduce or prevent feather pecking in laying hens: Any potential to intervene during rearing? *World's Poult. Sci. J.* **2020**, *76*, 591–610. [CrossRef]
116. Ross, M.; Rausch, Q.; Vandenberg, B.; Mason, G. Hens with benefits: Can environmental enrichment make chickens more resilient to stress? *Physiol. Behav.* **2020**, *226*, 113077. [CrossRef]
117. Martin, C.D.; Mullens, B. Housing and dustbathing effects on northern fowl mites (*Ornithonyssus sylviarum*) and chicken body lice (*Menacanthus stramineus*) on hens. *Med. Vet. Entomol.* **2012**, *26*, 323–333. [CrossRef] [PubMed]
118. Huber-eicher, B.; Wechsler, B. The effect of quality and availability of foraging materials on feather pecking in laying hen chicks. *Anim. Behav.* **1998**, *55*, 861–873. [CrossRef]
119. Tahamtani, F.M.; Brantsaeter, M.; Nordgreen, J.; Sandberg, E.; Hansen, T.B.; Nodtvedt, A. Effects of litter provision during early rearing and environmental enrichment during the production phase on feather pecking and feather damage in laying hens. *Poult. Sci.* **2016**, *95*, 2747–2756. [CrossRef] [PubMed]
120. Brantsæter, M.; Tahamtani, F.M.; Nordgreen, J.; Sandberg, E.; Hansen, T.B.; Rodenburg, T.B. Access to litter during rearing and environmental enrichment during production reduce fearfulness in adult laying hens. *Appl. Anim. Behav. Sci.* **2017**, *189*, 49–56. [CrossRef]
121. Schreiter, R.; Damme, K.; von Borell, E.; Vogt, I.; Klunker, M.; Freick, M. Effects of litter and additional enrichment elements on the occurrence of feather pecking in pullets and laying hens—A focused review. *Vet. Med. Sci.* **2019**, *5*, 500–507. [CrossRef]

Article

The Effect of Housing System and Gender on Relative Brain Weight, Body Temperature, Hematological Traits, and Bone Quality in Muscovy Ducks

Ondřej Krunt [1], Adam Kraus [1], Lukáš Zita [1,*], Karolína Machová [2], Eva Chmelíková [3], Stanislav Petrásek [4] and Petr Novák [4]

[1] Department of Animal Science, Faculty of Agrobiology, Food and Natural Resources, Czech University of Life Sciences Prague, 165 00 Prague, Czech Republic; krunt@af.czu.cz (O.K.); krausa@af.czu.cz (A.K.)
[2] Department of Genetics and Breeding, Faculty of Agrobiology, Food and Natural Resources, Czech University of Life Sciences Prague, Kamýcká 129, 165 00 Prague, Czech Republic; machovakarolina@af.czu.cz
[3] Department of Veterinary Sciences, Faculty of Agrobiology, Food and Natural Resources, Czech University of Life Sciences Prague, Kamýcká 129, 165 00 Prague, Czech Republic; chmelikova@af.czu.cz
[4] Department of Agricultural Machines, Faculty of Engineering, Czech University of Life Sciences Prague, Kamýcká 129, 165 00 Prague, Czech Republic; petrasek@tf.czu.cz (S.P.); novakpetr@tf.czu.cz (P.N.)
* Correspondence: zita@af.czu.cz; Tel.: +420-224-383-053

Simple Summary: Free access to water with the possibility of swimming has the potential to be a good alternative to intensive housing of Muscovy ducks. The effect of this housing type was studied concerning hematological parameters, body temperature, relative brain weight, and bone quality. Birds with the possibility of swimming (S group) were compared to birds housed on deep litter with natural conditions (D group). Moreover, the effect of gender (G) was also studied. The housing of the birds had a significant effect on some hematological traits, body temperature, and relative brain weight. On the other hand, fracture toughness was not affected. Regarding the gender effect, it was found out that drakes had higher relative brain weight, lower body temperature, and higher fracture toughness of bones. These results help us understand the physiological and anatomical functioning of individual categories of animals monitored by us from a higher perspective with possible impacts on welfare and health.

Abstract: The study was conducted during the summer season (June–August 2020). Two hundred sixty-four 5-week-old sexed Muscovy ducklings were randomly divided into four equal experimental groups by housing system and by gender. Each group had three replicates (22 birds/replicate) in a randomized design experiment. Regarding the hematological traits, the volume of leukocytes was higher in the D group (by 0.34×10^9/L; $p < 0.05$) than in the S group. Furthermore, body temperature was found to be higher in ducks (by 0.84 °C; $p < 0.05$) and in the D group (by 0.5 °C; $p < 0.05$) in comparison with drakes and birds from the S group. Considering relative brain weight, drakes had higher values than ducks (by 0.56 g; $p < 0.05$), and birds from the S group also manifested higher values (by 0.78 g; $p < 0.05$). In terms of bone quality, there were no differences in studied parameters of tibia and femur bones regarding housing systems. The results provide valuable evidence of differences in the fattening of intensively bred Muscovy ducks within the housing system but also regarding gender.

Keywords: alternative housing; body temperature; bone quality; relative brain weight

Citation: Krunt, O.; Kraus, A.; Zita, L.; Machová, K.; Chmelíková, E.; Petrásek, S.; Novák, P. The Effect of Housing System and Gender on Relative Brain Weight, Body Temperature, Hematological Traits, and Bone Quality in Muscovy Ducks. *Animals* **2022**, *12*, 370. https://doi.org/10.3390/ani12030370

Academic Editors: Lilong Chai and Yang Zhao

Received: 18 January 2022
Accepted: 1 February 2022
Published: 3 February 2022

1. Introduction

Pekin, Muscovy, and Mule ducks are commonly reared for meat production in Europe. There are large differences in housing systems, which are related to different behavior and levels of welfare [1]. The tendencies of improving a duck's welfare are more and more common. Specifically, although outdoor runs were commonly used in the past [2], nowadays, different strategies of approving access to outdoor water for ducks are made to improve their health status or well-being [3]. Water provision in the form of swimming ponds allows ducks to manifest their species-specific behavior, such as dabbling, bathing, and swimming. Moreover, water effectively solves hygiene problems with dirty feathers and increases preening as the comfortable behavior of ducks [1]. Rearing ducks outdoor during the summer season can affect their body temperature, which can increase during hot days, or their blood profile [3]. The heat stress causes a reduction in feed intake and appetite and therefore compromises ducks' welfare. These negative aspects can be eliminated by enriching the duck's environment with access to water, which increases bird comfort [4]. The authors of [5] defined enrichment as an improvement in the biological functioning of captive animals resulting from modifications of their environment. Enrichment should be used for reducing negative emotional states such as fear, the stress associated with exposure to novel stimuli or boredom, and apathy from inappropriate housing. Moreover, environmental stimuli were found to increase the brain weight of rabbit males [6], probably due to stimulating neurogenesis in the hippocampus [7] or the higher energy requirements of animals [8]. Changes in brain size can also be supported by the expensive-tissue hypothesis, which predicts that the higher the brain size, the lower the size of another costly organ, such as the gut or others [9]. Additionally, the ability to learn tasks is a stimulus of increasing brain size [10]. Another factor that influences the well-being of animals, is bone quality, which can be affected by movement. In the study of [11], it was found that hens that were housed in floor systems with an increased possibility of movement had higher fracture toughness than hens in cages. In addition, the importance of the gender effect should not be overlooked. The differences between genders were described in intensively reared broiler chickens [12,13] due to hormonal differences [14]. Our overarching hypothesis is that free access to swimming ponds during the summer season improves Muscovy ducks' welfare. Specifically, we hypothesize that outdoor runs with the possibility of swimming will increase the relative brain weight of birds. We also hypothesize the possibility of swimming will reduce body temperature, positively change blood profile, and improve the bone quality of birds. We further hypothesize that drakes will have higher relative brain weight, lower body temperature, changed blood profile, and higher fracture toughness of bones than ducks.

2. Materials and Methods

2.1. Animals and Husbandry

The present study was approved by the ethics committee of the Czech University of Life Sciences Prague (case number, 07/2020). The study was conducted during the summer season (June–August 2020). Two hundred and sixty-four 5-week-old sexed Muscovy ducklings were randomly divided into 4 equal experimental groups by housing system and by gender (female/deep litter, male/deep litter, female/swimming pond, male/swimming pond). Each group (66 birds per group and gender) had three replicates (22 birds/replicate). Birds were housed in a close-sided house on deep litter (D) with regard to gender and in an open-sided house with free access to a swimming pond (S) with regard to gender under natural conditions. On average, the length of the day was 16 h and that of the night 8 h. Moreover, the average temperatures were: 17.9 °C (June), 18.9 °C (July), and 20.3 °C (August). Housing systems for S groups included trees near the swimming ponds, which provided shadow during hot summer days. All animals were reared under the same conditions. Wheat straw was used as deep litter in every housing system. Each group was kept at density of 4 ducks per m^2. Moreover, group S had a swimming pond (10 m length $\times$ 6 m width $\times$ 3 m depth with concrete floor) at its disposal. Fresh water was provided into the pond from supply channels. All birds were fed ad libitum with a pelleted diet (20% CP and 11.2 MJ/kg ME) and had water fully available in their housing system. At the end of the experiment (14 weeks of age), all birds were slaughtered by jugular venesection after 12 h fasting.

2.2. Measurements of Hematological Parameters

Blood samples from 9 animals (14 weeks of age) from each group and replicates were taken during slaughtering in sterile syringes from the jugular vein. Samples were centrifuged at 2.328× *g* for 15 min at 4 °C to collect serum. After obtaining whole blood samples, blood films were made using the slide method of [15]. Blood films were stained using Pappenheim May–Grunwald Giemsa stain. A differential number of leukocytes was made of three horizontal edge fields followed by two fields towards the center. They were followed by two fields in a horizontal direction and after that by two fields in a vertical direction to obtain the edge again. The field takes a crisscross shape with right angles. Two hundred cells, with 100 cells on each edge of the film, were differentiated, and the percentages of heterophils and lymphocytes were calculated. Erythrocytes were determined manually by hemocytometer. Blood hemoglobin (Hb) was determined according to [16]. Mean corpuscular volume (MCV) and mean corpuscular hemoglobin concentration (MCHC) were calculated according to [17]. The H/L ratios were determined according to the formula:

$$H/L = \text{heterophiles}/\text{lymphocytes}.$$

2.3. Body Temperature

The body temperature from 9 animals from each group and replicate was recorded from rectum by thermometer (TH—802, OEM brands, CE ISO FDA, Guangdong, China) once a week on Wednesday from 12:00 until 12:30. All birds were gently treated, and the thermometer was tenderly inserted into the rectum to 2 cm depth. The temperature was recorded after the alarm signal (usually after 45 s). Each animal was processed for less than 1.5 min.

2.4. Relative Brain Weight

At 14 weeks of age, all birds (66 birds/group/gender) were slaughtered by cutting the jugular vein. All brains were removed according to the methods of Bozicovich et al. [6] by cutting the frontal bone with stainless steel scissors, and they were weighed on an analytical scale Ohaus (Model: Traveler TA502, Parsippany, NJ 07054) with 0.01 g precision. Housing system and gender averages were used in the analyses.

2.5. Bone Quality Characteristics, Sampling, and Analyses

The raw bones from 9 animals from each group and replicate were examined for weight, length, width and fracture toughness. Bones were weighed on an analytical scale Ohaus (Model: Traveler TA502, Parsippany, NJ, USA, 07054), and diameter was measured by dial caliper (±0.02 mm) at the mid-diaphysis, where the breaking point was. Femur and tibia bones of the left hind leg were taken at slaughter and individually packed in polyethylene bags and stored at −20 °C until the analysis, when they were thawed overnight. When fully defrosted, soft tissue was removed from the tibia and femur. The length of the tibia/femur was measured as the distance from tibia/femur spine to inferior articular surface by dial caliper (±0.02 mm). The tibia and femur were subsequently boiled for 15 min in 95 °C water, de-fleshed and de-fatted, and dried at 25 °C for 24 h. The breaking strength was determined with a three-point flexure test using a Instron® Model 3342 (Instron, Norwood, MA, USA), and the load rate was 30 mm/min. The space between the two fulcra points supporting the bones was 45 and 38 mm. The bones were continually oriented for examination with their natural convex shape downwards.

Statistical Analyses

The effect of gender and housing system on each hematological trait, body temperature, relative brain weight, and bone quality parameters was assessed by the mixed model using the MIXED procedure of SAS (SAS Institute Inc., Cary, NC, USA, 2011):

$$yijk = \mu + HSi + Gj + (HS \times G)\ ij + eijk,$$

where yijk is the value of trait, μ is the overall mean, HSj is the effect of the housing system, Gi is the effect of gender (HS × G) ij is the effect of the interaction between housing system and gender, and eijk is the random residual error. The significance of the differences among groups was tested by Duncan's multiple range test. The value of $p \leq 0.05$ was considered as significant for all measurements.

3. Results

3.1. Relative Brain Weight

The effect of housing system and gender of Muscovy ducks on relative brain weight is displayed in Figure 1. Considering relative brain weight, differences were between gender (by 0.56 g; $p < 0.05$) and housing system (by 0.78 g; $p < 0.05$).

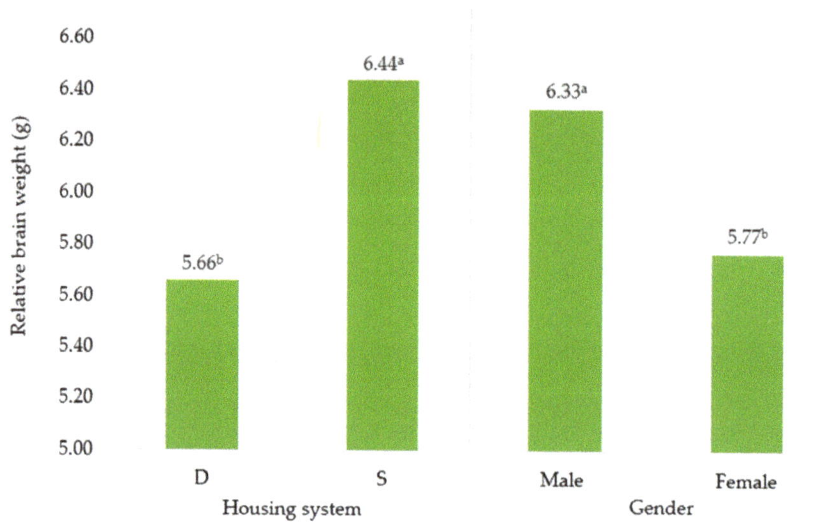

Figure 1. Effect of housing system and gender on relative brain weight (g). D = deep litter; S = swimming ponds. [a,b] Different superscripts indicate significant differences between means ($p \leq 0.05$).

3.2. Hematological Parameters and Body Temperature

The results concerning the hematological traits of birds are presented in Table 1, and body temperature values are displayed in Figure 2. Statistically significant interactions are discussed in detail in the text but not described in tables. Regarding the hematological traits, volume of leukocytes was higher in the D group (by $0.34 \times 10^9/L$; $p < 0.05$). Moreover, significant interaction between HS and G was found for leukocytes, where the highest values had drakes from the D group ($25.07 \times 10^9/L$; $p < 0.05$) and the lowest values had ducks from the S group, the D group, and drakes from the S group (24.58, 24.42, and $24.21 \times 10^9/L$; $p < 0.05$, respectively). Values of lymphocytes tended to be higher also in the D group in comparison with the S group). Furthermore, body temperature was found to be higher in ducks (by 0.84 °C; $p < 0.05$) and in the D group (by 0.5 °C; $p < 0.05$).

Table 1. The effect of housing system and gender on hematological traits.

Traits	Housing System (HS)		Gender (G)		SEM	*p*-Value		
	D	S	Male	Female		HS	G	HS × G
Hematocrit (%)	41.80	42.11	41.12	42.81	0.009	0.8620	0.3514	0.1102
Hemoglobin (g/L)	133.06	134.61	131.34	136.3	2.886	0.7896	0.3905	0.1135
Erythrocytes (10^{12}/L)	2.90	2.93	2.86	2.97	0.063	0.7813	0.3965	0.1160
Leukocytes (10^{9}/L)	24.74 ^{a}	24.40 ^{b}	24.64	24.51	0.096	0.0445	0.4113	0.0040
Heterophiles (10^{9}/L)	10.82	11.10	11.06	10.87	0.109	0.1897	0.3628	0.0888
Lymphocytes (10^{9}/L)	15.67	15.30	15.46	15.12	0.010	0.0512	0.7607	0.4484
H/L ratio	0.69	0.72	0.71	0.72	0.018	0.6541	0.4589	0.3461
MCV (fL)	144.21	143.65	143.67	144.18	0.213	0.1986	0.2440	0.8395
MCHC (g/L)	318.34	319.42	319.37	318.40	0.477	0.2709	0.3233	0.9631

D = deep litter; S = swimming ponds; MCV = mean corpuscular volume; MCHC = mean corpuscular hemoglobin concentration. ^{a,b} Different superscripts within a row indicate significant differences between means ($p \leq 0.05$).

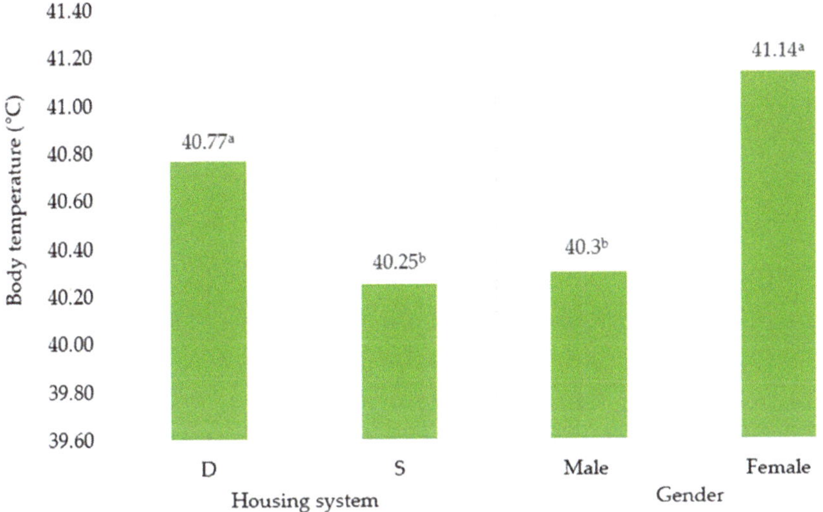

Figure 2. Effect of housing system and gender on body temperature (°C). D = deep litter; S = swimming ponds. ^{a,b} Different superscripts indicate significant differences between means ($p \leq 0.05$).

3.3. Bone Quality

In terms of bone quality (Table 2), there were no differences in studied parameters of tibia and femur bones regarding to housing systems. There was found just a tendency of higher weight in femur bones in favor of the D group (by 0.32 g; $p < 0.05$). On the contrary, a significant effect of gender was found in every evaluated parameter of both bones due to sexual dimorphism between the gender in favor of drakes.

Table 2. The effect of housing system and gender on some functional parameters of tibia and femur bones.

Traits	Housing System (HS)		Gender (G)		SEM	p-Value		
	D	S	Male	Female		HS	G	HS × G
tibia								
Length (mm)	112.32	114,09	124.99 [a]	102.21 [b]	2.125	0.2826	0.0001	0.2604
Width (mm)	8.03	8.05	9.20 [a]	6.95 [b]	0.220	0.7561	0.0001	0.5233
Weight (g)	10.64	10.63	14.65 [a]	6.88 [b]	0.721	0.3454	0.0001	0.5233
Fracture toughness (N/cm^2)	386.29	379.34	488.04 [a]	283.95 [b]	19.925	0.3564	0.0001	0.9206
femur								
Length (mm)	69.26	69.78	75.00 [a]	62.04 [b]	1.380	0.4303	0.0001	0.8159
Width (mm)	9.34	9.40	10.62 [a]	8.13 [b]	0.247	0.7647	0.0001	0.5152
Weight (g)	7.90	7.58	10.44 [a]	5.04 [b]	0.496	0.1008	0.0001	0.5542
Fracture toughness (N/cm^2)	372.22	361.63	450.17 [a]	283.68 [b]	17.716	0.5925	0.0001	0.3345

D = deep litter; S = swimming ponds. [a,b] Different superscripts within a row indicate significant differences between means ($p \leq 0.05$).

4. Discussion

4.1. Relative Brain Weight

Considering relative brain weight, differences between gender could seem expected due to sexual dimorphism between ducks and drakes, which are substantially larger than ducks [1]. However, the results may have a different reason. In humans, there was confirmation of the sex difference in adult brain size [18] or in 18-year-old students' brains [19]. Pakkenberg and Gundersen [20] explained the differences using a higher number (4 billion more) of cortical neurons in men. More exceptional are the differences in groups, which were housed differently. In total, birds housed in a system with a swimming pond had heavier brains than birds from the deep litter. According to the scientific literature, there are several studies that have found differences in relative brain weight among the birds. For example, parrots have larger brains as a response to higher seasonality and precipitation [21]. Passerine birds have larger brains when they experience higher environmental variation by migrating [22]. Unfortunately, there are no studies of intensively housed ducks or geese that considered relative brain weight as a possible aspect of physiological or mental state. The reason could be linked with higher interaction with the environment and connected exploratory behavior, since providing environmental stimuli influences brain weight in mice [10] and in male rabbits [6]. The authors of the latter study postulate that environmental enrichment promotes the development of specific regions in rabbits' brains. They base their claims on the conclusions of the study of [23], which found increased cortical thickness and enhanced dendritic ramification in the brains of rodents, which were exposed to the presence of environmental enrichment. In addition, our hypothesis could be supported by the findings of [24], which reported significant decreases in absolute and relative brain volume in captive-bred waterfowl compared to their wild counterparts. Additionally, the decrease in brain size in domesticated waterfowls in comparison with wild animals is generally attributed to a decrease in functional requirements resulting from the unnatural environment. Our results could indicate that the environment had a strong influence on brain development in Muscovy ducks.

4.2. Hematological Parameters and Body Temperature

The current study results show a decrease in leukocytes and lymphocytes in the S group compared to the D group, which could be attributed to potential non-specific immune response induced by heat [25], which probably acted more in deep litter housing. On the contrary, in the study of [3], it was reported that the highest volume of lymphocytes was in birds with the longest time with access to water. The body temperature of birds is also related to the time spent on the water. Animals from the S group had lower body temperature than those from the D group. This trend was also confirmed by [26] and by [3] with the explanation of cooling the body due to better evaporation. More interesting are the differences between ducks and drakes, which had lower body temperatures than ducks. Pis [27] mentioned the link between metabolic rate and body temperature in galliform birds, that body temperature is associated with the large variability because of gender, season, and measurement conditions and therefore resulted in "unpopularity". The effect of gender on body temperature was previously studied in mice. The study of [28] reported the importance of body temperature measurements at the same time of the day when it is performed. This is consistent with our methodology of measuring body temperature at the same time of the day. Findings of this study reported that female mice had higher body temperature than males, with a possible effect on lifespan. The activity of hormones is suggested as one possible explanation for the higher body temperature in females. In general, progesterone promotes less vasodilatation, heat conservation, and higher values of body temperature in women [29]. In poultry, the effect of gender on body temperature was also confirmed in Japanese quails. Female quails had higher body temperature than male quails [30]. It is very difficult to explain the differences between genders, but another possible connection to consider could be the basal metabolic rate, which was linked with increased capacity of heat production [31].

4.3. Bone Quality

In general, bone fracture toughness can reflect the welfare levels of animals in their housing system. Fractures of keel bones are a real problem in rearing systems of intensive laying hens [32]. There were also found to be differences in fracture toughness due to higher movement in tibia bones of hens that were housed in flat floors (these hens had bones more resistant to fracture) than of hens in cages [11]. Results of these studies should mean that ducks that will have a greater ability to move will also have bones that are more resistant to fracture. On the other hand, our results show that birds from both systems did not differ in these terms. In the end, this is good information, because we can summarize that no-swimming housing conditions did not decrease bone strength or vice versa. Considering the gender effect, which was significant in all parameters of tibia and femur bones, the length, width, and weight of these bones were expectably higher in drakes due to their weight dimorphism in general. The fracture toughness was probably also affected by the same factor, which was mentioned in the previous statement. It was higher in drakes due to greater width or weight, which should indicate the higher content of elements that affected bone strength [33].

5. Conclusions

In conclusion, free access to water with the possibility of swimming had an effect on leukocytes and a positive effect on body temperature. Moreover, relative brain weight was strongly influenced by housing, whereas bone quality did not differ. With regard to gender, no effect on hematological traits was found, whereas body temperature was significantly higher in ducks when compared to drakes. Additionally, ducks had a lower relative brain weight than drakes. Nevertheless, according to the bone quality analyses, drakes had higher values of every single parameter than ducks. Our results provide valuable evidence of differences in the fattening of intensively bred Muscovy ducks within the housing system, but also regarding gender. These results reveal the physiological and anatomical functioning of individual categories of animals monitored by us from a higher perspective with possible impacts on welfare and health.

Author Contributions: Conceptualization, O.K.; methodology, L.Z., O.K. and A.K.; software, L.Z.; validation, K.M. and S.P.; formal analysis, O.K. and A.K.; laboratory analysis., O.K., E.C. and A.K.; investigation, A.K., O.K. and L.Z..; resources, S.P. and P.N.; data curation, K.M., O.K. and A.K.; writing—original draft preparation, O.K.; writing—review and editing, O.K., A.K., L.Z., S.P., K.M., E.C. and P.N.; visualization, O.K. and A.K.; supervision, L.Z.; project administration, K.M..; funding acquisition, S.P. All authors have read and agreed to the published version of the manuscript.

Funding: This research was funded by an "S" grant from the Ministry of Education, Youth and Sports of the Czech Republic, Project No. SV21-6-21320 and IGA 2021:31160/1312/3116.

Institutional Review Board Statement: The welfare of the ducks was carefully considered during the whole experiment. The animals were not subjected to pain, suffering, distress, or lasting harm. Feed and water were provided ad libitum. The whole study was carried out in harmony with the guidelines of Act No. 246/1992, which directs the protection against animal cruelty. The present study was approved by the ethics committee of the Czech University of Life Sciences Prague (case number, 07/2020).

Informed Consent Statement: Not applicable.

Data Availability Statement: The data presented in this study are available by reasonable request from the corresponding author.

Acknowledgments: We would like to thank Richard Hardy for the English language corrections.

Conflicts of Interest: The authors declare no conflict of interest. The funders had no role in the design of the study; in the collection, analyses, or interpretation of data; in the writing of the manuscript; or in the decision to publish the results.

References

1. Rodenburg, T.B.; Bracke, M.B.M.; Berk, J.; Cooper, J.; Faure, J.M.; Guémené, D.G.U.Y.; Ruis, M.A.W. Welfare of ducks in European duck husbandry systems. *Worlds Poult. Sci. J.* **2005**, *61*, 633–646. [CrossRef]
2. Reiter, K.; Zernig, F.; Bessei, W. Effect of water bath and free range on behaviour and feathering in Pekin, Muscovy, and Mulard duck. In Proceedings of the 11th European Symposium on Waterfowl, Nantes, France, 8–10 September 1997; pp. 224–229.
3. Farghly, M.F.; Mahmoud, U.T. Access to outdoor swimming pond during summer season improved Muscovy ducks performance and health status. *Livest. Sci.* **2018**, *211*, 98–103. [CrossRef]
4. Blokhuis, H.J.; Van Niekerk, T.F.; Bessei, W.; Elson, A.; Guémené, D.; Kjaer, J.B.; Van De Weerd, H.A. The LayWel project: Welfare implications of changes in production systems for laying hens. *Worlds Poult. Sci. J.* **2007**, *63*, 101–114. [CrossRef]
5. Newberry, R.C. Environmental enrichment: Increasing the biological relevance of captive environments. *Appl. Anim. Behav. Sci.* **1995**, *44*, 229–243. [CrossRef]
6. Bozicovich, T.F.; Moura, A.S.A.; Fernandes, S.; Oliveira, A.A.; Siqueira, E.R.S. Effect of environmental enrichment and composition of the social group on the behavior, welfare, and relative brain weight of growing rabbits. *Appl. Anim. Behav. Sci.* **2016**, *182*, 72–79. [CrossRef]
7. Kempermann, G.; Kuhn, H.G.; Gage, F.H. More hippocampal neurons in adult mice living in an enriched environment. *Nature* **1997**, *386*, 493–495. [CrossRef]
8. Allman, J. *Evolving Brains*; Scientific American Library: New York, NY, USA, 2000.
9. Liu, Y.T.; Luo, Y.; Gu, J.; Jiang, S.; Liao, W.B. The relationship between brain size and digestive tract length do not support expensive-tissue hypothesis in *Hylarana guentheri*. *Acta Herpetol.* **2018**, *13*, 141–146. [CrossRef]
10. Van Praag, H.; Kempermann, G.; Gage, F.H. Neural consequences of enviromental enrichment. *Nat. Rev. Neurosci.* **2000**, *1*, 191–198. [CrossRef]
11. Qiaoxian, Y.; Hui, C.; Yingjue, X.; Chenxuan, H.; Jianzhong, X.; Rongyan, Z.; Lijun, X.; Han, W.; Ye, C. Effect of housing system and age on products and bone properties of Taihang chickens. *Poult. Sci.* **2020**, *99*, 1341–1348. [CrossRef]
12. Venäläinen, E.; Valaja, J.; Jalava, T. Effects of dietary metabolisable energy, calcium and phosphorus on bone mineralisation, leg weakness and performance of broiler chickens. *Brit. Poult. Sci.* **2006**, *47*, 301–310. [CrossRef]
13. Tasoniero, G.; Cullere, M.; Baldan, G.; Dalle Zotte, A. Productive performances and carcase quality of male and female Italian Padovana and Polverara slow-growing chicken breeds. *Ital. J. Anim. Sci.* **2018**, *17*, 530–539. [CrossRef]
14. Rath, N.C.; Balog, J.M.; Huff, W.E.; Huff, G.R.; Kulkarni, G.B.; Tierce, J.F. Comparative differences in the composition and bio-mechanical properties of tibiae of seven-and seventy-two-week-old male and female broiler breeder chickens. *Poult. Sci.* **1999**, *78*, 1232–1239. [CrossRef]
15. Schalm, O.W.; Jain, N.C.H.; Carroll, E.J. *Veterinary Hematology*; Lea & Febiger: Philadelphia, PA, USA, 1975.
16. Dukes, H.H.; Schwarte, L.H. The hemoglobin content of the blood of fowls. *Am. J. Physiol. Cont.* **1931**, *96*, 89–93. [CrossRef]
17. Al-Daraji, H.J.; Al-Hayani, W.K.; Al-Hassani, A.S. *Avian Hematology*; Ministry of Higher Education and Scientific Research: Cairo, Egypt; College of Agriculture, University of Baghdad: Baghdad, Iraq, 2008.
18. Good, C.D.; Johnsrude, I.S.; Ashburner, J.; Henson, R.N.A.; Friston, K.J.; Frackowiak, R.S.J. A voxel-based morphometric study of ageing in 465 normal adult brains. *NeuroImage* **2001**, *14*, 21–36. [CrossRef]
19. Ivanovic, D.M.; Leiva, B.P.; Castro, C.G.; Olivares, M.G.; Jansana, J.M.M.; Castro, V.G. Brain development parameters and intelligence in Chilean high school graduates. *Intelligence* **2004**, *32*, 461–479. [CrossRef]
20. Pakkenberg, B.; Gundersen, H.J.G. Neocortical neuron number in humans: Effects of sex and age. *J. Comp. Neurol.* **1997**, *384*, 312–320. [CrossRef]

21. Schuck-Paim, C.; Alonso, W.J.; Ottoni, E.B. Cognition in an ever-changing world: Climatic variability is associated with brain size in neotropical parrots. *Brain Behav. Evol.* **2008**, *71*, 200–215. [CrossRef]
22. Sol, D.; Lefebvre, L.; Rodríguez-Teijeiro, J.D. Brain size, innovative propensity and migratory behaviour in temperate Palaearctic birds. *Proc. R. Soc. B Biol. Sci.* **2005**, *272*, 1433–1441. [CrossRef]
23. Rosenzweig, M.R.; Bennett, E.L. Psychobiology of plasticity: Effects of training and experience on brain and behavior. *Behav. Brain Res.* **1996**, *78*, 57–65. [CrossRef]
24. Guay, P.J.; Iwaniuk, A.N. Captive breeding reduces brain volume in waterfowl (Anseriformes). *Condor* **2008**, *110*, 276–284. [CrossRef]
25. Kamel, N.N.; Ahmed, A.M.; Mehaisen, G.M.; Mashaly, M.M.; Abass, A.O. Depression of leukocyte protein synthesis, immune function and growth performance induced by high environmental temperature in broiler chickens. *Int. J. Biometeorol.* **2017**, *61*, 1637–1645. [CrossRef] [PubMed]
26. El-Shafaei, H.E.; Sharaf, M.M.; Rashed, R.R. The effect of different intervention strategies to alleviate heat stress on behavior, performance and some blood parameters of growing Muscovy ducks. *Alex. J. Vet. Sci.* **2016**, *48*, 69–76. [CrossRef]
27. Pis, T. The link between metabolic rate and body temperature in galliform birds in thermoneutral and heat exposure conditions: The classical and phylogenetically corrected approach. *J. Therm. Biol.* **2010**, *35*, 309–316. [CrossRef]
28. Sanchez-Alavez, M.; Alboni, S.; Conti, B. Sex-and age-specific differences in core body temperature of C57Bl/6 mice. *AGE* **2011**, *33*, 89–99. [CrossRef]
29. Charkoudian, N.; Hart, E.C.; Barnes, J.N.; Joyner, M.J. Autonomic control of body temperature and blood pressure: Influences of female sex hormones. *Clin. Auton. Res.* **2017**, *27*, 149–155. [CrossRef]
30. Gilbreath, J.C.; Ko, R.C. Sex differential for body temperature in Japanese quail. *Poult. Sci.* **1970**, *49*, 34–36. [CrossRef]
31. Chamane, S.C.; Downs, C.T. Seasonal effects on metabolism and thermoregulation abilities of the red-winged starling (*Onychognathus morio*). *J. Therm. Biol.* **2009**, *34*, 337–341. [CrossRef]
32. Casey-Trott, T.M.; Guerin, M.T.; Sandilands, V.; Torrey, S.; Widowski, T.M. Rearing system affects prevalence of keel-bone damage in laying hens: A longitudinal study of four consecutive flocks. *Poult. Sci.* **2017**, *96*, 2029–2039. [CrossRef]
33. Krunt, O.; Zita, L.; Kraus, A.; Volek, Z. How can housing system affect growth and carcass traits, meat quality and muscle fiber characteristics in biceps femoris and mineral content of tibia and femur bones in growing rabbits? *Livest. Sci.* **2021**, *249*, 104531. [CrossRef]

MDPI

Article

Impacts of Air Velocity Treatments under Summer Condition: Part I—Heavy Broiler's Surface Temperature Response

Suraiya Akter [1], Bin Cheng [1], Derek West [1], Yingying Liu [1,2], Yan Qian [1,2], Xiuguo Zou [1,2], John Classen [1], Hernan Cordova [3], Edgar Oviedo [3] and Lingjuan Wang-Li [1,*]

1 Department of Biological and Agricultural Engineering, North Carolina State University, Raleigh, NC 27695, USA; sakter@ncsu.edu (S.A.); chengbin0228@gmail.com (B.C.); drwest@ncsu.edu (D.W.); lyy@njau.edu.cn (Y.L.); qianyan@njau.edu.cn (Y.Q.); xiuguozou@gmail.com (X.Z.); classen@ncsu.edu (J.C.)
2 Department of Electrical Engineering, College of Engineering, Nanjing Agricultural University, Nanjing 210095, China
3 Prestage Poultry Science Department, North Carolina State University, Raleigh, NC 27695, USA; hacordovanoboa@gmail.com (H.C.); eooviedo@ncsu.edu (E.O.)
* Correspondence: lwang5@ncsu.edu; Tel.: +1-919-515-6762

Simple Summary: The surface temperature variation of heavy broilers (42–61 d age) under heat stress is an important indicator of thermal comfort, but it is not well studied and reported yet. This study examined the variation of surface temperatures of broilers through two dynamic air velocity treatments under hot summer conditions. It was discovered that the surface temperatures varied over age, daytime, and environmental factors (air temperature, relative humidity, and temperature humidity index). A simple linear regression model to predict the surface temperature of heavy broilers was developed. The findings from this study will enhance knowledge to understand the broilers' responses under heat stress, which will be helpful in providing necessary management decisions to create a comfortable thermal environment.

Abstract: Heavy broilers exposed to hot summer conditions experience fluctuations in surface temperatures due to heat stress, which leads to decreased performance. Maintaining a bird's homeostasis depends on several environmental factors (temperature, relative humidity, and air velocity). It is important to understand the responses of birds to environmental factors and the amount of heat loss to the surrounding environment to create thermal comfort for the heavy broilers for improved performances and welfare. This study investigates the variation in surface temperatures of heavy broilers under high and low air velocity treatments. Daytime, age and bird location's effect on the surface temperature variation was also examined. The experiment was carried out in the poultry engineering laboratory of North Carolina State University during summers of 2017, 2018, and 2019 as a part of a comprehensive study on the effectiveness of wind chill application to mitigate heat stress on heavy broilers. This live broiler heat stress experiment was conducted under two dynamic air velocity treatments (high and low) with three chambers per treatment and 44 birds per chamber. Surface temperatures of the birds were recorded periodically through the experimental treatment cycles (flocks, 35–61 d) with infrared thermography in the morning, noon, evening, and nighttime. The overall mean surface temperature of the broilers under two treatments was found to be 35.89 ± 2.37 °C. The variation in surface temperature happened due to air temperature, thermal index, air velocity, bird's age, daytime, and position of birds inside the experimental chambers. The surface temperatures were found lower under high air velocity treatment and higher under low air velocity treatment. During the afternoon time, the broilers' surface temperatures were higher than other times of the day. It was also found that the birds' surface temperature increased with age and temperature humidity indices. Based upon the experimental data of five flocks, a simple linear regression model was developed to predict surface temperature from the birds' age, thermal indices, and air velocity. It will help assess heavy broilers' thermal comfort under heat stress, which is essential to provide a comfortable environment for them.

Citation: Akter, S.; Cheng, B.; West, D.; Liu, Y.; Qian, Y.; Zou, X.; Classen, J.; Cordova, H.; Oviedo, E.; Wang-Li, L. Impacts of Air Velocity Treatments under Summer Condition: Part I—Heavy Broiler's Surface Temperature Response. *Animals* **2022**, *12*, 328. https://doi.org/10.3390/ani12030328

Academic Editors: Lilong Chai and Yang Zhao

Received: 12 December 2021
Accepted: 26 January 2022
Published: 29 January 2022

Keywords: heavy broiler; heat stress; surface temperature; air velocity; temperature humidity index

1. Introduction

As homoeothermic birds, broiler chickens maintain their body temperatures between 40.6–41.7 °C by dissipating heat produced from metabolism [1]. When the environmental temperature goes above a bird's thermoneutral zone, it becomes challenging for the bird to maintain its thermoregulatory status as it cannot release excess heat. As a result, heat stress occurs, which affects the chicken's behavioral, physiological and immunological properties [2]. Increased panting [3], mortality rate [4,5] feed conversion ratio [6,7], and decreased feed intake [7–9] and body weight gain [6,7,9] are some of the significant effects of heat stress on broiler. Moreover, heat stress hinders overall poultry and egg quality [10] leading to annual loss [11]. Most studies that investigating the broilers' performance or responses under heat stress conditions were conducted for 1–42-day-old birds. Producers are moving towards bigger broiler production with a market age of 63 d and a bodyweight of 3.75 kg to meet the continuous demand for poultry meat [12–14]. However, they are challenged to maintain these heavy birds' constant performance and welfare due to heat stress, especially in hot and humid summer seasons. Moreover, the existing production structures cannot provide adequate thermal comfort due to the increased incidence and severity of heatwaves in the broiler-producing region caused by climate change. To provide a comfortable thermal environment to the heavy broilers (i.e., 42–61 d), the assessment of the birds' response to heat stress is of utmost importance.

Some researchers suggested deep body temperature (DBT) as an indicator of stress as it is very responsive to different stress levels [15,16]. Although Hamrita [17] found DBT responses of birds of age 8.6 to 9.4 weeks under heat stress, the method of implementing biotelemetry to obtain DBT was exhaustive and stressful for the birds. Cangar [18] and Giloh [19] found that surface temperatures (Ts) are indicative of the broilers' comfort or thermal stress. The birds' Ts vary at different body parts and change with age [18]. Moreover, research has indicated that DBT has a strong correlation with Ts [19]. Besides physiology, the environment also impacts Ts [18–20]. Studies showed the Ts changes with air temperature (Ta) and relative humidity (RH). An increase in Ta and RH increases the Ts of broilers of age 1–35 d [21–23]. The Ts variation with all these environmental factors was assessed for the birds of age up to 42 d. Hence, although Ts can be used as an indicator of stress, the evaluations are not available for heavier and bigger birds of current market size.

Among various heat stress mitigation strategies, ventilation improvement was preferred by producers. Various studies were conducted to verify the impact of air velocity (AV) on the thermal comfort of broilers under stressful conditions [20,24–26]. Furlan [27] suggested that a 50% increase in AV will help maintain optimum weight gain and feed gain ratios for heavy broilers under hot weather. Moreover, they found an increase in AV decreases skin Ts, but those are for birds aged up to 35 d. As AV is found to be potential in mitigating heat stress, it is also important to know the responses of birds under different AVs.

Since Ts is useful in assessing the thermal stress of broiler chickens [18,21], several approaches were taken to build models and methods that can be easily used to practically predict the Ts of broilers [20,23,28]. However, limitations in Ts measurement, changes of Ts with the environment, and variations of Ts at various ages underestimated the Ts of broiler chicken. Moreover, all the previous models were established on the Ts found from broilers of age up to 42 d only. Thus, an updated Ts prediction model which can be used for current market-sized broiler chicken is needed.

It is not only the production house Ta that can determine the comfort alone as the birds' heat loss depends on the difference between body Ts and Ta [28,29]. Moreover, RH and AV strongly influence the thermal comfort of birds. Thermal indices including temperature and humidity govern the comfortable thermal environment of broilers. Hence, the assessment of the broilers' Ts at stressful environmental conditions is vital to design a comfortable ambiance for the birds and ensure their performances and welfare. Broiler Ts can be used as an indicator of thermoregulatory status. Moreover, thermal imaging can help assess the condition without touching the birds and also for the birds in commercial production houses [23]. The Ts variation with environmental and physiological factors was investigated through some research, but none of them were conducted for the heavy broilers of age 42–61 d, nor the whole flock of birds. This study was aimed to fill that gap. The experiment was designed to obtain temporal and spatial variations of Ts for different ages and environmental conditions under heat stress. The variation in Ts was investigated under two dynamic AV treatments—high and low. Additionally, a regression model between Ts and environmental factors (AV, T-RH-index) and age was established to help assess the thermal comfort of heavy broilers in commercial flocks under heat-stressed conditions and take necessary management decisions.

2. Materials and Methods

2.1. Experimental Unit

The experiment was carried out in the poultry engineering laboratory (PEL) of NCSU for three consecutive summers from 2017 to 2019. In 2017 and 2018, two flocks of birds were experimented each summer, but only one flock was studied in 2019. Hence, heat stress experiments were conducted on a total of five flocks of heavy broilers.

The PEL has six simulated poultry chamber systems with the core chambers in 2.44 m × 2.44 m × 2.44 m dimension. All these chambers are equipped with a nipple drinker line, four feeders, and an automated switch-timer soft lighting system (Figure 1). Each of the chambers was comprised with a blower house, a conditioning chamber, a turnaround, and an exhaust duct. A belt-driven blower controlled by a variable frequency drive (VFD) system provides various ventilation rates and desired airspeeds in the range of 0.9–4.6 m/s at birds' height according to birds' age and ambient condition. In each of the blower houses, at the outlet of the blower, an adjustable damper controls the amount of fresh air entering the system based on the chamber inlet temperature and blower's revolution per minute (RPM). More detailed descriptions of the PEL and operations are reported by Wang-Li [30].

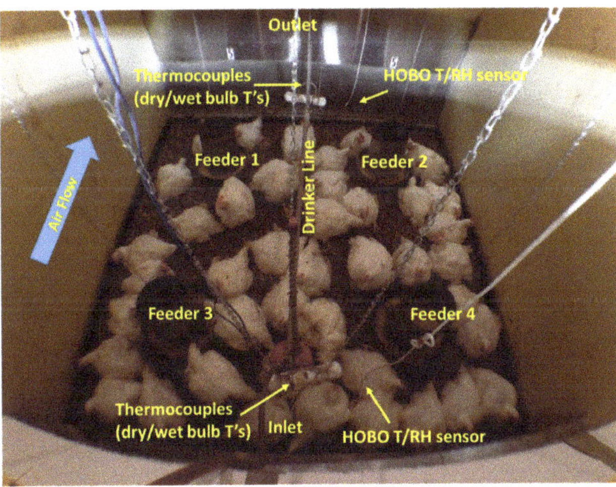

Figure 1. Broilers, feeders, drinker-line, and sensors in the core chambers.

2.2. Animals

Before each flock, a total of 400 male broilers ("Ross-708" for flocks 1–4, "COBB 500" for flock 5) were hatched and raised in the floor pans under similar conditions at the NCSU poultry unit. Then, 264 birds without leg defects were selected randomly to be placed in the six experimental chambers with 44 birds per chamber at the age of 28 d. They remained in the chamber until the age of 61 d. The final stocking density was ≤40 kg/m^2, followed by the animal welfare guideline.

2.3. Core Chamber Environmental Data Monitoring

In each core chamber where birds were housed, the Ta and RH were monitored continuously with a Thermocouple, and HOBO Pro v2 External T/RH Data Logger, Model U23-002 (Onset, Computer Corporation, MA, USA) placed at the airflow inlet and outlet of the chamber at birds' height. Calibrated thermocouples (range: −5 °C to 50 °C, and accuracy: ±0.002 °C) recorded temperature data continuously at 1 min interval. All the thermocouple measurements were integrated into 6 PLC control boards for data acquisition onsite readings. The HOBO sensors recorded both temperature and RH at 10 min intervals. Air velocities were tested and pre-determined under different blower frequencies in Hz displayed in the VFD system.

2.4. Air Velocity Treatments

Increasing AV is an effective way to improve broiler performance and welfare. However, there is no fixed range or value of AV established for heavy broilers yet since the AV requirements depend on weather conditions. Birds' body weight and age also impact the ventilation requirement. There are some guidelines about AV that can bring wind-chill effect on effective temperature [31,32]. On top of that, Czarick and Fairchild [33] reported that for heavier birds, air velocities 50% higher than usual (up to 3 m/s) would help maintain optimum weight gain and feed gain ratio during hot weather. Additionally, Yahav [26] found that the AV of 1.5 to 2.0 m/s is optimal to maintain the performance of broilers under harsh summer conditions. Two sets of dynamic AV treatments (high AV and low AV) were designed to test the broilers' response to AV for this study. The treatment velocities were designed by the research team based upon birds' possible responses to air temperature classes and birds' age. Tables 1 and 2 show the high and low AV treatment designs. The difference between the two treatments varied in range depending on birds' age and how far the measured chamber temperature differed from the optimal thermal condition. It is important to note that compared to the previous broiler studies [25,26,34] with static AV treatments, this study implements dynamic AV treatment levels based on the age of the birds and temperature of the air around the birds, which is highly recommended and widely used in the broiler industry.

The AV treatment started on the birds at the age of 35. As shown in Tables 1 and 2, when the Ta was below optimal temperature, there was no AV differences between the treatments to avoid cold stress to complicate the experiments. From 35–61 d of age, high AV treatment was applied to chambers 1, 3, 5 and the low AV treatment on chambers 2, 4, 6. Change in AV was achieved by changing variable speed blowers' frequency of the VFD.

Table 1. High AV treatment design.

Broiler Age (Days)	AV (m/s) below Optimum T	Temp °C (°F)	AV (m/s) around Optimum T	Temp °C (°F)	AV (m/s) above Optimum T (Moderate)	Temp °C (°F)	AV (m/s) above Optimum T (Severe)	Temp °C (°F)	AV (m/s) above Optimum T (Life Threatening)	Temp °C (°F)	AV (m/s) above Optimum T (Warning)	Temp °C (°F)
28–34	0.90	<26.0 (78.8)	1.23	26.0–27.8 (78.8–82)	1.33	27.8–28.9 (82–84)	1.48	28.9–32.2 (84–90)	1.64	32.2–33.3 (90–93)	1.75	>37.8 (93)
35–40	0.90	<22.2 (71.0)	1.23	22.0–26.0 (72–78)	2.02	26.0–30.0 (78–86)	2.77	30.0–33.0 (86–92)	3.45	33.0–37.8 (92–100)	3.95	>37.8 (100)
41–42	1.48	<21.1 (70.0)	1.48	22.0–26.0 (72–78)	2.02	26.0–30.0 (78–86)	2.77	30.0–33.0 (86–92)	3.45	33.0–37.8 (92–100)	3.95	>37.8 (100)
43–52	1.48	<20.6 (69.0)	1.75	22.0–26.0 (72–78)	2.02	26.0–30.0 (78–86)	2.77	30.0–33.0 (86–92)	3.95	33.0–37.2 (92–99)	4.33	>37.2 (99)
53–54	1.48	<19.4 (67.0)	1.75	21.1–25.0 (70–77)	2.43	25.0–29.4 (77–85)	3.02	29.4–32.7 (85–91)	3.95	32.7–35.6 (91–97)	4.33	>36.1 (97)
55–56	1.48	<19.4 (67.0)	1.75	21.1–25.0 (70–77)	2.43	25.0–29.4 (77–85)	3.02	29.4–32.7 (85–91)	3.95	32.7–35.6 (91–96)	4.33	>35.6 (96)
57–58	1.48	<18.9 (66.0)	1.75	20.6–25.0 (69–77)	2.43	25.0–29.4 (77–85)	3.02	29.4–32.2 (85–90)	3.95	32.2–35.6 (90–96)	4.33	>35.6 (95)
59–60	1.48	<18.9 (66.0)	2.43	20.6–24.4 (69–76)	3.02	24.4–28.9 (76–84)	3.45	28.9–31.7 (84–89)	4.33	31.7–35 (89–95)	4.43	>35.0 (95)
61	1.48	<18.3 (65.0)	2.43	20.0–23.9 (68–75)	3.02	24.4–28.9 (76–84)	3.45	28.9–31.7 (84–89)	4.33	31.1–33.9 (88–93)	4.60	>33.9 (93)

Table 2. Low AV treatment design.

AV (m/s) below Optimum T	Temp °C (°F)	AV (m/s) around Optimum T	Temp °C (°F)	AV (m/s) above Optimum T (Moderate)	Temp °C (°F)	AV (m/s) above Optimum T (Severe)	Temp °C (°F)	AV (m/s) above Optimum T (Life Threat-ening)	Temp °C (°F)	AV (m/s) above Optimum T (Warning)	Temp °C (°F)
0.90	<26.0 (78.8)	1.23	26.0–27.8 (78.8–82)	1.33	27.8–28.9 (82–84)	1.48	28.9–32.2 (84–90)	1.64	32.2–33.3 (90–93)	1.75	>37.8 (93)
0.90	<22.2 (71.0)	1.23	22.0–26.0 (72–78)	1.48	26.0–30.0 (78–86)	2.02	30.0–33.0 (86–92)	2.77	33.0–37.8 (92–100)	3.45	>37.8 (100)
1.48	<21.1 (70.0)	1.48	22.0–26.0 (72–78)	1.48	26.0–30.0 (78–86)	2.02	30.0–33.0 (86–92)	2.77	33.0–37.8 (92–100)	3.45	>37.8 (100)
1.48	<20.6 (69.0)	1.48	22.0–26.0 (72–78)	1.75	26.0–30.0 (78–86)	2.43	30.0–33.0 (86–92)	3.02	33.0–37.2 (92–99)	3.65	>37.2 (99)
1.48	<19.4 (67.0)	1.48	21.1–25.0 (70–77)	1.75	25.0–29.4 (77–85)	2.43	29.4–32.7 (85–91)	3.02	32.7–35.6 (91–97)	3.65	>36.1 (97)
1.48	<19.4 (67.0)	1.48	21.1–25.0 (70–77)	1.75	25.0–29.4 (77–85)	2.43	29.4–32.7 (85–91)	3.02	32.7–35.6 (91–96)	3.65	>35.6 (96)
1.48	<18.9 (66.0)	1.48	20.6–25.0 (69–77)	1.75	25.0–29.4 (77–85)	2.43	29.4–32.2 (85–90)	3.02	32.2–35.6 (90–96)	3.65	>35.6 (95)
1.48	<18.9 (66.0)	1.75	20.6–24.4 (69–76)	2.43	24.4–28.9 (76–84)	2.77	28.9–31.7 (84–89)	3.65	31.7–35 (89–95)	3.80	>35.0 (95)
1.48	<18.3 (65.0)	1.75	20.0–23.9 (68–75)	2.43	24.4–28.9 (76–84)	2.77	28.9–31.7 (84–89)	3.65	31.1–33.9 (88–93)	3.95	>33.9 (93)

2.5. Thermal Image Capturing

Infrared thermography has been widely used to estimate the surface temperature of chickens [18,19,35–37]. Birds' head surface temperature was considered surface temperature (Ts) as it was easier to get the skin temperature from the head than the other body parts with many feathers. Top-view thermal images of birds were captured using handheld infrared cameras, FLIR T400 (Teledyne FLIR LLC, Wilsonville, OR, USA) for flocks 1–4 and FLIR E8 (Teledyne FLIR LLC, Wilsonville, OR, USA) for flock 5. The opening of the door was kept minimal so that the birds were assumed undisturbed during image capture. For every flock, images were captured on randomly selected days between 42–60 d at different times of the day and at different ages. Each core chamber was virtually divided into the following six segments (see Figures 1 and 2 for relative locations): 3 in the outlet area (feeder-1, back-middle, feeder 2); 3 in the inlet area (feeder-3, front-middle, feeder-4). Six images (one image per segment) were taken in each core chamber during every image taking time at early morning (5:00–7:00), morning (7:00–10:00), late morning (10:00–12:00), noon (12:00), early afternoon (12:00–14:00), late afternoon (14:00–16:00), evening (17:00–19:00), and night (19:00–22:00), respectively. Figure 2 illustrates representative images of six segments in one chamber. Images were downloaded by "FLIR" software (Teledyne FLIR LLC, Wilsonville, OR, USA), which allowed reading temperatures in °C. All six images' total countable head temperature were averaged and considered as birds' surface temperature in the given chamber.

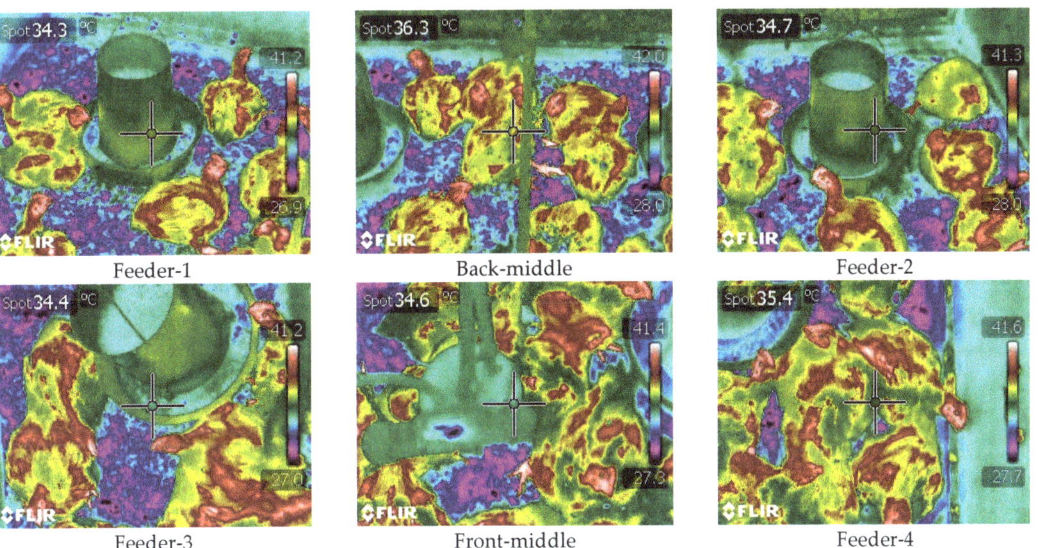

Figure 2. Thermal images by segments in a chamber. Feeder-1, back-middle and feeder-2 were captured in the outlet of the chamber and feeder-3, front-middle and feeder-4 were captured from the inlet of the chamber (see Figure 1 for feeder's relative locations).

2.6. Temperature Humidity Index

Since there is no equation yet established to calculate heavy broilers' temperature humidity index (THI) directly from Ta and RH, this study used the following equation established by [38]:

$$THI = 0.8 \times Ta + RH\,(Ta - 14.3)/100 + 46.3 \quad (1)$$

Ta = air dry-bulb temperature (°C)
RH = relative humidity of air (%)

The calculated indices were then used to identify the thermal comfort for the broilers, according to Table 3. The thermal environments were classified into five comfort/discomfort conditions (Table 3) following the method by [38]. The classification was adopted by Moraes [38] from the combination of average values of temperature, and relative humidity recommended to commercial broilers and laying chickens by several other researchers mentioned in the literature, hence we assumed this could be applied in the current study.

Table 3. Broiler comfort levels under different THI.

THI	Birds Comfort
≤72	Absolute comfort
73–76	Light discomfort
77–80	Moderate discomfort
81–84	Severe discomfort
≥85	Life-threatening

2.7. Data Analysis

Each of the five flock experiments was conducted in a completely randomized experimental design with two treatments (high AV and low AV) and in subplots with three replications for each treatment. So, while checking the difference between two treatments within a flock, we used Student's t-test when there was only one factor. The number of samples used was $n = 3$ since we had three replicates under each treatment within a flock. While checking the differences for multiple factors, the data were evaluated through analysis of covariance (ANCOVA) and the means compared using the Tukey test at the level of 5% probability. When investigating the significant difference for the high and low treatment for the overall flocks, the total samples were 15 for each treatment. The descriptive analysis such as mean, median, quantile, standard deviation of any parameter was performed using the statistical software RStudio (version 1.0.143) (RStudio, Boston, MA, USA). Statistical analysis was also conducted with the same software.

3. Results

3.1. Environmental Conditions

All the five flocks' experiments were conducted under summer conditions. Table 4 represents the average environmental conditions at the core chamber inlets while the thermal images were captured. There were no significant differences in Ta, RH, and THI between treatments in any flock. The AV treatments were designed to be dynamic, i.e., changed according to the inlet Ta and age of the birds (Tables 1 and 2). The AV during data collection time was significantly different under two treatments ($p < 0.05$) in each flock (Table 4). The AV under high treatment was always higher than those of low.

Table 4. Average environmental conditions in chambers during the image data collection times.

Flock	AV Treatment	Ta (°C)	RH (%)	THI	AV **
1	High	30.38 ± 3.35	67.86 ± 11.67	81.17 ± 3.25	2.51 ± 0.65 ^{a}
	Low	29.54 ± 3.09	69.14 ± 12.10	80.14 ± 2.95	1.68 ± 0.51 ^{b}
2	High	27.09 ± 3.54	72.62 ± 18.85	77.01 ± 4.48	1.93 ± 0.85 ^{a}
	Low	26.37 ± 3.68	73.85 ± 14.12	76.20 ± 5.20	1.22 ± 0.46 ^{b}
3	High	27.75 ± 4.61	72.09 ± 16.15	77.49 ± 4.30	1.88 ± 0.44 ^{a}
	Low	27.39 ± 4.23	72.79 ± 15.06	77.15 ± 4.09	1.43 ± 0.34 ^{b}
4	High	27.26 ± 3.59	77.98 ± 11.52	77.84 ± 4.02	1.61 ± 0.26 ^{a}
	Low	27.01 ± 3.52	78.32 ± 11.08	77.50 ± 4.04	1.25 ± 0.09 ^{b}
5	High	28.56 ± 3.63	62.64 ± 12.71	77.68 ± 3.79	1.99 ± 0.53 ^{a}
	Low	27.93 ± 3.43	63.62 ± 12.85	76.95 ± 3.68	1.43 ± 0.37 ^{b}

Means within flocks with different letter superscripts are significantly different at ($p < 0.05$); ** the AV during the image collection time.

The experiment was designed to obtain Ts variation under heat stress conditions. The thermal index was "severe discomfort" on average during the first flock. The rest of the flocks were "moderate discomfort". Table 5 represents the percentage of time the birds were under heat stress during the experiment. The data presented in Table 5 reflects only the time of image capture, which is around 1–3% of the overall experiment period for each flock. During the image capture time, the chamber environment (inlet Ta and RH) exceeded the optimal growth condition 98.6%, 47.9%, 55.6%, 50%, and 71.7% time of flocks 1 through 5 consecutively (Table 5). Flocks 1 and 5 were more stressed than flocks 2, 3, and 4.

Table 5. Distribution of AV treatments implemented from 35 d to 61 d during the image-capturing time.

	% of Occurrences					
Flock	**AV below Optimal**	**AV around Optimal**	**AV Moderate**	**AV Severe**	**AV Life-Threatening**	**AV Warning**
1	0.0	1.4	48.6	30.6	19.4	0.0
2	12.5	39.6	29.2	8.3	10.4	0.0
3	0.0	44.4	30.6	8.3	16.7	0.0
4	0.0	50.0	26.2	11.9	11.9	0.0
5	0.0	28.3	26.8	24.6	16.7	3.6

3.2. *Average Surface Temperature Variation in Flocks*

In investigation of the Ts variation under AV treatment over the experiment, ANCOVA test was conducted considering AV Treatment and Flock as main factors (Table 6). It was observed that both AV and Flock significantly impacted ($p < 0.05$) the birds' average Ts. The differences among flocks (Table 7) were found by Tukey's HSD test.

Table 6. Results of analysis of covariance test for differences in surface temperature under AV treatment.

Source	Degrees of Freedom	Type III Sum of Squares	Mean Square	F-Value	$p > F$
AV Treatment	1	4.08	4.082	40.05	1.43×10^{-6} ***
Flock	4	34.94	8.74	85.69	7.53×10^{-14} ***

*** Numbers with asterisk represent the significant effects (confidence interval of 95%).

Table 7. Mean surface temperatures of heavy broilers under two AV treatments in the summer condition.

Flock	AV Treatment	Average Ts (°C)	
1	High	36.68 ± 1.86	37.02 ± 1.84 [A]
	Low	37.36 ± 1.78	
2	High	33.89 ± 1.69	34.21 ± 1.60 [C]
	Low	34.53 ± 1.47	
3	High	34.98 ± 2.39	35.35 ± 2.29 [B]
	Low	35.72 ± 2.17	
4	High	34.04 ± 2.59	34.64 ± 2.54 [BC]
	Low	35.26 ± 2.34	
5	High	36.35 ± 2.34	36.56 ± 2.27 [A]
	Low	36.76 ± 2.21	

[A–C] Means followed by the different letter within flock differ significantly ($p < 0.05$).

The average Ts for heavy broilers from all five flocks is summarized in Table 7. Flocks 1 and 5 had significantly higher Ts ($p < 0.05$) than the other flocks (Table 7). Flocks 2 and 4 observed the lowest average Ts among all the flocks. Mean Ts in the third flock was higher than that of the second flock. The first and fifth flock observed more stressful times than the other three flocks and, hence higher mean Ts for those flocks were reasonable.

3.3. *Ts Variation with Ta*

The variation of Ts with Ta was investigated by building a simple linear relationship between them. It was discovered that the Ts had a positive co-relationship with Ta (Figure 3).

This relationship was significant at a significance level of 0.05. Although there was no interaction effect for Ta and AV treatment at α = 0.05 level, Ts was found lower for high AV treatment than that of the low AV treatment for all flocks. Besides Ta and AV, age was also found to impact Ts for these heavy birds significantly.

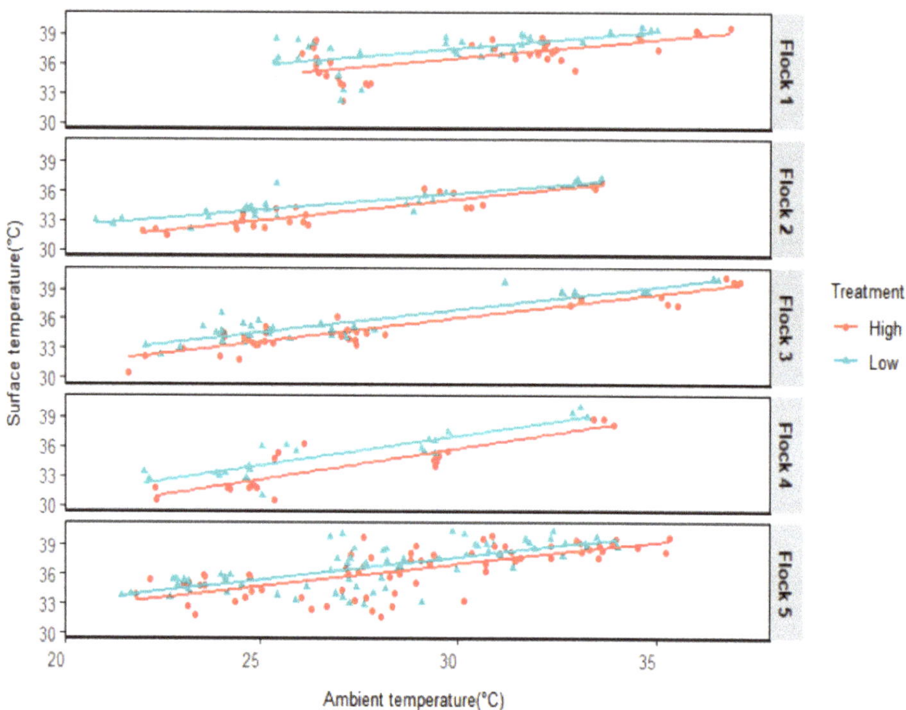

Figure 3. Changes in heavy broilers' surface temperature with environmental temperature.

3.4. Surface Temperature Variation with Temperature Humidity Index

Like the Ta, the broilers' thermal comfort depends on RH as birds' respiration also works as a pathway to lose heat under heat stress. Hence, the variation of Ts was checked under various indices for all flocks. Figure 4 indicates that the Ts vary over different THI classes. The THI significantly changed the Ts in any flock. The Ts primarily increase when the THI condition changes from comfortable to life-threatening. The high AV treatment had a significant positive effect on Ts during the second, third, and fourth flocks. Interaction between THI and AV did not change the Ts during any flock.

3.5. Surface Temperature Variation over the Time of the Day

The time of the day was divided into six periods. Although the data represents all six time periods, not all flocks had each interval's representative data. According to Figure 5, the Ts of broilers was highest in the afternoon (14:00–18:00) at any flock and lowest in the early morning (5:00–8:00). ANOVA analysis suggested time of the day significantly impacted ($p < 0.05$) the variation in Ts. Additionally, AV treatments had a significant effect on Ts at any time; the Ts under high AV was consistently lower than that under low AV (Figure 5). The interaction between AV and daytime had no significant effect on Ts variation in any flock.

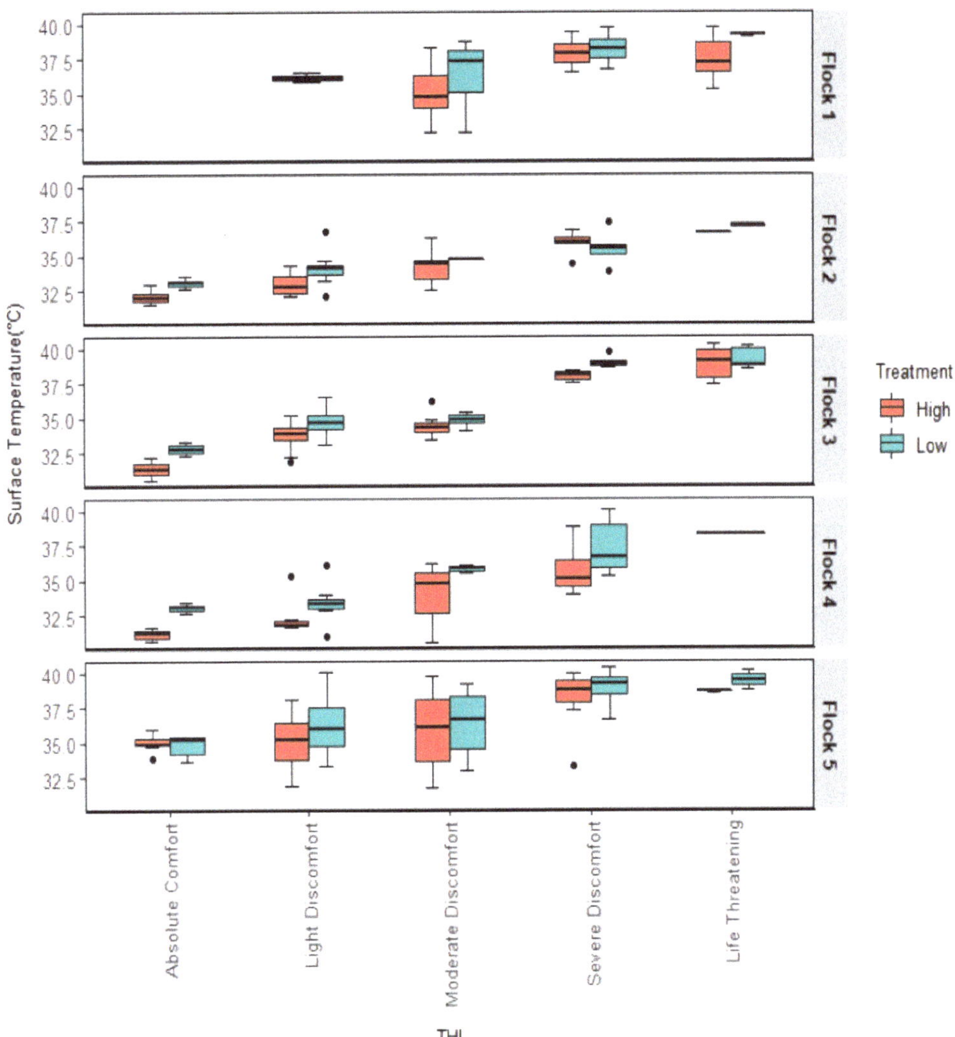

Figure 4. Variation in Ts for different THIs in different flocks under two AV treatments.

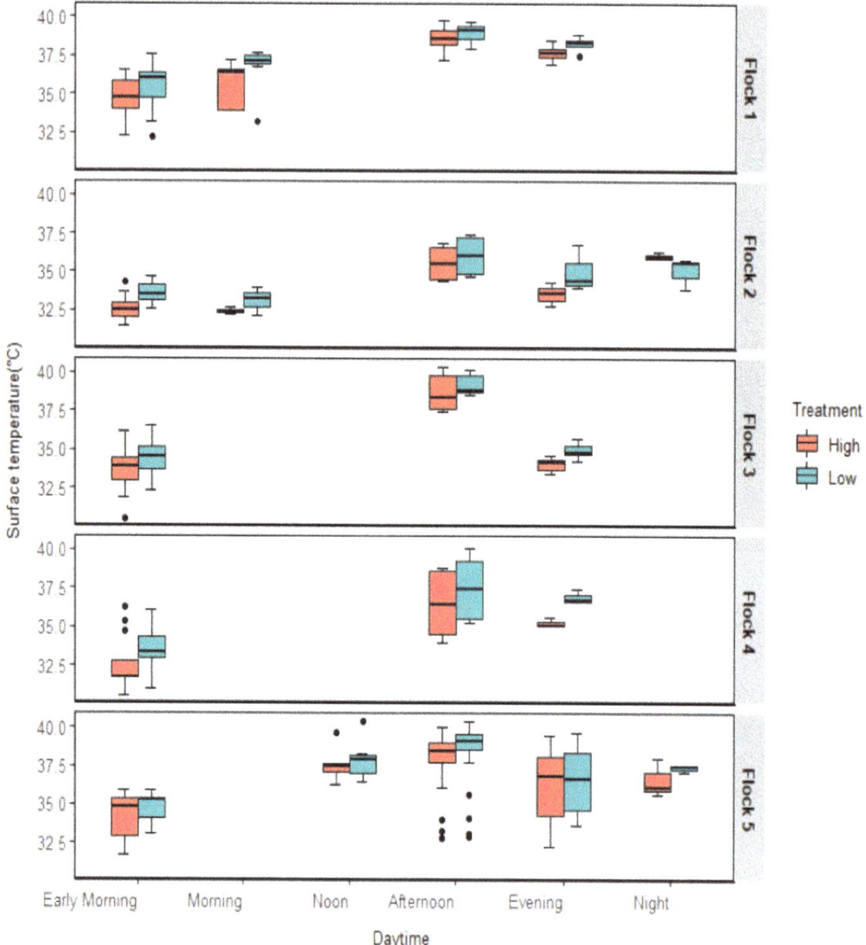

Figure 5. Diurnal variation of heavy broilers' surface temperature variation under two AV treatments.

3.6. Ts Variation with Age

The Ts variation of broilers aged from 6th to 9th weeks under both AV treatment for all flocks is presented in Table 8. Under high AV, the Ts decreased significantly ($p < 0.05$) over the week during 2nd and 4th flocks. Under low AV treatments, Ts changes with age during 4th flock. There was no significant difference in Ts under low AV treatment at any week.

A two-way ANCOVA test was conducted to test the effects of age (in week) and the two AV treatments on the Ts of chicken. The main effect of age in week revealed a significant ($p < 0.05$) impact on the broilers' Ts (Table 9). A significant ($p < 0.05$) positive correlation between Ts and age for the overall flocks was found. The AV treatment also significantly affected the broilers' Ts. However, the interaction between age and AV treatment was not significant ($p < 0.05$). Although we considered flock as a blocking factor, the ANCOVA test revealed it had a significant impact on the variation of the broilers' AV. According to Table 4, there were no statistical differences observed in the environmental condition of five flocks. Hence, we considered flock as a blocking factor, and the effects due to flock were not discussed later on.

Table 8. Heavy broilers' surface temperatures variation with age.

Flock	Treatment	Surface Temperature °C (Mean ± SD)			
		6th Week	7th Week	8th Week	9th Week
1	High	NA	37.10 ± 0.60 [a]	37.23 ± 0.35 [a]	35.74 ± 0.61 [a]
	Low	NA	37.71 ± 0.61 [a]	37.84 ± 0.27 [a]	36.53 ± 1.09 [a]
2	High	34.70 ± 0.57 [a]	NA	33.07 ± 0.36 [b]	NA
	Low	35.07 ± 0.57 [a]	NA	33.97 ± 0.71 [a]	NA
3	High	NA	35.35 ± 0.61 [a]	34.69 ± 0.62 [a]	34.90 ± 0.71 [a]
	Low	NA	35.89 ± 0.47 [a]	35.37 ± 0.46 [a]	35.89 ± 0.51 [a]
4	High	NA	35.25 ± 0.35 [a]	32.43 ± 0.39 [b]	NA
	Low	NA	36.37 ± 0.36 [a]	33.77 ± 0.69 [b]	NA
5	High	NA	NA	35.49 ± 0.79 [a]	36.70 ± 0.94 [a]
	Low	NA	NA	35.94 ± 0.68 [a]	37.08 ± 0.61 [a]

[a,b] Means under each treatment within Flocks with different superscripts are different at ($p < 0.05$).

Table 9. Results of analysis of covariance test for testing age effect on broiler surface temperature.

Source	Degrees of Freedom	Type III Sum of Squares	Mean Square	F-Value	$p > F$
Week	3	112.9	37.62	8.496	1.82×10^{-5} ***
AV Treatment	1	39.1	39.14	8.838	0.00315 **
Flock	4	336.3	84.97	18.984	3.54×10^{-14} ***
Week × AV Treatment	3	1	0.33	0.075	0.97345

The asterisk (**,***) represent the significance at different level (0.01, 0.001).

3.7. Surface Temperature Variation at Inlets and Outlets

Table 10 reflects the differences in Ts at the inlets and outlets of the chambers in all flocks. Under high AV treatments, the average Ts at the outlet was not significantly different from that at the inlet for all flocks. This was the same for low AV treatment. Air entered the chamber through the inlet, and so the birds under heat stress wanted to stay and hang around more at that side. The entered air helped birds cool down, so the Ts were lower at this side. On the other hand, air exited through the outlet. The heat released by the birds exited the atmosphere through the outlet, so the air at the outlet was warmer than the inlet. Hence, the birds experienced higher surface temperature at the outlet.

Table 10. Difference in surface temperatures at the inlets and outlets of the chambers.

Flock	AV Treatment	Ts (Mean ± SD)	
		Outlet	Inlet
1	High	36.84 ± 1.82	36.75 ± 2.01
	Low	37.76 ± 1.77	37.22 ± 1.95
2	High	33.97 ± 1.76	33.79 ± 1.83
	Low	34.49 ± 1.49	34.58 ± 1.78
3	High	35.09 ± 2.39	34.86 ± 2.50
	Low	35.82 ± 2.08	35.61 ± 2.32
4	High	34.02 ± 2.76 [a]	34.06 ± 2.71
	Low	35.56 ± 2.30 [b]	34.95 ± 2.72
5	High	36.66 ± 2.37	36.04 ± 2.52
	Low	37.02 ± 2.33	36.67 ± 2.25

Different superscript in a column under each flock indicates values were significantly different ($p < 0.05$) in columns.

3.8. Regression Modeling

Since the results indicated the heavy broilers' Ts were affected by age and several environmental factors (Ta, RH, THI, AV), a regression modeling was conducted to predict Ts from the multiple factors. A correlation matrix was first built and analyzed to identify which factors had a significant impact on Ts.

The broilers' Ts had a strong positive correlation with Ta (0.76), THI (0.68), and AV (0.43) (Figure 6). On the other hand, Ts was negatively correlated with RH (0.71). The Ts was positively correlated with age, although the correlation factor (0.06) was not high. Hence, a regression model was built to predict Ts from all these factors (Ta, RH, THI, AV, and Age). Primarily, a simple linear regression model was built. Then, two-way interaction between all the predictor variables was incorporated into the first model. However, a large variance inflation factor (VIF) was found in both models. The THI is strongly correlated to Ta and RH, so the model had a large variance inflation factor. Hence, the issue was resolved by excluding the model's correlated factors Ta and RH. Since THI was calculated from Ta and RH, hence the thermal stress condition for chicken can be explained by THI solely.

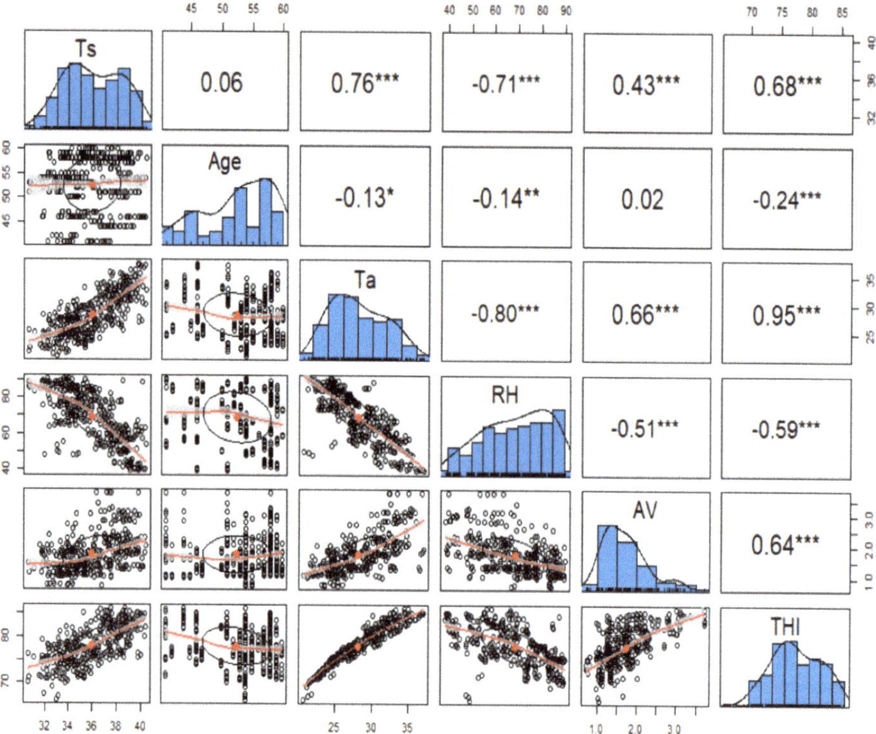

Figure 6. Correlation matrix for the dependent and independent factors. The asterisks ("***", "**", "*") indicate the significance levels of the correlations at different significance level (0.001, 0.01, 0.05).

The proposed model to predict Ts is given below:

$$\text{Ts} = -4.40 + 0.11 \times \text{Age} + 0.45 \times \text{THI} - 0.29 \times \text{AV} \quad (2)$$

where, Ts = Surface temperature of broiler (°C)
Age = Age of the chicken (day)
THI = Temperature humidity index
AV = Air velocity (m/s)

According to the model (2), the Ts increases with age and THI. Additionally, Ts is inversely related to AV, i.e., if AV decreases then Ts increase. This model is a significant model at significance value 0.05. The R^2 value of the above-mentioned model is 0.512.

4. Discussion

As heat stress is one of the biggest hindrances in poultry production, mitigation of heat stress is a must-needed step to ensure well-performing birds. Consumers are not only concerned with meat, but the animal welfare is also of a great concern. Both performance and welfare depend on a suitable environment to live in throughout the growth period of birds. Understanding the bird's response to environmental changes is a necessary action as the environment cannot solely tell the birds' comfort. The birds' body adjusts to the environment through different changes in extreme environmental conditions. So, it is crucial to recognize the birds' responses to environmental changes besides the sensors put in place.

Through a set of live broiler heat stress experiments under summer conditions, it was discovered that the average Ts of birds aged 41–62 d was 35.89 ± 2.37 °C regardless of the flock and AV treatment. This study indicated the effect of age on the broilers' Ts. Cangar [18] observed a decrease in mean Ts from 1st to 6th week of birds. During 2nd and 4th flock, high AV caused a significant decrease in Ts with age. Although it was hypothesized that the Ts increases with age under heat stress, more investigation is required to establish this hypothesis.

Furlan [27], Cangar [18], and Naas [39] measured head Ts of birds under a wide range of Ta (23–32 °C) up to 42 d, and the measured average value were between 27–36.2 °C. This study found that the mean head Ts of the birds from age 42–61 d was 35.89 ± 2.37 °C under heat-stressed conditions. Since the birds were not able to release excess body heat under heat-stressed conditions due to less activity for bigger bodies, the mean Ts was found to be higher than those from the cited studies. While measuring Ts with thermal imaging, measuring from the different body parts is critical. Hence, the head surface, which is exposed to the environment at a bigger age, can be used as a representative area to measure heavier birds' Ts.

Heavy broilers' Ts was found to be varied over the changes in Ta. A positive relationship between mean Ts and Ta for young birds (1–7 d of age) under three controlled Ta of 20, 25, and 35 °C treatments was found by Malherois [20]. Nascimento [35] also found the mean Ts increased with the Ta but independent of age. These studies did not verify the Ta effect on Ts for bigger birds of age more than 42 d. In accordance with these studies, current study found the Ts can be impacted by Ta for broilers beyond 42 d. Moreover, age also influences Ts. Since the birds become heavier and their activity decreases with the body weight gain under stressful condition, their heat dissipation rate also lessen. As a result, body Ts increases at a later age.

Air temperature Ta is not only indicative of the broilers' thermal comfort, but the assessment also needs to be conducted under THI. At any age, the birds' Ts increases when the THI goes beyond optimal environmental condition, Ts are higher under high THI values. It was also observed that birds were panting more under high THI stressed conditions (Table 3), producing a different vocal signal, indicating that birds need more attention during the higher THI times. Under discomfort to life-threatening thermal conditions, bigger weight of birds made them less active and hence could not actively release excessive heat. This leads to stress and death ultimately.

The diurnal variation of Ts was observed in every flock under heat-stressed conditions. In the morning, the birds remain more active than in the afternoon However, as the Ta rose during the daytime, the birds suffered more from heat stress and became less active, releasing less heat from their body surface. So, higher Ts were observed during noon to afternoon than in the morning period. Additionally, during the experimental period, every morning from 7:00 to 9:00 the screens at the inlet and outlet of each chamber were cleaned, which pushed the birds to move around more and helped release more heat from their body surfaces.

The proposed linear regression model from this study can be used to predict birds' Ts under heat stress by using only age, THI and AV. Moghbeli [40] also found a mean Ts increasing pattern with age up to 6 weeks due to the feathering index of the head surface. In agreement with Moghbeli [40], the current Ts model indicates that even after 6 weeks of age, the birds' Ts may increase with age. Under the heat stress conditions, birds became less active and hence were could not release heat from the body surface, so their Ts might increase. Nascimento [23] also established a regression model to predict Ts from feathered and unfeathered areas, but that model was only applicable for the birds of only 42 d. The proposed Ts model upgraded the prediction level for birds of more than 6 weeks. Therefore, it can be used to predict the Ts of heavy broilers. This model developed a positive relationship with THI; hence, the birds' Ts increases with THI. When the environmental condition exceeds the birds' thermal comfort zone, it tends to increase the body temperature. Moreover, beyond the thermal comfort zone, birds are no more capable of releasing their metabolic energy as the higher THI does not allow adequate heat transfer anymore. Hence, their body temperature keeps rising. Birds will suffer more at this point. They cannot perform well, i.e., their body weight gain, water intake, and feed intake decline more than usual. The proposed Ts model suggests that the birds' Ts change inversely with AV. The increase in AV will decrease the Ts of the broiler. Although the time of day had a significant impact on Ts, it was not included in the model as all the thermal image data were captured during the lightning period.

5. Conclusions

Broiler surface temperature can be used to indicate their thermoregulatory status under heat stress conditions. Surface temperature varies according to the birds' age and environmental factors such as Ta, THI, and AV. Through five flocks (35–61 d) of live broiler experiments under summer conditions and two AV treatments, it was discovered that the broiler's Ts changed temporally and spatially. The variation in Ts happened due to Ta, THI, AV, birds' age, daytime, and position of birds inside the experimental chambers. The Ts were found lower under high AV treatment. During afternoon time, the broilers' Ts were higher than other times of the day. It was also found that Ts increased with age and THI. Based upon the experimental data of five flocks, a simple linear regression model was developed to predict Ts from the birds' age, THI, and AV. Thermal imaging can be used to easily detect surface temperature of broilers under commercial settings. It will help assess heavy broilers' thermal comfort under heat stress. Producers can take the necessary steps to mitigate heat stress impact by observing surface temperature changes at different ages of birds and environmental conditions.

Author Contributions: Conceptualization and methodology, L.W.-L., E.O. and S.A.; formal analysis, S.A.; investigation, S.A.; data curation, S.A., B.C., L.W.-L., D.W., Y.L., Y.Q., X.Z. and H.C.; writing—original draft preparation, S.A.; writing—review and editing, L.W.-L., J.C. and S.A.; supervision, L.W.-L.; project administration, L.W.-L. and E.O.; funding acquisition, L.W.-L., E.O. and J.C. All authors have read and agreed to the published version of the manuscript.

Funding: This research was funded by USDA-NIFA-AFRI, grant number 2017-67021-26329.

Institutional Review Board Statement: Experimental use of birds was approved by the Animal Care and Use Committee and conducted in compliance with the Guidelines for Care and Use of Laboratory Animals at North Carolina State University (NCSU) (IACUC#16-279-A).

Informed Consent Statement: Not applicable.

Data Availability Statement: Data presented in this study can be available upon request from the corresponding author.

Acknowledgments: We acknowledge the funding agency and all the other contributors for their enormous support. North Carolina State University contributed to this research through general faculty and graduate research support. C. Sayde, M. Adcock, P. Harris, R. Currin, D. Carroll and S. Hall helped building the research facility and installing sensors and other controlling devices. Graduate student Yijia Dietrich, eight undergraduate research assistants and three high school interns assisted with data collection.

Conflicts of Interest: The authors declare no conflict of interest.

References

1. University of Kentucky College of Agriculture and Kentucky Poultry Federation. Chapter 7 Ventilation Principals. Poultry Production Manual. 2014. Available online: https://afs.ca.uky.edu/files/chapter7.pdf (accessed on 9 September 2020).
2. Lara, L.J.; Rostagno, M.H. Impact of Heat Stress on Poultry Production. *Animals* **2013**, *3*, 356–369. [CrossRef] [PubMed]
3. Mack, L.A.; Felver-Gant, J.N.; Dennis, R.L.; Cheng, H.W. Genetic variations alter production and behavioral responses following heat stress in 2 strains of laying hens. *Poult. Sci.* **2013**, *92*, 285–294. [CrossRef] [PubMed]
4. Kang, S.; Kim, D.-H.; Lee, S.; Lee, T.; Lee, K.-W.; Chang, H.-H.; Moon, B.; Ayasan, T.; Choi, Y.-H. An Acute, Rather Than Progressive, Increase in Temperature-Humidity Index Has Severe Effects on Mortality in Laying Hens. *Front. Vet. Sci.* **2020**, *7*. [CrossRef] [PubMed]
5. Warriss, P.; Pagazaurtundua, A.; Brown, S. Relationship between maximum daily temperature and mortality of broiler chickens during transport and lairage. *Br. Poult. Sci.* **2005**, *46*, 647–651. [CrossRef] [PubMed]
6. Awad, E.A.; Najaa, M.; Zulaikha, Z.A.; Zulkifli, I.; Soleimani, A.F. Effects of heat stress on growth performance, selected physiological and immunological parameters, caecal microflora, and meat quality in two broiler strains. *J. Anim. Sci.* **2016**, *33*, 778–787. [CrossRef] [PubMed]
7. Sohail, M.U.; Hume, M.E.; Byrd, J.A.; Nisbet, D.J.; Ijaz, A.; Sohail, A.; Shabbir, M.Z.; Rehman, H. Effect of supplementation of prebiotic mannan-oligosaccharides and probiotic mixture on growth performance of broilers subjected to chronic heat stress. *Poult. Sci.* **2012**, *91*, 2235–2240. [CrossRef] [PubMed]
8. Mashaly, M.M.; Hendricks, G.L.; Kalama, M.A.; Gehad, A.E.; Abbas, A.O.; Patterson, P.H. Effect of Heat Stress on Production Parameters and Immune Responses of Commercial Laying Hens. *Poult. Sci.* **2004**, *83*, 889–894. [CrossRef]
9. Quinteiro-Filho, W.M.; Gomes, A.V.S.; Pinheiro, M.L.; Ribeiro, A.; Ferraz-De-Paula, V.; Astolfi-Ferreira, C.S.; Ferreira, A.J.P.; Palermo-Neto, J. Heat stress impairs performance and induces intestinal inflammation in broiler chickens infected with *Salmonella* Enteritidis. *Avian Pathol.* **2012**, *41*, 421–427. [CrossRef]
10. Ranjan, A.; Sinha, R.; Devi, I.; Rahim, A.; Tiwari, S. Effect of Heat Stress on Poultry Production and their Managemental Approaches. *Int. J. Curr. Microbiol. Appl. Sci.* **2019**, *8*, 1548–1555. [CrossRef]
11. St-Pierre, N.; Cobanov, B.; Schnitkey, G. Economic Losses from Heat Stress by US Livestock Industries. *J. Dairy Sci.* **2003**, *86*, E52–E77. [CrossRef]
12. Chicken Checkin, What is the Difference between Faster- and Slower-Growing Chicken? National Chicken Council. 2021. Available online: https://www.chickencheck.in/faq/difference-faster-slower-growing-chicken/ (accessed on 10 August 2021).
13. Torrey, S.; Mohammadigheisar, M.; dos Santos, M.N.; Rothschild, D.; Dawson, L.C.; Liu, Z.; Kiarie, E.G.; Edwards, A.M.; Mandell, I.; Karrow, N.; et al. In pursuit of a better broiler: Growth, efficiency, and mortality of 16 strains of broiler chickens. *Poult. Sci.* **2020**, *100*, 100955. [CrossRef] [PubMed]
14. National Chicken Council. US Broiler Performance, 1925 to Present. National Chicken Council. 2020. Available online: https://www.nationalchickencouncil.org/about-the-industry/statistics/u-s-broiler-performance/ (accessed on 10 November 2021).
15. Dadgar, S.; Lee, E.S.; Leer, T.L.V.; Burlinguette, N.; Classen, H.L.; Crowe, T.G.; Shand, P.J. Effect of microclimate temperature during transportation of broiler chickens on quality of the pectoralis major muscle. *Poult. Sci.* **2010**, *89*, 1033–1041. [CrossRef] [PubMed]
16. Kettlewell, P.; Mitchell, M.; Meeks, I. An implantable radio-telemetry system for remote monitoring of heart rate and deep body temperature in poultry. *Comput. Electron. Agric.* **1997**, *17*, 161–175. [CrossRef]
17. Hamrita, T.K.; Conway, R.H. Effect of air velocity on deep body temperature and weight gain in the broiler chicken. *J. Appl. Poult. Res.* **2017**, *26*, 111–121. [CrossRef]
18. Cangar, J.M.; Aerts, J.-M.; Buyse, J.; Berckmans, D. Quantification of the Spatial Distribution of Surface Temperatures of Broilers. *Poult. Sci.* **2008**, *87*, 2493–2499. [CrossRef] [PubMed]
19. Giloh, M.; Shinder, D.; Yahav, S. Skin surface temperature of broiler chickens is correlated to body core temperature and is indicative of their thermoregulatory status. *Poult. Sci.* **2012**, *91*, 175–188. [CrossRef]
20. Malheiros, R.D.; Moraes, V.M.B.; Bruno, L.D.G.; Malheiros, E.B.; Furlan, R.L.; Macari, M. Environmental Temperature and Cloacal and Surface Temperatures of Broiler Chicks in First Week Post-Hatch. *J. Appl. Poult. Res.* **2000**, *9*, 111–117. [CrossRef]
21. Bahuti, M.; Abreu, L.H.P.; Junior, T.Y.; de Lima, R.R.; Campos, A.T. Performance of fuzzy inference systems to predict the surface temperature of broieler chickens. *Eng. Agric.* **2018**, *38*, 813–823.
22. Lin, H.; Zhang, H.F.; Jiao, H.C.; Zhao, T.; Sui, S.J.; Gu, X.H.; Zhang, Z.Y.; Buyse, J.; Decuypere, E. Thermoregulation responses of broiler chickens to humidity at different ambient temperatures. I. One week of age. *Poult. Sci.* **2005**, *84*, 1166–1172. [CrossRef]
23. Nascimento, S.T.; Da Silva, I.J.O.; Maia, A.S.C.; De Castro, A.C.; Vieira, F.M.C. Mean surface temperature prediction models for broiler chickens—A study of sensible heat flow. *Int. J. Biometeorol.* **2014**, *58*, 195–201. [CrossRef]
24. Mitchell, M.A. Effects of air velocity on convective and radiant heat transfer from domestic fowls at environmental temperatures of 20° and 30 °C. *Br. Poult. Sci.* **1985**, *26*, 413–423. [CrossRef] [PubMed]
25. Simmons, J.D.; Lott, B.D.; May, J.D. Heat loss from broiler chickens subjected to various air speeds and ambient temperatures. *Appl. Eng. Agric.* **1997**, *13*, 665–669. [CrossRef]
26. Yahav, S.; Straschnow, A.; Vax, E.; Razpakovski, V.; Shinder, D. Air Velocity Alters Broiler Performance Under Harsh Environmental Conditions. *Poult. Sci.* **2001**, *80*, 724–726. [CrossRef] [PubMed]

27. Furlan, R.L.; Macari, M.; Secato, E.R.; Guerreiro, J.R.; Malheiros, E.B. Air Velocity and Exposure Time to Ventilation Affect Body Surface and Rectal Temperature of Broiler Chickens. *J. Appl. Poult. Res.* **2000**, *9*, 1–5. [CrossRef]
28. Richards, S.A. The significance of changes in the temperature of changes in the temperature of the skin and body core of the chicken in the regulation of heat loss. *J. Physiol.* **1971**, *216*, 1–10. [CrossRef] [PubMed]
29. Baracho, M.; Naas, I.; Nascimento, G.; Cassiano, J.; Oliveira, K. Surface temperature distribution in broiler houses. *Rev. Bras. Cienc. Avic.* **2011**, *13*, 177–182. [CrossRef]
30. Wang-Li, L.; Xu, Y.; Shivkumar, A.P.; Williams, M.; Brake, J. Effect of dietary coarse corn inclusion on broiler live performance, litter characteristics, and ammonia emission. *Poult. Sci.* **2020**, *99*, 869–878. [CrossRef]
31. Aviagen, Ross Broiler Pocket Guide. 2018. Available online: https://en.aviagen.com/assets/Tech_Center/Ross_Broiler/Ross-Broiler-Pocket-Guide-2020-EN.pdf (accessed on 15 January 2020).
32. COBB.com. COBB Broiler Management Guide. 2018. Available online: https://www.cobb-vantress.com/assets/5c7576a214/Broiler-guide-R1.pdf (accessed on 10 February 2021).
33. Czarick, M.; Fairchild, B.D. Poultry housing for hot climates. In *Poultry Production in Hot Climates*, 2nd ed.; CABI: Wallingford, UK, 2008.
34. Simmons, J.; Lott, B.; Miles, D. The effects of high-air velocity on broiler performance. *Poult. Sci.* **2003**, *82*, 232–234. [CrossRef]
35. Nascimento, G.R.; Nääs, I.; Pereira, D.; Baracho, M.S.; Garcia, R.G. Assessment of broiler surface temperature variation when exposed to different air temperatures. *Rev. Bras. Cienc. Avic.* **2011**, *13*, 259–263. [CrossRef]
36. Xiong, X.; Lu, M.; Yang, W.; Duan, G.; Yuan, Q.; Shen, M.; Berckmans, D. An Automatic Head Surface Temperature Extraction Individual Broiler. *Sensors* **2019**, *19*, 5286. [CrossRef]
37. Abreu, L.H.P.; Junior, T.Y.; Campos, A.T.; Bahuti, M.; Fassani, J. Cloacal and Surface Temperatures of Broilers Subject to Thermal Stress. *Eng. Agric.* **2017**, *37*, 877–886. [CrossRef]
38. de Moraes, S.R.P.; Júnior, T.Y.; de Oliveira, A.L.R.; Yanagi, S.D.M.; Café, M.B. Classification of the temperature and humidity index (THI), aptitude of the region, and conditions of comfort for broilers and layer hens in Brazil. In Proceedings of the International Conference of Agricultural Engineering, XXXVII Brazilian Congress of Agricultural Engineering, International Livestock Environment Symposium-ILES VIII, Iguassu Falls City, Brazil, 31 August–4 September 2008.
39. Nääs, I.D.A.; Romanini, C.E.B.; Neves, D.P.; Nascimento, G.R.D.; Vercellino, R.D.A. Broiler surface temperature distribution of 42 day old chickens. *Sci. Agric.* **2010**, *67*, 497–502. [CrossRef]
40. Damane, M.M.; Barazandeh, A.; Mokhtari, M.S.; Esmaeilipour, O.; Badakhshan, Y. Evaluation of body surface temperature in broiler chickens during the rearing period based on age, air temperature and feather condition. *Iran. J. Appl. Anim. Sci.* **2018**, *8*, 499–504.

Article

Impacts of Air Velocity Treatments under Summer Conditions: Part II—Heavy Broiler's Behavioral Response

Suraiya Akter [1], Yingying Liu [1,2], Bin Cheng [1], John Classen [1], Edgar Oviedo [3], Dan Harris [4] and Lingjuan Wang-Li [1,*]

1 Department of Biological and Agricultural Engineering, North Carolina State University, Raleigh, NC 27695, USA; sakter@ncsu.edu (S.A.); lyy@njau.edu.cn (Y.L.); chengbin0228@gmail.com (B.C.); classen@ncsu.edu (J.C.)
2 Department of Automation, College of Artificial Intelligence, Nanjing Agricultural University, Nanjing 210095, China
3 Prestage Poultry Science Department, North Carolina State University, Raleigh, NC 27695, USA; eooviedo@ncsu.edu
4 Department of Statistics, North Carolina State University, Raleigh, NC 27695, USA; doharris@ncsu.edu
* Correspondence: lwang5@ncsu.edu; Tel.: +1-919-515-6762

Simple Summary: Behavioral changes are one of the mechanisms for broilers to adjust their body temperature under heat stress conditions. However, the behavioral responses of heavy broilers to environmental changes have not yet been studied well. Therefore, this research investigated the behavioral changes of broilers under two dynamic air velocity treatments (high and low) under summer conditions. Video data collected from a heat stress experiment conducted on broilers aged 42–54 days were used to investigate variations in the number of chickens feeding, drinking, standing, walking, sitting, wing flapping, and leg stretching. The results indicated that the high air velocity treatments increased the number of chickens feeding, standing, and walking. In addition, age significantly affected the number of birds feeding, drinking, panting, and sitting, while the time of the day also affected the number of chickens drinking and panting. This study reveals the thermal stress of heavy broilers from their behavior under summer conditions to help manage the performance and welfare of birds under environmental stress.

Abstract: Broiler chickens exposed to heat stress adapt to various behavioral changes to regulate their comfortable body temperature, which is critical to ensure their performance and welfare. Hence, assessing various behavioral responses in birds when they are subjected to environmental changes can be essential for assessing their welfare under heat-stressed conditions. This study aimed to evaluate the effect of two air velocity (AV) treatments on heavy broilers' behavioral changes from 43 to 54 days under summer conditions. Two AV treatments (high and low) were applied in six poultry growth chambers with three chambers per treatment and 44 COBB broilers per chamber from 28 to 61 days in the summer of 2019. Three video cameras placed inside each chamber (2.44 m × 2.44 m × 2.44 m in dimension) were used to record the behavior of different undisturbed birds, such as feeding, drinking, resting, standing, walking, panting, etc. The results indicate that the number of chickens feeding, drinking, standing, walking, sitting, wing flapping, and leg stretching changed under AV treatments. High AV increased the number of chickens feeding, standing, and walking. Moreover, a two-way interaction with age and the time of day can affect drinking and panting. This study provides insights into heavy broilers' behavioral changes under heat-stressed conditions and AV treatments, which will help guide management practices to improve birds' performance and welfare under commercial conditions in the future.

Keywords: heavy broiler; heat stress; air velocity; behavior

Citation: Akter, S.; Liu, Y.; Cheng, B.; Classen, J.; Oviedo, E.; Harris, D.; Wang-Li, L. Impacts of Air Velocity Treatments under Summer Conditions: Part II—Heavy Broiler's Behavioral Response. *Animals* **2022**, *12*, 1050. https://doi.org/10.3390/ani12091050

Academic Editors: Lilong Chai and Yang Zhao

Received: 31 March 2022
Accepted: 15 April 2022
Published: 19 April 2022

1. Introduction

Broiler chickens are now bred to reach their market size weight of about 2.3 to 4.5 kg at the age of 42 to 63 days due to the demand for deboned meat compared to the whole bird [1,2]. Faster growth and heavier body weight in a confined facility often challenge these birds' performance and welfare [3]. Moreover, global warming and climate change cause more heatwaves in the summertime. Consequently, birds are experiencing heat stress more often in summer due to increased temperature and relative humidity. Heat stress increases mortality rate [4,5] and feed conversion ratio [6,7], and decreases feed intake [7–9] and body weight gain [6,7,9]. This not only leads to economic loss but also compromises animal welfare.

Chickens adapt to heat stress by adopting several behavioral changes to maintain their homeostasis [10,11]. For example, birds eat less and drink more water [12,13]. Moreover, birds tend to sit, elevate their wings, and pant to dissipate excess heat produced from metabolism [7,14]. Therefore, when heat-stressed birds cannot release their body heat to the environment, they try to transfer heat to the environment through various activities and behavioral changes. In other words, these kinds of behavioral changes are indicators of their discomfort, and they provide further evidence of compromised welfare. Hence, it is essential to understand how birds respond under thermal stresses to provide them with the necessary support and means to ensure performance and welfare.

Several studies have been conducted to help understand broiler chickens' behavioral changes under different environmental conditions and other management strategies, such as through the use of dietary manipulation or the addition of supplements [11,15,16]. Adding different levels of propolis in feed to heat-stressed broilers increased walking but did not change feeding, drinking, wing elevation, or preening in 15–42-day-old chickens [11]. However, synbiotic-fed 15–42-day-old broilers showed less panting and wing spreading and more standing, sitting, walking, feeding, and preening [16]. The increased light intensity significantly affected 35-day-old broilers in behaviors such as lying, eating, drinking, standing, walking, preening while lying, wing/leg stretching, sleeping, dozing, vocalization, and idling [17,18]. Since diet manipulation does not always impact broiler behavior, these approaches do not consistently reduce heat stress. Moreover, none of these approaches were investigated for current market-sized broilers.

Controlling the inside environment of broiler grow-out houses is being recommended by several researchers [14,19] to reduce heat stress impact on broilers. Various studies have been conducted to verify the impact of air velocity (AV) on the thermal comfort of broilers under stressful conditions [20–23]. Since producers are growing heavier birds, it is even more crucial to understand the behavioral responses of bigger birds to provide a comfortable environment. However, no researchers have yet studied birds' behavioral changes pattern from 49 to 61 days under various air velocity treatments. Hence, the objective of this study was to investigate heavy broilers' behavioral responses to AV treatments under heat stress conditions. The effect of AV on feeding, drinking, standing, walking, sitting, panting, wing flapping, and leg stretching was investigated. The variation in these behaviors according to time of day and the age of the broilers was also examined.

2. Materials and Methods

2.1. Experimental Unit

The experiment was conducted in the poultry engineering laboratory (PEL) of North Carolina State University (NCSU) in the summer of 2019 from 29 May to 1 July. The PEL has six simulated poultry chamber systems with core chambers with dimensions of 2.44 m × 2.44 m × 2.44 m for the birds' stay (Figure 1). All these chambers are equipped with a nipple drinker line, four feeders, and an automated switch-timer soft lighting system. A belt-driven blower controlled by a variable frequency drive (VFD) system provides various ventilation rates and desired airspeeds to all the chambers in the range of 0.9–4.6 m/s at birds' heights according to their age and the ambient temperature. More detailed descriptions of the PEL and its operations are reported by Wang-Li [24], West [25], and Shivkumar [26].

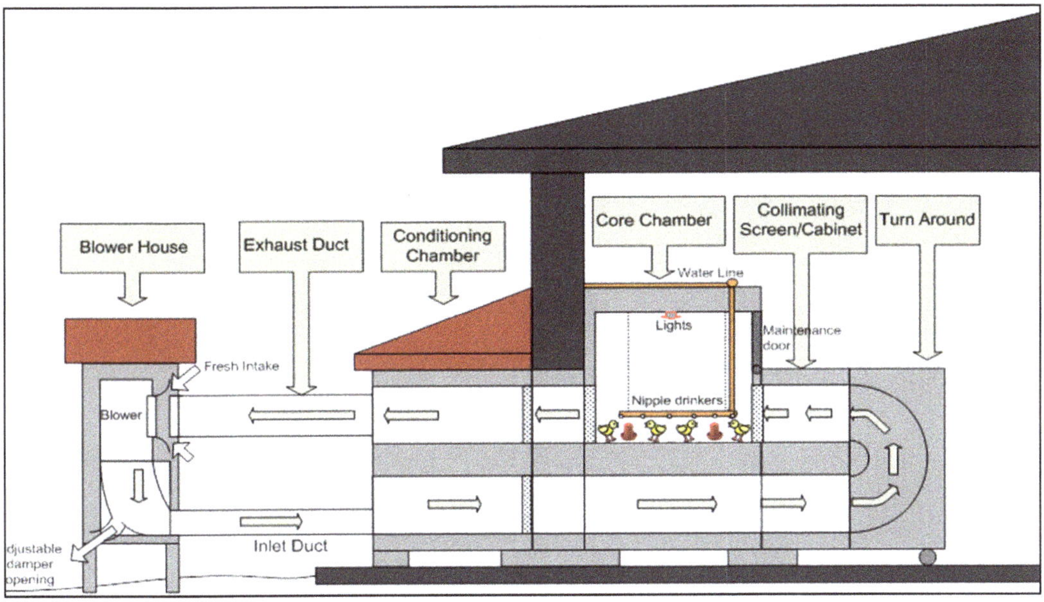

Figure 1. Cross section of the poultry chamber systems (from Shivkumar [26], used with permission).

2.2. *Animals*

A total of 400 male COBB 500 broilers were hatched and raised in the floor pens under similar conditions at the NCSU poultry unit. Then, 264 birds without leg defects were randomly selected to be placed in the six experimental chambers, with 44 birds per chamber after 28 days. They remained in the chamber until reaching the age of 61 days. The final stocking density was $\leq$40 kg/m^2, following the animal welfare guideline.

2.3. *Core Chamber Environmental Data Monitoring*

Each chamber's air temperatures (Ta) were monitored with a thermocouple, and a HOBO Pro v2 External T/RH Data Logger, Model U23-002 (Onset, Computer Corporation, MA, USA), was placed at the airflow inlet and outlet of each chamber at the birds' height. Calibrated thermocouples (range: −5 °C to 50 °C, and accuracy: ±0.002 °C) recorded temperature data continuously at 1 min intervals, while the HOBO logged both Ta and RH at 10 min intervals. The hourly average Ta values from both sensors were then averaged to obtain the hourly average temperature at the inlet.

2.4. *Air Velocity Treatments*

Two sets of dynamic AV treatments (high AV and low AV) were designed depending on inlet Ta and bird age. The AV treatment design criteria are detailed in [27]. As shown in Table 1, high and low AVs were designed for each of the six following growth condition classes: below optimum, around optimum, above optimum (moderate), above optimum (severe), above optimum (life-threatening), above optimum (warning). Changes in AV were achieved with programs written for VFD. It is important to note that, unlike previous broiler studies [21,22,28] using static AV treatments, this study implemented dynamic AV treatment levels based on the age of the birds and the air temperature, a procedure which is highly recommended and widely used in the broiler industry.

Table 1. High and low AV treatment design.

Treatment	Broiler Age (Days)	Temp °C	AV (m/s) Below Optimum T	Temp °C	AV (m/s) around Optimum T	Temp °C	AV (m/s) above Optimum T (Moderate)	Temp °C	AV (m/s) above Optimum T (Severe)	Temp °C	AV (m/s) above Optimum T (Life-Threatening)	Temp °C	AV (m/s) above Optimum T (Warning)
High Low	28–34 *	<26.0	0.9 0.9	26.0–27.8	1.23 1.23	27.8–28.9	1.33 1.33	28.9–32.2	1.48 1.48	32.2–33.9	1.64 1.64	>33.9	1.75 1.75
High Low	35–40	<21.7	0.9 0.9	21.7–26.0	1.23 1.23	26.0–30.0	2.02 1.48	30.0–33.0	2.77 2.02	33.3–37.8	3.45 2.77	>37.8	3.95 3.45
High Low	41–42	<21.1	1.48 1.48	21.1–26.0	1.48 1.48	26.0–30.0	2.02 1.48	30.0–33.0	2.77 2.02	33.3–37.8	3.45 2.77	>37.8	3.95 3.45
High Low	43–52	<20.6	1.48 1.48	20.6–26.0	1.75 1.75	26.0–30.0	2.02 1.75	30.0–33.0	2.77 2.43	33.3–37.2	3.95 3.02	>37.8	4.33 3.65
High Low	53–54	<19.4	1.48 1.48	19.4–25.0	1.75 1.48	25.0–29.5	2.43 1.75	29.4–32.7	3.02 2.43	32.7–36.1	3.95 3.02	>36.1	4.33 3.65
High Low	55–56	<19.4	1.48 1.48	19.4–25.0	1.75 1.48	25.0–29.5	2.43 1.75	29.4–32.7	3.02 2.43	32.7–35.6	3.95 3.02	>35.6	4.33 3.65
High Low	57–58	<18.9	1.48 1.48	18.9–25.0	1.75 1.48	25.0–29.5	2.43 1.75	29.9–32.2	3.02 2.43	32.2–35.6	3.95 3.02	>35.6	4.33 3.65
High Low	59–60	<18.9	1.48 1.48	18.9–24.4	2.43 1.75	24.4–28.9	3.02 2.43	28.9–31.7	3.45 2.77	31.7–35.0	4.33 3.65	>35.0	4.43 3.8
High Low	61	<18.3	1.48 1.48	18.3–23.9	2.43 1.75	23.9–28.9	3.02 2.43	28.9–31.7	3.45 2.77	31.1–33.9	4.33 3.65	>33.9	4.6 3.95

* A non-treatment period (in the first week) allowed the broilers to acclimate to their new environment.

The AV treatments began on the birds aged 35 days after the birds moved into the chamber for a week. As shown in Table 1, when the Ta was below optimal temperature, there were no AV differences between the two treatments in order to avoid cold stress, which would have complicated the experiments. For broilers aged 35–61 days, high AV treatments were applied to chambers 1, 3, and 5 and low AV treatments to chambers 2, 4, and 6. The difference in AV varied depending on bird age and the extent to which the measured chamber Ta differed from its value under optimal thermal conditions.

2.5. Behavioral Data Collection

Three video cameras (DVR-4580, Swann Communications, Santa Fe Springs, CA, USA) were installed in each chamber on the left- and right-side walls and on the celling to capture videos of undisturbed bird activity. The video recordings were saved in an external hard drive connected to each camera. Unfortunately, the hard drive that stored all the recordings of the cameras in chamber 2 and the ceiling cameras of all six chambers was destroyed; hence, only the two side wall video recordings of five chambers were available. Segmented videos were selected based on video quality, availability from both side cameras for the same period, and undisturbed bird appearance, as in Figure 2. The available videos lasted 10–20 min for various ages (43, 44, 49, 51, and 54 days) and times of day (early morning, morning, noon, afternoon, evening, and night); 8–10 min spans of these videos were watched manually to count the number of chickens for various behavioral poses. The classification of the time of day was as follows: early morning: 5:00–8:00, morning: 8:00–11:00, noon: 11:00–13:00, afternoon: 13:00–17:00, evening: 17:00–20:00 and night: 20:00–24:00.

Figure 2. Snap shots of videos from the same time from both (**left**) and (**right**) cameras in a chamber, showing 2 chickens drinking, 4 feeding, 12 panting, 1 leg stretching, 0 wing flapping, 1 standing, and 24 sitting.

The number of chickens feeding, drinking, walking, and standing (up on their feet but not feeding, drinking, or walking), panting, resting, stretching legs, or wing flapping was counted manually by an observer for each individual video. The ethogram in Table 2 [11,29] was used for observing various behaviors.

No specific chickens were marked or colored for observation; hence, only the number of chickens engaging in any behavior listed in Table 2 was counted from the videos. Only one chicken from chambers 3 and 4 died on the 53rd day. Hence, the total number of chickens in each chamber was 44 for the video monitoring periods selected.

Table 2. Bird behavior ethogram.

Behavior	Definition
Feeding	The bird's head is located inside the feeder.
Drinking	The bird's beak is in contact with the drinker.
Panting	The bird is breathing hard and quickly with a wide-open mouth and constantly shallow respiration.
Standing or Walking	Both feet are in contact with the floor; no other body part is in contact with floor.
Walking	The bird is in the process of taking at least 2 steps, including scratching the litter.
Sitting	Most of the ventral region of the bird's body is in contact with the floor. No space is visible between the floor and the bird.
Wing flap	Flapping wings so that space can be seen between the bird's wings and its body.
Leg stretching	Stretching one leg, often together with the wing of the same side, but the leg may also be stretched alone while sitting or standing.
	The behaviors were mutually exclusive.

2.6. Statistical Analysis

The data were analyzed using Rstudio (version 1.0.143) (RStudio, Boston, MA, USA). The number of birds having different behavior was the average from the replicated chambers under the two treatment AVs. A two-way ANOVA test was used to analyze the effect of treatment, age, time of day, and interactions on the number of chickens engaging in various behaviors. The main effects and the interactions were considered significant at $p < 0.05$. If any factor had main effects, the Tukey HSD test was performed to check the differences at the level of that variable. The replicated chambers were considered blocking factors and the number of chickens was considered the experimental unit.

3. Results

3.1. Environmental Conditions

The hourly averages of Ta and RH varied by time of day (Figure 3). There was no significant difference in Ta and RH in high and low AV treatment chambers. The hourly Ta values for inlets at 49, 51, and 54 days were higher than for those at 43 and 44 days as the later days were warmer. The average hourly Ta during the video recording days was 24.84 ± 4.16 °C, while the average hourly RH was 68.37 ± 15.42%. Higher hourly Ta values and lower RH values were observed during the afternoon as compared with other times of day. In general, RH was higher in the early morning.

The experiment was designed to obtain the behavioral changes of broiler chickens due to AV treatments under heat stress conditions. Figure 4 represents the time distribution of the AV treatments under the different growth conditions defined in Table 1. The birds were under heat stress conditions when the Ta was in one of the four growth conditions (i.e., moderate, severe, life-threatening, and warning). The data presented in Figure 4 reflect only the five days of video recording. The inlet Ta exceeded the optimal growth condition 35% of the time during those days. The Ta was below or around its optimum value during the 43rd and 44th days. On the 49th day, AV was primarily moderate; however, 26% of the time, the condition was severe. The growth condition never reached the AV warning condition during the observation period. There was a 22% life-threatening growth condition on the 54th day, when there was also an 18% severe condition. During these five days, there were no occurrences of the warning condition. The severe and life-threatening conditions mainly occurred during the afternoon and evening.

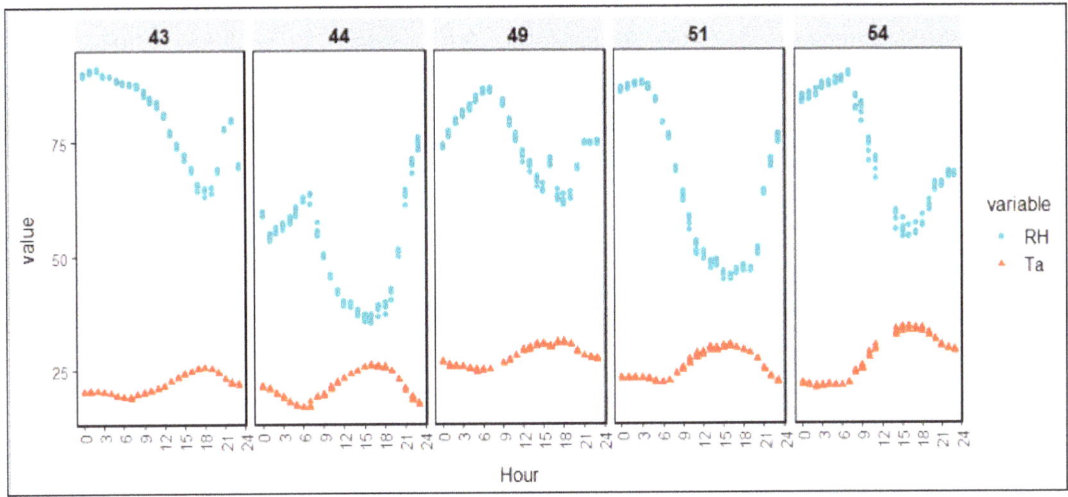

Figure 3. Average hourly Ta values and RH values by treatment during video observation periods.

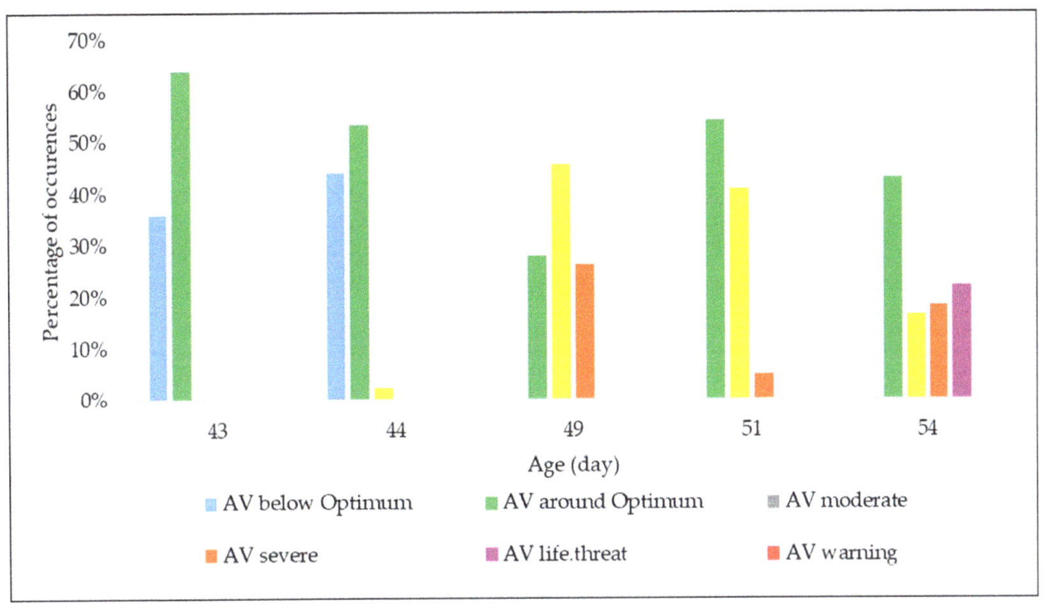

Figure 4. Time distribution of AV treatments implemented in all chambers only during video capture times.

3.2. Effect of the AV Treatments on Behavior

3.2.1. Feeding

Treatment and age significantly affected the number of chickens engaging in feeding behavior (Table 3). The number of birds feeding decreased with age (Figure 5). The number of chickens feeding was significantly higher under high AV treatment ($p < 0.05$) than low AV treatment (Table 4). Although the time of the day did not affect the number of birds feeding, the interaction with the treatment affected the number of chickens (Table 3).

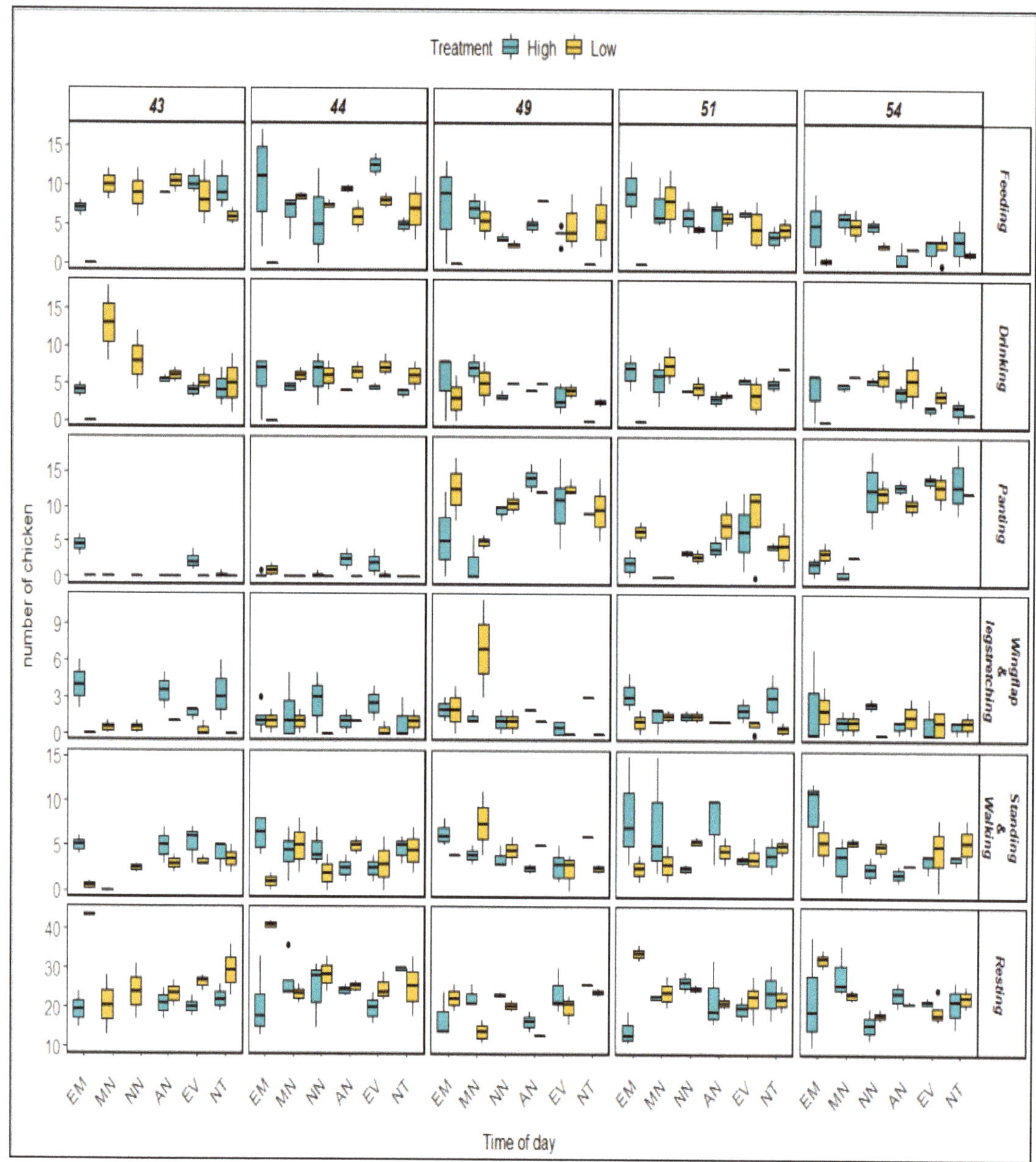

Figure 5. Broilers' various behaviors separated by age and time of day (EM = early morning; MN = morning; NN = noon; AN = afternoon; EV = evening; NT = night) under high and low AV treatments.

Table 3. Results of ANOVA test for differences in behavior according to AV treatment, age, and time of day.

Behavior	Effects (p-Value)						
	AV Treatment	Time of Day	Age	AV Treatment × Time of Day	AV Treatment × Age	Time of Day × Age	Chamber
Feeding	0.0171 *	0.1145	1.33×10^{-9} ***	1.10×10^{-5} ***	0.8334	0.4446	0.0205 *
Drinking	0.26698	4.69×10^{-5} ***	0.00353 **	2.26×10^{-6} ***	0.65882	0.01318 *	1.03×10^{-5} ***
Panting	0.593863	4.46×10^{-13} ***	$<2 \times 10^{-16}$ ***	0.285085	0.068111	2.30×10^{-6} ***	0.000473 ***
Standing and Walking	0.0478 *	0.1784	0.1844	0.0380 *	0.0175 *	0.6045	0.1844
Sitting	0.01777 *	0.14526	0.00945 **	1.38×10^{-7} ***	0.03087 *	0.27508	0.15007
Wing flapping and leg stretching	0.00493 **	0.2688	0.75742	0.17656	0.01473 *	0.68147	0.37366
df	1	5	4	20	4	20	4

Different asterisks represent different levels of significance (*** $p < 0.001$; ** $p < 0.01$; * $p < 0.05$, $p < 0.1$).

Table 4. Differences in the average number of chickens with different behavior according to various factors.

Behavior	Number of Chickens (Mean ± Std)												
	AV		Age (Days)					Time of Day *					
	High	Low	43	44	49	51	54	EM	MN	NN	AN	EN	NT
Feeding	6.1 ± 3.8 a	4.9 ± 3.0 b	8.2 ± 3.5 a	7.3 ± 4.1 ab	4.5 ± 3.3 cd	5.6 ± 3.2 bc	2.9 ± 2.4 d	4.8 ± 3.4 a	6.9 ± 2.6 a	5.1 ± 3.1 a	5.9 ± 3.4 a	5.8 ± 3.6 a	4.9 ± 3.5 a
Drinking	4.2 ± 2.2 a	4.6 ± 3.3 a	5.4 ± 3.9 a	5.1 ± 2.5 a	4.0 ± 2.4 ab	4.6 ± 2.5 ab	3.5 ± 2.4 b	3.4 ± 3.5 c	6.3 ± 3.2 a	5.3 ± 2.5 ab	4.5 ± 1.8 abc	4.0 ± 1.9 bc	3.7 ± 2.4 c
Panting	4.8 ± 5.5 a	5.1 ± 5.5 a	0.7 ± 1.6 c	0.5 ± 1.1 c	9.2 ± 4.7 a	4.3 ± 4.0 b	9.3 ± 5.7 a	3.4 ± 4.2 c	1.1 ± 1.9 d	5.6 ± 5.5 abc	6.3 ± 5.7 ab	7.6 ± 5.9 a	5.1 ± 5.9 bc
Standing and Walking	5.6 ± 2.9 a	3.7 ± 2.3 b	3.3 ± 2.1 a	1.0 ± 2.4 a	3.9 ± 2.1 a	5.1 ± 3.5 a	4.4 ± 2.8 a	5.3 ± 3.7 a	4.5 ± 3.6 a	3.7 ± 1.8 a	4.1 ± 2.5 a	3.5 ± 1.9 a	4.2 ± 1.8 a
Sitting	22.4 ± 5.9 b	24.5 ± 7.0 a	24.8 ± 7.9 ab	25.9 ± 6.9 a	20.8 ± 4.5 b	22.9 ± 5.9 ab	22.7 ± 6.3 ab	25.2 ± 11.1 a	23.5 ± 5.7 a	23.1 ± 4.6 a	21.9 ± 4.7 a	22.1 ± 3.9 a	24.7 ± 5.4 a
Wing Flapping and Leg Stretching	1.7 ± 1.6 a	0.9 ± 1.6 b	1.5 ± 1.9 a	1.3 ± 1.5 a	1.5 ± 2.1 a	1.5 ± 1.3 a	1.2 ± 1.6 a	1.9 ± 2.0 a	1.7 ± 2.4 a	1.3 ± 1.4 a	1.3 ± 1.1 a	87 ± 1.1 a	1.3 ± 1.3 a

a–d Different letters within a row under the same factor mean a significant difference at level 0.05; * EM = early morning; MN = morning; NN = noon; AN = afternoon; EV = evening; NT = night.

3.2.2. Drinking

The number of chickens drinking did not vary under AV treatment (Table 3). Although the treatment had no main effect on the number of chickens drinking, the interaction with the time of day was significant ($p < 0.05$) (Table 3). The number of chickens drinking significantly changed according to time of day, age, and their interaction (Table 3).

3.2.3. Standing and Walking

The numbers of chickens standing and walking were analyzed together, since it was difficult to distinguish between the two behaviors as there was no marking on the birds' bodies. The numbers of birds standing and walking were affected by AV treatments and their interaction with the time of day (Table 3). More birds were standing and walking under high AV treatment than low AV treatment (Figure 5). The time of day did not have main effects, but its interaction with the time of day significantly changed the number of chickens standing or walking. Age had no effect on the number of birds standing or walking.

3.2.4. Panting

Panting is one of the expected behaviors for maintaining thermoregulation under heat stress. The AV treatment had no main effect on the number of birds panting (Table 3). However, time of day, age, and their interaction significantly affected the number of panting birds. The number of birds panting increased with age (Figure 5). More birds were panting on day 49, 51, and 54 than on day 43 and 44 (Table 4). From noon to evening, more birds were panting than at night, in the early morning, or in the morning (Table 4).

3.2.5. Sitting

The AV treatment significantly affected the number of chickens sitting (Table 3). More birds were sitting under low AV treatment (Table 4). Age had main effects on the number of birds sitting (Table 3). More chickens were resting on day 44 than on day 49. Time of day had no main effect on the number of chickens sitting, but the interaction with AV treatment significantly affected the number of birds sitting. The interaction between age and AV treatment also changed the number of birds sitting.

3.2.6. Wing Flapping and Leg Stretching

Wing flapping and leg stretching are two common behaviors under heat stress. This study discovered that the AV treatment significantly changed the number of birds flapping their wings and stretching their legs (Table 3). The number of birds flapping their wings or stretching their legs was higher under high AV (Figure 5). Age and the time of day did not have any effect on these behaviors, but the interaction between AV and age significantly changed the number of birds flapping their wings or stretching their legs under heat stress conditions.

3.3. *Bird Sitting Location*

The chickens moved around the chamber for various purposes. For example, they moved to the feeder location for feeding and moved to the middle of the chamber for drinking. While resting, they tended to sit at various locations. The number of chickens sitting varied at the inlet and the outlet (Table 5). Under both AV treatments, the number of chickens sitting near the inlets was significantly higher ($p < 0.05$) than the number of those sitting near the outlets for chickens of any age and at any time of day (Table 5 and Figure 6). Time of day and age did not impact the number of chickens sitting at the inlet or the outlet.

Table 5. Results of ANOVA test for the effect of testing location on broilers' sitting behavior.

AV	Factors	Degrees of Freedom	Type III Sum of Squares	Mean Square	F-Value	$p > F$
High	Location (inlet/outlet)	1	441.6	441.6	104.587	$<2 \times 10^{-0.5}$ ***
	Time of Day	5	21	4.2	0.994	0.424
	Age	4	4.5	1.1	0.267	0.899
Low	Location (inlet/outlet)	1	473.5	473.5	127.459	$<2 \times 10^{-0.5}$ ***
	Time of Day	5	9.8	2	0.527	0.755
	Age	4	5	1.2	0.333	0.855

*** $p < 0.001$.

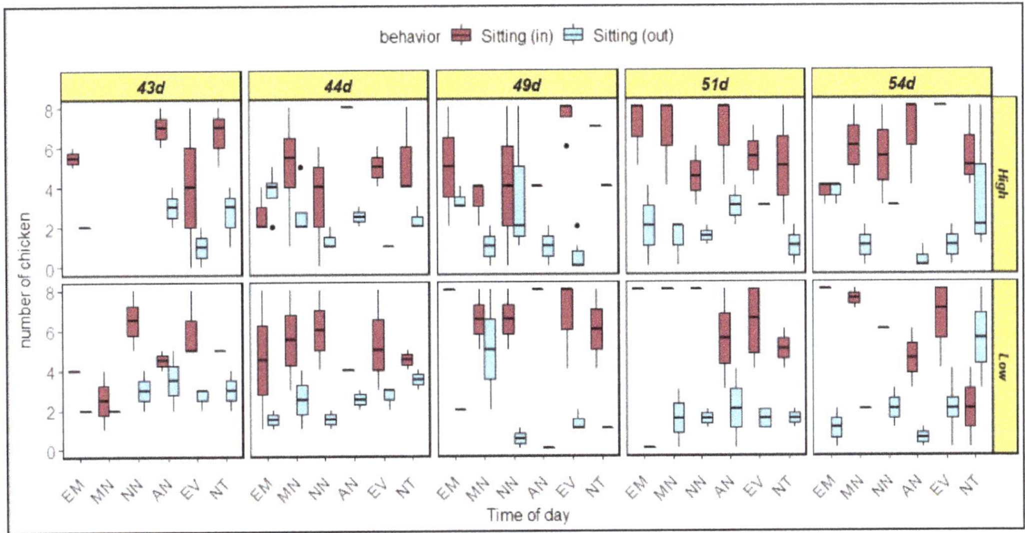

Figure 6. Difference in the number of chickens sitting at the inlets and the outlets of the chambers under both AV treatments during different times of the day (EM = early morning; MN = morning; NN = noon; AN = afternoon; EV = evening; NT = night.

4. Discussion

Under heat stress, birds usually reduce their feed intake according to their age up to 42 days [9,10]. Although the time budget for feeding and feed intake was not determined in this study, the decreased number of chickens feeding implies that decreased feeding occurs at later ages (42–54 days). A heavier body weight at a later age lessened birds' activity under heat stress. As a result, they could not release enough metabolic heat to bring their body temperature to the thermoneutral zone. Hence, their coping mechanism was to eat less in this context. The high AV treatment helped more chickens to eat under heat-stressed conditions. Hence, increased AV and proper mixing of AV can help regulate broiler performance even under faster growth rates and environmental stresses.

Irrespective of AV treatment, drinking declined with broiler age. Bizeray [10] and Jacobs [30] found increased drinking behavior in broilers up to six weeks of age, while Newberry [31] found a decreased water intake from the sixth to the ninth week under summer conditions. Newberry [31] found no time-of-day effect on drinking behavior, since the Ta was controlled and remained the same throughout the experiment. In this study, the AV treatments were dynamically changed with changes in Ta and bird age. Hence, it was reasonable to expect variations in the number of chickens drinking water according to time of day. Birds often drink more under heat stress [32], but due to their heavier body weight in this study, the birds tended to walk less, which might be another reason for the decrease in the number of chickens drinking. An interesting but not statistically significant observation is that during the early morning on any day, at noon and in the morning on the 49th day, and in the evening of the 51st day and the night at 54th day, more birds were drinking under high AV treatment (Figure 4). The collected videos from the low AV chambers were all during the lights-off condition for the early morning. Hence, the birds had almost zero activity in those chambers, leading to fewer chickens drinking. The same condition is applicable for the 54th day's night observation. On the 49th day, the AVs in both treatments were from moderate to severe all day long, except for the early morning. Hence, under both treatments, the birds tended to drink more.

The birds selected for this experiment were all without leg defects. During the experimental period, the birds' activity was more involved in either feeding or drinking. The heavier broilers were found to be less likely to walk or stand. It was even observed that they barely made more than two or three steps unless it was required to reach the drinker or feeder. The heavier body weight and stressful weather condition made them sit more often than walk or stand. Li [33] found restless walking behavior under heat stress in 21-day-old birds, but during this experiment, less walking and standing were observed among birds aged 42–54 days. As the birds aged, they sometimes resorted immediately to standing. Moreover, lameness or laziness was prominent when the thermal environment exceeded severe conditions. Less walking and standing indicate heat stress behavior for heavier birds. Since the number of chickens standing and walking increased under high AV treatment, increasing AV might be a good management strategy under heat-stressed conditions to cool down the birds and make them comfortable.

The absence of sweat glands and the presence of feathers increase broilers' panting under higher environmental temperature to release excessive body heat. The AV treatments in this study did not significantly impact panting behavior. Hence, the number of chickens panting under the two treatments did not differ significantly. Panting increased with age (Figure 4). On the 49th, 51st, and 54th days, the number of chickens panting was higher than on the 43rd and 44th days. Under both treatments, the growth condition exceeded the optimum condition 72%, 46%, and 57% of the time on days 49, 51, and 54, respectively (Figure 3). Hence, the birds panted more often on those days. The number of birds panting was higher during noon, afternoon, and evening while the environment was warmer, which is consitent with the results of Lott et al. [34], who found that 4-to-6-week-old birds panted more often during warmer periods of the day while sitting on the floor pen at 0.25 m/s AV. A lower number of panting birds was observed in the early morning and morning. The growth condition was around or below optimum in the early morning. In the morning, the condition was mainly optimum or moderate, and the birds did not experience significant heat stress during those times; there was therefore less of a necessity to pant to release heat. At a later age, even in the nighttime, many birds were panting. This occurred because the growth condition reached severe and moderate on those days. Lott et al. [34] found decreased panting in birds aged 4 to 6 weeks under tunnel ventilation with AV in 2.08 m/s. Hence, the effect of AV requires further investigation to help heavy broilers manage under stressful conditions.

This study suggests that the number of birds sitting increases with age. Li [33] found that the duration of lying down increased with age up to 21 days in order to decrease basal metabolic rate and resist heat stress. Although our study did not analyze the duration of lying down or sitting, the higher number of birds indicates that they were more likely to sit at a later age. At an older age, the birds' bodies were heavier, and hence they were less likely to walk or stand. Even activities for releasing heat, such as panting, wing flapping, or leg stretching, mostly occurred while the birds were sitting. Feeding and drinking activity also decreased with age, so that the number of birds sitting was higher under every treatment. Fewer birds were sitting under high AV treatment because high AV helped them release some heat and thereby feel comfortable enough to move for feeding, drinking, or other activities. Although the time of day did not impact the number of birds resting, its interaction with AV treatment impacted the number. Tao and Xin [35] suggest that broiler chickens' core body temperature responds to the cumulative action of dry-bulb temperature, dew point temperature, and air velocity. Hence, in this experiment, the birds' core body temperature also changed at different times of the day under different temperatures and corresponding air velocities, which led to changes in the number of birds sitting or resting.

Although the number of chickens engaging in wing flapping or leg stretching was small, the occurrences were observed under stressed conditions. On average, only a couple of chickens flapped their wings or stretched their legs, but at a later age, when the Ta was higher, at most 11 birds in the low AV treatment chamber were found engaging in this activity. They might have been trying to cope with the stressful environment with this behavior.

According to the ANOVA test, the blocking factor chamber had a significant impact on the number of chickens feeding, drinking, and panting (Table 3). Although the replicated chambers under high AV treatments were not significantly different according to environmental conditions (Figure 3), surprisingly, the number of chickens feeding, drinking, and panting was significantly higher in chambers 1 and 3 than in chamber 5. On days 43, 44, and 49, early morning and nighttime videos from chamber 5 were taken while the lights were off. Therefore, the number of chickens with different activities was very low at those times, which might have impacted the results. Hence, a balanced design and careful data collection processes are recommended for future investigation.

The birds moved around the chamber floor primarily to drink or eat. They sometimes walked simply to find a better location in the chamber. During warmer periods, the birds tended to sit more often near the inlet to experience a higher speed of air flow passing over their bodies. Since the air entered the chamber through the inlet and the air speed was higher there, the birds under heat stress wanted to sit more often at the inlet side compared to the outlet, where upwind birds may block some flow to reduce the air speed in the downwind zone. From this study, it was found that the heavier birds tended to stay closer to the location where air entered when conditions were warmer or stressful. Hence, it is important to make sure that all the birds in a facility can experience uniform air velocity passing over them to keep them comfortable. Bringing down air flow to birds' height and properly mixing AV throughout the housing facility could be a good strategy for pacifying birds under heat stress. A collective understanding of these behavioral changes will help in understanding the potential issues associated with broilers' welfare in the flock in order to take necessary management actions to maintain and/or enhance performance and welfare.

5. Conclusions

These results indicate that high AV treatment significantly changed the number of chickens engaging in feeding, standing, walking, sitting, wing flapping, and leg stretching behavior under heat stress. The numbers of chickens feeding and panting increased with age, but the number of those drinking and sitting declined. The number of chickens drinking and panting varied significantly according to the time of the day. The applied AV did not directly affect drinking and panting, but its two-way interaction with age and time of day significantly altered the prevalence of these behaviors under thermal stress. In general, heavy broilers changed their typical behaviors, such as feeding, drinking, walking, or resting, under heat-stressed conditions in order to adapt to the stressor. Air velocity can enhance the heat release activity of larger and older broiler chickens, ensuring their growth performance and welfare. The findings from this study will help to identify thermal stress through behavioral changes in the birds themselves. This will help producers make necessary management decisions to keep their birds healthy and happy under conditions of environmental stress.

Author Contributions: Conceptualization and methodology, L.W.-L., E.O. and S.A.; formal analysis, S.A., D.H.; investigation, S.A.; data curation, S.A., B.C., L.W.-L. and Y.L.; writing—original draft preparation, S.A.; writing—review and editing, L.W.-L., J.C. and S.A.; supervision, L.W.-L.; project administration, L.W.-L. and E.O.; funding acquisition, L.W.-L., E.O. and J.C. All authors have read and agreed to the published version of the manuscript.

Funding: This research was funded by USDA-NIFA-AFRI, grant number 2017-67021-26329.

Institutional Review Board Statement: The experimental use of birds was approved by the Animal Care and Use Committee and conducted in compliance with the Guidelines for Care and Use of Laboratory Animals at North Carolina State University (NCSU) (IACUC#16-279-A).

Data Availability Statement: The data presented in this study are available upon request from the corresponding author.

Acknowledgments: We acknowledge the funding agency and all the other contributors for their enormous support. North Carolina State University contributed to this research through general faculty and graduate research support. C. Sayde, M. Adcock, P. Harris, R. Currin, D. Carroll, and S. Hall helped with building the research facility and installing sensors and other controlling devices. D. Harris helped with statistical analysis. Three graduate students, Derek West, Yijia Dietrich, and Blake Stratton, eight undergraduate research assistants, and three high school interns assisted with data collection.

Conflicts of Interest: The authors declare no conflict of interest.

References

1. Guidefreak. Expected Broiler Weight per Week and How to Get It! 2019. Available online: https://guidefreak.com/broiler-weight-per-week/amp/ (accessed on 10 August 2021).
2. National Chicken Council. Chicken Checkin, What is the Difference between Faster- and Slower-Growing Chicken? 2021. Available online: https://www.chickencheck.in/faq/difference-faster-slower-growing-chicken/ (accessed on 10 August 2021).
3. Lange, K. Super-Size Problem. The Humane Society of the United States. 2017. Available online: https://www.humanesociety.org/news/super-size-problem-broiler-chickens (accessed on 15 September 2021).
4. Kang, S.; Kim, D.-H.; Lee, S.; Lee, T.; Lee, K.-W.; Chang, H.-H.; Moon, B.; Ayasan, T.; Choi, Y.-H. An Acute, Rather Than Progressive, Increase in Temperature-Humidity Index Has Severe Effects on Mortality in Laying Hens. *Front. Veter. Sci.* **2020**, *7*, 853. [CrossRef] [PubMed]
5. Warriss, P.; Pagazaurtundua, A.; Brown, S. Relationship between maximum daily temperature and mortality of broiler chickens during transport and lairage. *Br. Poult. Sci.* **2005**, *46*, 647–651. [CrossRef]
6. Awad, E.A.; Najaa, M.; Zulaikha, Z.A.; Zulkifli, I.; Soleimani, A.F. Effects of heat stress on growth performance, selected physiological and immunological parameters, caecal microflora, and meat quality in two broiler strains. *J. Anim. Sci.* **2016**, *33*, 778–787. [CrossRef] [PubMed]
7. Sohail, M.U.; Hume, M.E.; Byrd, J.A.; Nisbet, D.J.; Ijaz, A.; Sohail, A.; Shabbir, M.Z.; Rehman, H. Effect of supplementation of prebiotic mannan-oligosaccharides and probiotic mixture on growth performance of broilers subjected to chronic heat stress. *Poult. Sci.* **2012**, *91*, 2235–2240. [CrossRef] [PubMed]
8. Mashaly, M.M.; Hendricks, G.L.; Kalama, M.A.; Gehad, A.E.; Abbas, A.O.; Patterson, P.H. Effect of heat stress on production parameters and immune responses of commercial laying hens. *Poult. Sci.* **2004**, *83*, 889–894. [CrossRef]
9. Quinteiro-Filho, W.M.; Gomes, A.V.S.; Pinheiro, M.L.; Ribeiro, A.; Ferraz-De-Paula, V.; Astolfi-Ferreira, C.S.; Ferreira, A.J.P.; Palermo-Neto, J. Heat stress impairs performance and induces intestinal inflammation in broiler chickens infected with Salmonella Enteritidis. *Avian Pathol.* **2012**, *41*, 421–427. [CrossRef]
10. Bizeray, D.; Estevez, I.; Leterrier, C.; Faure, J.M. Effects of increasing environmental complexity on the physical activity of broiler chickens. *Appl. Anim. Behav. Sci.* **2002**, *79*, 27–41. [CrossRef]
11. Mahmoud, U.T.; Abdel-Rahman, M.A.M.; Darwish, M.H.A.; Applegate, T.J.; Cheng, H.W. Behavioral changes and feathering score in heat stressed broiler chickens fed diets containing different levels of propolis. *Appl. Anim. Behav. Sci.* **2015**, *166*, 98–105. [CrossRef]
12. Mack, L.A.; Felver-Gant, J.N.; Dennis, R.L.; Cheng, H.W. Genetic variations alter production and behavioral responses following heat stress in 2 strains of laying hens. *Poult. Sci.* **2013**, *92*, 285–294. [CrossRef]
13. Saeed, M.; Abbas, G.; Alagawany, M.; Kamboh, A.A.; Abd El-Hack, M.E.; Khafaga, A.F.; Chao, S. Heat stress management in poultry farms: A comprehensive overview. *J. Therm. Biol.* **2019**, *84*, 414–425. [CrossRef]
14. Daghir, N.J. *Poultry Production in Hot Climates*, 2nd ed.; Cab International: Oxfordshire, UK, 2008; pp. 11–30.
15. Shakeri, M.; Cottrell, J.J.; Wilkinson, S.; Le, H.H.; Suleria, H.A.R.; Warner, R.D.; Dunshea, F.R. Growth performance and characterization of meat quality of broiler chickens supplemented with betaine and antioxidants under cyclic heat stress. *Antioxidants* **2019**, *89*, 336. [CrossRef] [PubMed]

16. Mohammed, A.A.; Jacobs, J.A.; Murugesan, G.R.; Cheng, H.W. Effect of dietary synbiotic supplement on behavioral patterns and growth performance of broiler chickens reared under heat stress. *Poult. Sci.* **2018**, *97*, 1101–1108. [CrossRef] [PubMed]
17. Calvet, S.; van den Weghe, H.; Kosch, R.; Estellés, F. The influence of the lighting program on broiler activity and dust production. *Poult. Sci.* **2009**, *88*, 2504–2511. [CrossRef] [PubMed]
18. Senaratna, D.; Samarakone, T.S.; Gunawardena, W.W.D.A. Red color light at different intensities affects the performance, behavioral activities and welfare of broilers. *Asian-Australas. J. Anim. Sci.* **2016**, *29*, 1052–1059. [CrossRef]
19. Renaudeau, D.; Collin, A.; de Basilio Yahav, S.; Gourdine, J.L.; Collier, R.J. Adaptation to hot climate and strategies to alleviate heat stress in livestock production. *Animal* **2012**, *6*, 707–728. [CrossRef]
20. Mitchell, M.A. Effects of air velocity on convective and radiant heat transfer from domestic fowls at environmental temperatures of 20° and 30 °C. *Br. Poult. Sci.* **1985**, *26*, 413–423. [CrossRef]
21. Simmons, J.D.; Lott, B.D.; May, J.D. Heat loss from broiler chickens subjected to various air speeds and ambient temperatures. *Appl. Eng. Agric.* **1997**, *13*, 665–669. [CrossRef]
22. Yahav, S.; Straschnow, A.; Vax, E.; Razpakovski, V.; Shinder, D. Air velocity alters broiler performance under harsh environmental conditions. *Poult. Sci.* **2001**, *80*, 724–726. [CrossRef]
23. Furlan, R.L.; Macari, M.; Secato, E.R.; Guerreiro, J.R.; Malheiros, E.B. Air velocity and exposure time to ventilation affect body surface and rectal temperature of broiler chickens. *J. Appl. Poult. Res.* **2000**, *9*, 1–5. [CrossRef]
24. Wang-Li, L.; Shivkumar, A.P.; Xu, Y.; Munilla, R.D.; Adcock, M.E.; Tutor, J.C.; Brake, J.; Williams, C.M. Performance of a poultry engineering chamber complex for animal environment and welfare studies. In Proceedings of the International Symposium on Animal Environment and Welfare, Chongqing, China, 19–22 October 2013.
25. West, D.R. Litter Characteristics, Ammonia Emissions, and Leg Health of Heavy Broilers as Impacted by Air Velocity Treatments: A Chamber Study. Master's Thesis, North Carolina State University, Raleigh, NC, USA, 2020.
26. Shivkumar, A.P.; Wang-Li, L.; Shah, S.B.; Stikeleather, L.; Fuentes, M. Performance analysis of a poultry engineering chamber complex for animal environment, air quality, and welfare studies. *Trans. ASABE* **2016**, *59*, 1371–1381.
27. Akter, S.; Cheng, B.; West, D.; Liu, Y.; Qian, Y.; Zou, X.; Classen, J.; Cordova, H.; Oviedo, E.; Wang-Li, L. Impacts of air velocity treatment under summer condition: Part I —heavy broilers' surface temperature response. *Animals* **2022**, *12*, 328. [CrossRef] [PubMed]
28. Simmons, J.D.; Lott, B.D.; Miles, J.D. The effects of high-air velocity on broiler performance. *Poult. Sci.* **2003**, *82*, 232–234. [CrossRef] [PubMed]
29. Pichova, K.; Nordgreen, J.; Leterrier, C.; Kostal, L.; Moe, R.O. The effects of food-related environmental complexity on litter directed behaviour, fear and exploration of novel stimuli in young broiler chickens. *Appl. Anim. Behav. Sci.* **2016**, *174*, 83–89. [CrossRef]
30. Jacobs, L.; Melick, S.; Freeman, N.; Garmyn, A.; Tuyttens, F.A.M. Broiler chicken behavior and activity are affected by novel flooring treatments. *Animals* **2021**, *11*, 2841. [CrossRef] [PubMed]
31. Newberry, R.C.; Hunt, J.R.; Gardiner, E.E. Influence of light intensity on behavior and performance of broiler chickens. *Poult. Sci.* **1988**, *67*, 1020–1025. [CrossRef] [PubMed]
32. Bruno, L.D.G.; Maiorka, A.; Macari, M.; Furlan, R.L.; Givisiez, P.E.N. Water intake behavior of broiler chickens exposed to heat stress and drinking from bell or and nipple drinkers. *Rev. Bras. Cienc. Avic.* **2011**, *13*, 147–152. [CrossRef]
33. Li, M.; Wu, J.; Chen, Z. Effects of heat stress on the daily behavior of wenchang chickens. *Braz. J. Poult. Sci.* **2015**, *17*, 559–566. [CrossRef]
34. Simmons, J.D.; Lott, B.D.; May, J.D. Air Velocity and High Temperature Effects on Broiler Performance. *Poult. Sci.* **1998**, *77*, 391–393. [CrossRef]
35. Tao, X.; Xin, H. Acute synergistic effects of air temperature, humidity, and velocity on homeostasis of market–size broilers. *Trans. ASAE* **2003**, *46*, 491. [CrossRef]

MDPI

Article

Survival of *Escherichia coli* in Airborne and Settled Poultry Litter Particles

Xuan Dung Nguyen [1], Yang Zhao [1,*], Jeffrey D. Evans [2], Jun Lin [1] and Joseph L. Purswell [2]

1 Department of Animal Science, The University of Tennessee, Knoxville, TN 37996, USA; ndung@vols.utk.edu (X.D.N.); jlin6@utk.edu (J.L.)
2 Poultry Research Unit, Agriculture Research Service, United States Department of Agriculture (USDA), Mississippi State, MS 39762, USA; jeff.evans@usda.gov (J.D.E.); joseph.purswell@usda.gov (J.L.P.)
* Correspondence: yzhao@utk.edu; Tel.: +1-865-974-6466

Simple Summary: Airborne transmission is recognized as an important mechanism of disease spreading in livestock and poultry production, yet is far from being fully understood. Evaluating the impact of airborne transmission requires information of the microbial survivability. We determined the survivability of the *E. coli*—a common microbial species found in poultry environment—in airborne particles, settled dust, and poultry litter under laboratory environmental conditions. The poultry litter which contained mainly manure mixed with fresh wood shavings was collected from a commercial farm. Results of the study showed that the half-life time of airborne *E. coli* was 5.7 ± 1.2 min. The half-life time of *E. coli* in poultry litter and settled particles was 15.9 ± 1.3 h and 9.6 ± 1.6 h, respectively. The findings of this study will help better estimate the impact of airborne transmission of *E. coli* in poultry production.

Abstract: Airborne *Escherichia coli* (*E. coli*) in the poultry environment can migrate inside and outside houses through air movement. The airborne *E. coli*, after settling on surfaces, could be re-aerosolized or picked up by vectors (e.g., caretakers, rodents, transport trucks) for further transmission. To assess the impacts of airborne *E. coli* transmission among poultry farms, understanding the survivability of the bacteria is necessary. The objective of this study is to determine the survivability of airborne *E. coli*, settled *E. coli*, and *E. coli* in poultry litter under laboratory environmental conditions (22–28 °C with relative humidity of 54–63%). To determine the survivability of airborne *E. coli*, an AGI-30 bioaerosol sampler (AGI-30) was used to collect the *E. coli* at 0 and 20 min after the aerosolization. The half-life time of airborne *E. coli* was then determined by comparing the number of colony-forming units (CFUs) of the two samplings. To determine the survivability of settled *E. coli*, four sterile Petri dishes were placed on the chamber floor right after the aerosolization to collect settled *E. coli*. The Petri dishes were then divided into two groups, with each group being quantified for culturable *E. coli* concentrations and dust particle weight at 24-h intervals. The survivability of settled *E. coli* was then determined by comparing the number of viable *E. coli* per milligram settled dust collected in the Petri dishes in the two groups. The survivability of *E. coli* in the poultry litter sample (for aerosolization) was also determined. Results show that the half-life time of airborne *E. coli* was 5.7 ± 1.2 min. The survivability of *E. coli* in poultry litter and settled *E. coli* were much longer with the half-life time of 15.9 ± 1.3 h and 9.6 ± 1.6 h, respectively. In addition, the size distribution of airborne *E. coli* attached to dust particles and the size distribution of airborne dust particles were measured by using an Andersen impactor and a dust concentration monitor (DustTrak). Results show that most airborne *E. coli* (98.89% of total *E. coli*) were carried by the dust particles with aerodynamic diameter larger than 2.1 μm. The findings of this study may help better understand the fate of *E. coli* transmitted through the air and settled on surfaces and evaluate the impact of airborne transmission in poultry production.

Keywords: airborne *E. coli*; settled *E. coli*; survivability; airborne transmission; poultry

Citation: Nguyen, X.D.; Zhao, Y.; Evans, J.D.; Lin, J.; Purswell, J.L. Survival of *Escherichia coli* in Airborne and Settled Poultry Litter Particles. *Animals* **2022**, *12*, 284. https://doi.org/10.3390/ani12030284

Academic Editor: Colin Scanes

Received: 23 December 2021
Accepted: 19 January 2022
Published: 24 January 2022

1. Introduction

The United States of America (USA) is one of the leading countries in poultry production. Poultry products originating in the USA primarily consist of meat from broilers and turkeys and eggs from layers. According to the USDA report [1], the combined value of production from these products in 2020 exceeded USD 35 billion. These products provide important and affordable sources of dietary protein to the domestic population. In addition, approximately 18% of the USA poultry products are exported and poultry production in the USA was estimated to provide over 1 million jobs. However, the outbreak of infectious diseases is one of the biggest challenges for the poultry industry. For example, the Highly Pathogenic Avian Influenza (HPAI) outbreaks in the USA in 2015 resulted in losses of over 50 million birds and 3.3 billion dollars [2].

Escherichia coli (*E. coli*) is a member of the *Enterobacteriaceae* family and is commonly associated with the intestinal tract of warm-blooded animals and the environment in which these animals reside. In poultry, *E. coli* primarily inhabits the lower gastrointestinal tract as an indicator for the poultry environmental quality and exists there as an important commensal species. Typically, *E. coli* are harmless, but some *E. coli* strains may be pathogenic in nature and their virulence may lead to losses in the poultry industry. Pathogenic *E. coli* strains in poultry are commonly referred to as avian pathogenic *E. coli* (APEC) [3]. The APEC causes the systemic disease colibacillosis in broilers. The severity of APEC disease depends on the health status of the host, virulence characteristics of the *E. coli* strain, and other predisposing factors such as stress. Approximately 30% of broiler flocks in the U.S are infected by subclinical colibacillosis [4].

E. coli can be abundant in poultry house with concentrations up to 4 $\log_{10}$ CFU m^{-3} in the air [5], 3 $\log_{10}$ CFU g^{-1} in feeds [6], and 7 $\log_{10}$ CFU g^{-1} in poultry litter [7]. To reduce the economic losses caused by *E. coli*, antibiotics, such as tetracyclines and trimethoprim sulfamethoxazole, have been widely utilized in poultry feed [8]. However, the widespread use of antibiotics can cause the emergence and re-emergence of antibiotic resistant bacterial strains. Thus, the use of antibiotics has been limited and many bacteria, including *E. coli*, have reemerged as significant threats to poultry production. Some alternatives were developed to reduce *E. coli* contamination of the farm microclimate such as probiotics [9] and UV lights [10]. These methods do not rely on the use of antibiotics and are relatively effective in reducing microbial contamination in poultry houses. However, these studies have not mentioned the effectiveness of reducing airborne bacteria which attach to dust particles. Therefore, further studies on airborne *E. coli* attached to dust particles such as their survivability or size distribution which directly affects the effectiveness of the methods are needed to investigate.

The litter is a major reservoir of microorganisms in the poultry environment [11]. The dry matter contents can be about 70–80% of litter mass and it can contain abundant biological organisms and compounds that can affect the quality of the poultry environment [12]. Dust particles are aerosolized because of bird activity, as such, the poultry environment is highly dusty.

Air in the poultry houses may contain abundant microorganisms such as *E. coli* [13]. *E. coli* from manure first deposit into poultry litter and are then aerosolized through bird activities [14]. Ventilation systems can drive their migration across a poultry house or even from barn to barn. Airborne *E. coli* were shown to account for 2–6% of the total airborne bacteria in poultry houses [5]. With the high concentration of *E. coli* and the possibility of barn-to-barn transmission, the airborne *E. coli* can harm the entire wide range of environment outside the poultry houses, and they can deposit on surfaces near the poultry houses. The barn-to-barn airborne transmission of avian influenza was investigated in a study conducted in 2019 [15]. The probability of airborne infection is affected by several factors including farm type, flock size, and distance of transmission where the survivability of the pathogen is among the key factors for the modeling accuracy. Moreover, the survivability of *E. coli* on stainless steel under refrigeration conditions and room temperature was reported to exceed 28 days [16]. Therefore, it is also possible that *E. coli* can persist for a long time on various surfaces in the poultry production environment. With such a long survival period on the surface, they can spread to larger areas through vectors. These all raise the question of how long the airborne *E. coli*, carried by poultry litter particles, can survive in the air and on the physical surfaces when settled.

To determine the survivability of airborne and settled *E. coli* in laboratory, a proper aerosolization method that may mimic the fate of *E. coli* in the commercial poultry production environment is required. The wet aerosolization method such as nebulization was widely used to study the survivability of airborne *E. coli* [17]. However, the airborne *E. coli* in poultry houses are aerosolized from dried litter by bird activities, such as dust bathing [14]. So, the results of the study based on wet aerosolization cannot apply to the actual situation in the poultry house. In addition, the survivability of settled *E. coli* after going through the dry aerosolization process has never been investigated. Therefore, a study to determine the survivability of airborne *E. coli* and settled *E. coli* after being aerosolized based on dry aerosolization method needs to be done.

Size distribution of airborne *E. coli* attached to dust particles could affect the survivability of airborne *E. coli*. In a study conducted by Zuo et al. [18], the authors mentioned that carrier particle size had a significant effect on the survivability of airborne viruses. Lighthart et al. [19] also reported that test bacterial survivability increased directly with droplet size. However, most of the studies used droplets as aerosol particles to carry bacteria and viruses. The dry dust particles may yield different results compared to droplets. So, the size distribution of airborne *E. coli* attached to dry dust particles also needed to be investigated.

This study aimed to investigate the survivability of airborne and settled *E. coli* via dry aerosolization under room thermal conditions. In addition, the survivability of *E. coli* in poultry litter was also investigated as a reference parameter.

2. Materials and Methods

To investigate the survivability of the airborne *E. coli* and the settled *E. coli*, experiments were run in a test chamber in a Biosafety Level 2 (BSL-2) laboratory. The survivability test of *E. coli* in poultry litter was conducted in Biosafety Level 1 (BSL-1) laboratory. Both laboratories are located at the Animal Science Department, University of Tennessee, Knoxville, TN 37996, USA.

2.1. *Microorganism and System Descriptions*

2.1.1. Preparation of *E. coli* Solution

The *E. coli* strain used in this study was *Escherichia coli* GFP (ATCC® 25922GFP™) which was purchased from American Type Culture Collection (ATCC, Manassas, VA, USA). *E. coli* strain was cultured at 37 °C, 150 rpm for 24 h in ATCC® Medium 2855 (Tryptic Soy Broth 'TSB' with 100 mcg mL^{-1} Ampicillin and Tryptic Soy Agar 'TSA'). The bacterial concentrations of *E. coli* in the solution after 24 h were from 8 to 9 $\log_{10}$ colony-forming units ($\log_{10}$ CFU) mL^{-1}.

2.1.2. Litter Preparation

Litter from the commercial broiler farm was first collected and stored in a container. It was then brought back to the BSL-1 laboratory to analyze the dry matter content. After that, the litter was autoclaved at 121 °C in 20 min and divided into identical-size aluminum boxes with the amount of 6 kg per box. The autoclaved poultry litter was used as a source of organic matter to simulate the biological conditions in poultry environment [20]. The sterilization was confirmed to demonstrate a state of freedom from microbial contamination. The boxes were sealed by aluminum foils and covered by plastic caps to avoid contamination. They were stored in a 4 °C fridge until being used.

It was important to prepare litter so that the bacteria were evenly distributed. To do that, 240 g of litter needed for the survivability test of airborne *E. coli* and settled *E. coli* experiment were equally distributed into 40 ceramic cups (6 g litter per cup). The amount of airborne dust that can be generated using a mixer was determined in a previous experiment [21], and the results showed that 240 g of litter would produce dust concentrations ranging from 0.9 to 1.1 mg m^{-3} which was within a typical range of dust concentration in commercial poultry farm [22]. To prepare litter inoculated with *E. coli*, litter in each of the 40 cups was mixed with 6 mL of *E. coli* solution. The 6 mL bacteria solution was sprayed evenly onto the litter in each cup. In the meantime, an aluminum spoon was used to gently mix the litter and *E. coli* solution. The mixtures then went through a process of drying at 22 °C and 52–67% relative humidity (RH) for 48 h until the dry matter content (DMC) of the mixture reached about 70%. The *E. coli* concentration in each cup was approximately 4 $\log_{10}$ CFU mg^{-1} litter after the drying process. The litter containing *E. coli* was then transferred from 40 ceramic cups to a metal bowl of the mixer for aerosolization. In the bowl, the litter was gently mixed up again before aerosolization.

2.1.3. Test Chamber

Aerosolization was performed in an acrylic chamber. This chamber (2100 series, Cleatech, Orange, CA, USA) was a non-vacuum unit with two internal access doors with stainless steel frame, and a removable fully gasketed back wall. The dimension of the test chamber was 1.5 mL × 0.6 mW × 0.6 mH. The chamber was well sealed to prevent dust-laden particles from spilling out. It was also equipped with a temperature and RH sensor for continuously monitoring the inside thermal environment.

In the settled *E. coli* experiment, the chamber was modified to create a highly dusty environment in order to collect adequate settle dust for analysis. Initial results showed that the aerosolization space of the entire chamber was too large which led to the low concentration of airborne *E. coli* and dust particles. Thus, the chamber was modified by halving the aerosolization space using a partition acrylic film. The aerosolization space after modification was 0.75 mL × 0.6 mW × 0.6 mH.

2.1.4. Aerosolization System

A stand mixer (model DCSM350GBRD02, New York, NY, USA) was used for dry aerosolization of airborne *E. coli* in this study. The dimension of the mixer was 0.3 mL × 0.2 mW × 0.3 mH with a 3.3 L stainless steel bowl. They operated at the highest speed to ensure the bacteria concentration in the air was high enough. A stir fan was also used to distribute the airborne *E. coli* in the chamber evenly.

2.1.5. Dust Concentration Monitor

To monitor the dust concentration throughout the experiment, a dust concentration monitor (DustTrak DRX aerosol monitor 8533, TSI Inc., Shoreview, MN, USA) was used to provide data on the mass concentration of dust particles with different sizes. DustTrak was capable of measuring dust particles of PM 1, PM 2.5, PM 4.7, and PM 10. In this study, the dust concentration and particle size were recorded, and the results indicated that the particle concentration was relatively stable between experimental events.

2.1.6. Air Samplers

To evaluate the survivability of the airborne *E. coli*, the AGI-30 impinger (AGI-30) was used to collect *E. coli*-laden dust particles in a test chamber in a BSL-2 laboratory. The AGI-30 operates at 12.5 L min^{-1}. The airborne compounds were sucked through a fine nozzle in which the particles were accelerated and then impacted directly into the 20 mL TSB. The AGI-30 was proven to have the highest performance among three commonly used samplers (Andersen impactor, AGI-30 impinger, and BOBCAT ACD-200) for collecting airborne *E. coli* [21].

2.2. Experimental Design and Procedures

2.2.1. Bacterial Size Distribution and Viable *E. coli* Recovering in the Airborne *E. coli* Survivability Test

An Andersen impactor was used to monitor the bacterial size distribution. The Andersen impactor is designed as an aerodynamic classifying system for airborne particles. It operates at 28.3 L min^{-1}. Its six stages are designed to sort dust particles with different sizes of >7 µm, 4.7–7 µm, 3.3–4.7 µm, 2.1–3.3 µm, 1.1–2.1 µm, 0.65–1.1 µm, corresponding to stage 1 to stage 6. The dust particles carrying *E. coli*, after being aerosolized, were sucked in the intake on top of the Andersen impactor; then, the particles continuously went through 6 stages. For each stage, dust particles with sizes corresponding to each stage were collected on TSA agar plates.

In the process of sampling with the Andersen impactor, the stages of the sampler were often overloaded due to the excessive number of bacteria collected in each stage. Therefore, counting bacteria on agar plates directly was not possible. To overcome this problem, the agar plate washing method was applied. Bacteria, after being collected on agar plates, were immediately taken to the laboratory for analysis. Each agar plate was rinsed with 2 mL of TSB solution with the aid of a glass spreader, and then 1 mL of solution was collected by pipette. The 1 mL of this solution went through a traditional serial dilution process to determine the total *E. coli* in the solution. The agar plates, after washing, were also placed in an incubator letting the remaining *E. coli* on the plate grow. During the air sampling process, the agar plates in the Andersen impactor were dried by air flow in the sampler. Thus, the remaining 1 mL of solution in the washing process was mostly reabsorbed into the agar plates. However, to make sure that there is no residual solution that could affect the test results, the agar plates that have been washed instead of being turned upside down (due to traditional culture process) will be left right side up. The total *E. coli* on each stage was the combination of total *E. coli* collected from washing and total *E. coli* remaining on agar plates.

2.2.2. Dry Matter Content Measurement

The moisture content is one variable affecting the survivability of bacteria [23]. The dry matter content (DMC), which is the inverse term of moisture content, was measured over time in the experiment. The DMC measurement of poultry litter is the ratio of the litter mass before and after the litter is completely dried. To determine DMC, the process was divided into two stages. First, the litter mass (m_1) was weighted before going through a 48-h drying process until the litter mass was totally dried. After being dried at 105 °C, the litter mass (m_2) was weighted again. The DMC was then calculated by the litter mass m_2 divided by the litter mass m_1.

2.2.3. Sample Collection for Airborne *E. coli*

Two hundred and forty grams (240 g) of litter which contained ~4 $\log_{10}$ CFU mg^{-1} litter of *E. coli* were prepared and placed in the mixer. The mixer was placed in the center of the chamber to help evenly distribute the dust particles carrying *E. coli*. The mixer was fixed to the chamber surface by means of suckers, preventing it from moving during the running process. The stir fan was placed at the corner of the chamber to aid in distributing airborne particles. The AGI-30 was placed near the steel bowl of the mixer.

Each test lasted a total of 50 min. The first 20 min of the test was the aerosolization process of airborne *E. coli* using the mixer and stir fan. After the 20-min aerosolization, airborne *E. coli* was collected using the AGI-30 for 10 min and the dust concentration was determined using DustTrak. The second sampling of airborne *E. coli* and dust followed the same protocol but was performed 10 min after the first sampling. This test procedure was repeated 7 times.

2.2.4. Sample Collection for Settled *E. coli*

Two mixers were used for aerosolization. Two hundred and forty grams (240 g) of litter which contained about 4 $\log_{10}$ CFU mg^{-1} litter of *E. coli* were mixed gently and divided into two parts with 120 g for each mixer. The stir fan was operated during the aerosolization to improve the distribution of airborne *E. coli* in the chamber. Four Petri dishes were placed on both sides of the mixers to collect particles settled from the air. To avoid the position confounding effect, the Petri dishes were arranged randomly in a total of 4 experiment events. Each event started with 15 min aerosolization. After the aerosolization, the four Petri dishes were covered with caps and sealed by parafilm. Two Petri dishes were immediately analyzed to quantify viable *E. coli* via traditional culture technique. The remaining two Petri dishes were left at laboratory temperature at 20 °C, RH at 60% for 24 h. After that, they were quantified for viable *E. coli* by the same culture technique. The weight of each Petri dish was determined before and after aerosolization to determine the settle dust weight. The airborne dust concentration during the mixer running time was also monitored by DustTrak.

2.2.5. Viable *E. coli* Counting for *E. coli* Survivability Test in Poultry Litter

Fifteen (15) ceramic cups, each with six grams (6 g) of poultry litter, were prepared to determine the survivability of *E. coli* in the litter. The six grams of poultry litter were spread in each ceramic cup so that the thickness of the litter was uniform and without large lumps. Then, 6 mL of *E. coli* solution was added to the litter by using a pipette. The solution was sprayed onto the litter, ensuring that the bacterial fluid was distributed as evenly as possible. After that, the mixture of litter and bacterial solution were mixed gently by using an aluminum spoon. The cup was then placed in the BSL-1 under laboratory conditions. The viable *E. coli* in the litter were determined at 0, 12, 24, 48, and 72 h after litter samples were prepared in the ceramic cups. At each time point, three cups of samples were used. In addition, two cups of litter added with the TSB solution instead of the bacteria solution were used as a control for *E. coli* analysis and DMC measurement.

To determine the viable *E. coli* counts, TSB was added in each cup so that the total volume of the mixture reached 15 mL. The mixture was mixed evenly. Then, 0.1 mL of the solution (litter-bacteria mixture mixed with TSB) was taken out and transferred to 0.9 mL of TSB. After that, the solution went through a serial dilution process to determine the counts of viable *E. coli*. By doing back-calculation, the bacterial concentration in poultry litter was calculated.

2.2.6. Determining *E. coli* Concentration in Poultry Litter

To determine the viable *E. coli*, the *E. coli* concentrations were calculated in logarithm colony-forming units per gram ($\log_{10}$ CFU mg^{-1}) using Equation (1).

$$C = \log_{10}\left(\frac{N \times 10^n}{V_P} \times V_s \times \frac{1}{m_a}\right), \tag{1}$$

where C = the bacteria concentration, $\log_{10}$ CFU mg^{-1}; N = the number of colonies on a countable plate (30 to 300 colonies); n = serial dilution factor (n = 0 for undiluted sample, n = 1 for 10-fold diluted sample, etc.); V_P = the sample volume plated, mL (V_P = 0.1 mL in this study); V_s = the total volume of the original liquid sample, mL; m_a = the total poultry litter weight in each ceramic cup at the test time, mg.

2.2.7. Determining Airborne *E. coli* Concentration

Each air sample collected by AGI-30 in liquid form (in TSB medium) was used to quantify viable *E. coli* via traditional culture techniques. After vortexing for 5 s, a 0.1 mL subsample, after going through the serially diluted (1:10) process, was plated onto TSA agar plates. In each experimental event, the subsample was uniformly repeated 3 times to ensure the accuracy of the experiment. The plates were aerobically incubated at 37 °C for 24 h. The visible *E. coli* colonies formed on plates (30 to 300 colonies) were determined. Based on the culture results and the sampled air volume, airborne *E. coli* concentrations were calculated in logarithm colony-forming units per cubic meter ($\log_{10}$ CFU m^{-3}) using Equation (1). The parameter m_a converted to V_a which is the total air volume sampled using the bioaerosol samplers, m^3.

2.2.8. Determining Settled *E. coli* Concentration

Each settled sample on an empty Petri dish was used to quantify viable settled *E. coli*. After adding 10 mL of TSB medium (the culture medium) in each Petri dish, the Petri dish was gently shaken to wash the Petri dish surface and draw settled *E. coli* into TSB solution. After that, 0.1 mL of the solution containing *E. coli* was taken by using a pipette and went through a serial dilution process to count viable *E. coli*. Then, the viable *E. coli* was determined as the Equation (1). The parameter m_a was the mass of settled dust collected in each dish in each experiment, mg.

2.3. *Calculation of Half-Life Time*

The half-life time is the time interval needed for bacteria to decrease by half [24]. The bacterial concentrations throughout the experiments would be homogenized and normalized to the dust concentration (CFU mg^{-1}). In the survivability of the airborne *E. coli* test, the airborne *E. coli* concentration was calculated based on airborne *E. coli* concentration collected in the air (CFU m^{-3}) divided by total dust concentration (mg m^{-3}). In the survivability of the settled *E. coli* test, the settled *E. coli* concentration was calculated based on the settled *E. coli* concentration collected on each Petri dish (CFU mg^{-1}). The half-life time, then, was calculated by the following Equation (2).

$$t_{1/2} = \frac{\log_{10} 2 \times T}{\log_{10}\left(C_{\text{viable bacteria}}/C'_{\text{viable bacteria}}\right)}, \quad (2)$$

where $t_{1/2}$: half-life time (min or h); T = 20 (min) for airborne *E. coli* and 24 (h) for settled *E. coli* test; $C_{\text{viable bacteria}}$: *E. coli* concentration for the first sampling event, CFU mg^{-1}; $C'_{\text{viable bacteria}}$: *E. coli* concentration for the second sampling event, CFU mg^{-1}.

Linear simple regression was performed to calculate the half-life time of *E. coli*. The half-life time of *E. coli* in poultry litter was calculated based on the *E. coli* death over time by the linear Equation (3) [25]:

$$t_{1/2} = \frac{\text{constant} - \log_{10}\left(\frac{C_{\text{viable bacteria}}}{2}\right)}{k}, \quad (3)$$

where $C_{\text{viable bacteria}}$: the *E. coli* concentration at 0 h, CFU mg^{-1}; constant: intercept of the linear regression model, $\log_{10}$ CFU mg^{-1}; k: the death rate, [$\log_{10}$ CFU mg^{-1}] h^{-1}; and $t_{1/2}$: half-life time, h.

2.4. *Statistical Analysis*

Means and standard deviations for all experiments were calculated by using Rstudio (Rstudio, open-source license, Rstudio, Boston, MA, USA). Total 7 replicates for airborne *E. coli* experiment and 4 replicates for settled *E. coli* yielded decent statical analysis for calculating the half-life time. The conditions such as dust concentration among experiments were tested with the T-test to make sure there was no significant difference in terms of experimental conditions. The *t*-test significance level was 0.05 ($p < 0.05$). For the survivability of *E. coli* in poultry litter, at every time point, the concentration of *E. coli* in poultry litter was tested repeatedly 3 times for reliable viable *E. coli* data.

The half-life time of airborne *E. coli*, settled *E. coli* and *E. coli* in poultry litter were compared, and the differences between the survivability of *E. coli* under different conditions were tested by using a *t*-test run on Rstudio. The *t*-test was used to determine if the means of three sets of data (*E. coli* in poultry litter, airborne *E. coli*, and settled *E. coli*) are significantly different from each other. The *t*-test significance level was 0.05 ($p < 0.05$).

3. Results

3.1. Conditions for E. coli Survivability Test

Table 1 shows the litter DMC, initial litter *E. coli* concentration and environmental conditions during the experiments for determining survivability of airborne *E. coli*, settled *E. coli* and the *E. coli* in poultry litter. The DMC of litter, *E. coli* concentration and RH in the litter were kept stable throughout the experiments. In the test for settled *E. coli* survivability, instead of using one mixer, two mixers were used. Therefore, the heat generated in the two mixers caused the temperature in the test for settled *E. coli* survivability to be slightly higher than the two other tests.

Table 1. Conditions (Mean ± SD) for the *E. coli* in survivability test.

E. coli Concentration and Environmental Conditions	Test for Airborne *E. coli* Survivability	Test for Settled *E. coli* Survivability	Test for *E. coli* in Poultry Litter Survivability
DMC [1] of litter (%)	71 ± 5	72 ± 1	- [2]
E. coli concentration in litter (log_{10} CFU mg^{-1})	4.4 ± 0.6	4.0 ± 0.5	- [2]
Relative humidity (%)	54 ± 5	63 ± 7	36 ± 4
Temperature (°C)	22.1 ± 1.4	27.7 ± 5.1	20.5 ± 0.3

[1] Dry matter content, [2] DMC and bacteria concentration varied over 72 h.

3.2. Size Distribution of E. coli and Dust for the Airborne E. coli Survivability Test

The size distribution of airborne *E. coli* attached to dust particles and the size distribution of airborne dust particles were tested. The size distribution of airborne *E. coli* attached to dust particles during the 20-min aerosolization process is shown in Figure 1. The most *E. coli* were found in the particles larger than 7 μm with a percentage of 47.58%. The second large portion of *E. coli* was those attached to particles in the range of 4.7 to 7 μm, accounting for 27.34%. *E. coli* attached to dust particles in the ranges of 3.3–4.7 μm and 2.1–3.3 μm accounted for 14.05% and 9.92% of the total culturable *E. coli*, respectively. The least *E. coli* were found in particles smaller than 2.1 μm which accounted for 1.11% of the total culturable *E. coli*.

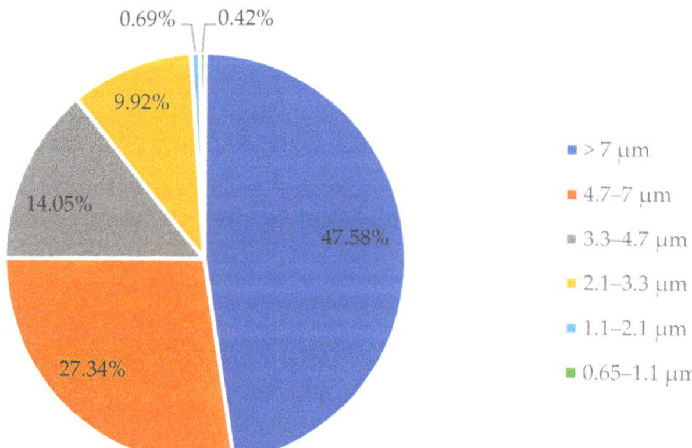

Figure 1. Size distribution of the airborne *E. coli* attached to dust particles in the airborne *E. coli* survivability test measured by an Andersen impactor.

The size distribution of airborne dust particles during the 20-min aerosolization process was monitored by the DustTrak and shown in Table 2. Most dust particles have the size smaller than 1 μm with a concentration of 0.678 ± 0.108 mg m^{-3}. The rest of the dust particles have size range of 1.0–2.5 μm, 2.5–4.7 μm, 4.7–10.0 μm and larger than 10.0 μm, with a concentration of 0.014 ± 0.001 mg m^{-3}, 0.016 ± 0.005 mg m^{-3}, 0.235 ± 0.042 mg m^{-3} and 0.232 ± 0.032 mg m^{-3}, respectively. The total dust concentration was about 1.176 ± 0.120 mg m^{-3}. As shown in Table 2 and Figure 1, although most dust particles were smaller than 1 μm, the size distribution of bacteria attached to dust particles was mainly larger than 2.1 μm, accounting for 98.89%. This indicates that when it comes to airborne *E. coli*, most are attached to dust particles with the size larger than 2.1 μm.

Table 2. Dust size distribution (Means ± SD) in the airborne *E. coli* survivability test.

<1.0 μm (mg m^{-3})	1.0–2.5 μm (mg m^{-3})	2.5–4.7 μm (mg m^{-3})	4.7–10.0 μm (mg m^{-3})	>10.0 μm (mg m^{-3})	Total (mg m^{-3})
0.678 ± 0.108 (57.60%) [1]	0.014 ± 0.001 (1.20%) [1]	0.016 ± 0.005 (1.40%) [1]	0.235 ± 0.042 (20.00%) [1]	0.232 ± 0.032 (19.80%) [1]	1.176 ± 0.120 (100.00%) [1]

[1] Percentage of the total for each size range.

3.3. *E. coli* Survivability in Poultry Litter

The survivability of *E. coli* in poultry litter was determined in a 72-h test under laboratory conditions and delineated in Figure 2. The temperature and RH remained stable throughout the test at 20.5 ± 0.3 °C and 36 ± 4%. The DMC of litter (containing *E. coli*) changed throughout the test and was presented in Figure 2. The *E. coli* concentration decreased from 4.5 $\log_{10}$ CFU mg^{-1} to 2.4 $\log_{10}$ CFU mg^{-1} over 72 h. The DMC increased from 38% to 82% due to moisture evaporation. The half-life time of *E. coli* in poultry litter calculated based on the linear regression was 15.9 ± 1.3 h.

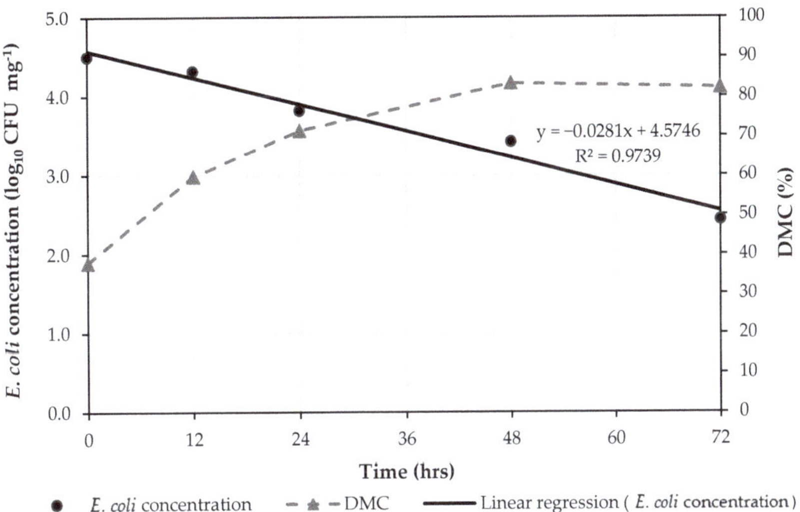

Figure 2. *E. coli* concentration and dry matter content (DMC) in poultry litter in a 72-h exposure under laboratory environmental condition (20.5 °C and 36%).

3.4. Airborne *E. coli* Survivability

The data collected from the first sampling and the second sampling to calculate the half-life time of *E. coli* were listed in the Table 3. As shown in Figure 1, most of the airborne *E. coli* were attached to dust particles larger than 2.1 μm, while only a small amount of total *E. coli* (1.11%) attached to dust particles smaller than 2.1 μm. Therefore, when calculating the concentration of *E. coli* in dust, we only considered the concentration of dust particles larger than 2.1 μm. The DustTrak was able to monitor the dust particles having size range of 1.0–2.5 μm, 2.5–4.7 μm, 4.7–10.0 μm and larger than 10.0 μm. In this study, we assumed that the amount of dust particles larger than 2.1 μm were equivalent to the amount of dust particles larger than 2.5 μm. The half-life time of the airborne *E. coli* based on dust with size > 2.5 μm was 5.7 ± 1.2 min.

Table 3. Concentrations (Mean ± SD) of dust particles with size larger than 2.5 μm, airborne *E. coli* and airborne *E. coli*-to-dust ratio during air sampling for survivability test of airborne *E. coli*. The 2nd sampling was performed 20 min after 1st sampling.

Concentrations of Dust Particles and Airborne *E. coli*	1st Sampling	2nd Sampling
Dust concentration with size > 2.5 μm (mg m^{-3})	0.032 ± 0.022	0.016 ± 0.012
Airborne *E. coli* concentration (log_{10} CFU m^{-3})	7.1 ± 0.7	5.7 ± 1.0
Airborne *E. coli* concentration carried by dust concentration with size > 2.5 μm (log_{10} CFU mg^{-1})	8.7 ± 0.7	7.5 ± 0.9

3.5. Settled E. coli Survivability

The survivability of settled *E. coli* was tested over 24 h. In 24 h, the concentration of settled *E. coli* declined from 3.7 ± 0.1 to 3.0 ± 0.2 log_{10} CFU mg^{-1}, yielding a half-life time of 9.6 ± 1.6 h for settled *E. coli*.

4. Discussion

The aim of this study was to determine the survivability of airborne and settled *E. coli* in laboratory under dry aerosolization conditions. Survivability of *E. coli* was determined using half-life time as the indicator. To calculate the half-life time, concentrations of airborne *E. coli* and settled *E. coli* collected at two different time points after the dry aerosolization process were measured and compared. The survivability of *E. coli* in poultry litter that was used for dry aerosolization was also determined in a 72-h test under laboratory conditions (20.5 ± 0.3 °C and 36 ± 4%). The results show that half-life times of airborne *E. coli*, settled *E. coli*, and *E. coli* in poultry litter were 5.7 ± 1.2 min, 9.6 ± 1.6 h, and 15.9 ± 1.3 h, respectively.

In the airborne *E. coli* survivability test, the mean half-life time of the bacteria based on dust particles with size larger than 2.5 μm was 5.7 min. Hoeksma et al. [26] tested survivability of airborne *E. coli* under wet aerosolization conditions at 20 °C and 40–60%. Their results showed that the half-life time of airborne *E. coli* under wet aerosolization conditions was about 2 min, which was much shorter than the half-life time calculated in the present study. The difference between the half-life time of airborne *E. coli* under wet aerosolization conditions and dry aerosolization conditions could be explained by inactivation due to evaporation. After being aerosolized, the wet aerosols lost their water film due to evaporation and become sensitive to ambient influences [26]. Moreover, the difference in preparation of *E. coli* for aerosolization between the two studies could be another reason of the discrepancy in survivability results. In the current study, the *E. coli* were prepared in poultry litter and exposed at laboratory conditions over 48 h before aerosolization. As such, the *E. coli* had already gone through a dehydration process before aerosolization, which might leave only dehydration-resistant *E. coli* for following dry aerosolization. In the study by Hoeksma et al. [26], the *E. coli* were aerosolized immediately after preparation via the wet aerosolization. In addition, the autoclaving process of poultry litter could affect the quality of poultry litter and produce Maillard reaction product. The Maillard reaction products were proven to inhibit growth of bacteria [27]. However, the effect of the preparation procedure was not well-studied in the present study. Therefore, the effect still needs further investigation.

Survivability and transmission range of airborne *E. coli* may be affected by the size of particles that *E. coli* attached to. Zuo et al. [18] reported that the carrier particle size had a significant influence in the transmission and survivability of airborne virus. In their study, the authors mentioned that the survivability of virus attached to larger particles was much longer than that attached to smaller particles. The possible explanation presented by

Zuo et al. [18] was the shielding effect. In other words, compared with viruses existing as a singlet or attaching to small particles, the virus attached to larger particles could be better protected from changes of ambient environment [28]. The concentration of *E. coli* should be proportional to the weight of airborne dust in the entire size spectrum, assuming a uniform mixture of *E. coli* and poultry litter. However, most of dust particles were smaller than 1.0 μm (accounted for 57.60%) and the majority of airborne *E. coli* were found to attach to dust particles larger than 2.1 μm (98.89%). This contradiction could be explained again by the shielding effect. While *E. coli* attached to large particles could be protected from ambient influences, *E. coli* attached to small particles received less protection effect. It led to a rapid death of the *E. coli* attached to small particles during the aerosolization and sampling.

The half-life time of settled *E. coli* in this study was about 9.6 h. Wilks et al. [16] tested the survivability of *E. coli* on metal surfaces at laboratory conditions at 20 °C. In their study, the total number of viable *E. coli* dropped by 1 log after the first 3 h, translating into an approximate 0.9 h half-life time. This discrepancy can be explained by differences in *E. coli* preparation methods, surfaces, and substrate (litter vs. liquid solution). As mentioned above, the *E. coli* preparation procedure in our study may affect the *E. coli* quality. Another possible explanation was metal surfaces used by Wilks et al. [16]. While the present study used regular plastic Petri dishes to collect settled *E. coli*, Wilks et al. [16] applied *E. coli* directly onto metal surfaces. This different material of surfaces could yield different survivability of *E. coli*. Ketkar et al. [29] indicated that stainless steel had antimicrobial effects. Further, different substrates (litter vs. liquid solution) used might have yielded different survivability of *E. coli*. While factors like pH and nutrient in poultry litter includes many affecting the survivability of bacteria [30,31], liquid solution used by Wilks et al. [16] for culturing *E. coli* was designed as a substrate for bacterial growth.

In the test of *E. coli* survivability in poultry litter, the half-life time was reported to be 15.9 ± 1.3 h. Compared with the half-life time of settled *E. coli* (9.6 h) and airborne *E. coli* (5.7 min), the half-life time of *E. coli* in poultry litter was significantly longer. A possible explanation was that the *E. coli* in the poultry litter did not go through the aerosolization process which negatively affect the *E. coli* survivability [32]. While settled *E. coli* and airborne *E. coli* were aerosolized, *E. coli* in the poultry litter were not aerosolized. In addition, degree of sample exposure to the environment could affect the survivability of *E. coli* as well. Ruiz et al. [33] reported that bacterial survival was highly influenced by ambient influences. The airborne *E. coli* were scattered in the air and the settled *E. coli* were prepared in thin layers where *E. coli* were exposed to ambient environment and were more susceptible to microenvironment changes [34,35], as compared to *E. coli* in the poultry litter. In contrast, the *E. coli* in poultry litter existed in a chuck form could be more protected from microenvironmental effects [34–36].

5. Conclusions

The study determined the survivability of airborne, settled, and poultry litter *E. coli* under dry aerosolization conditions in laboratory. Based on the results, we conclude that (1) most *E. coli* could be carried by the dust particles with aerodynamic diameter >2.1 μm, (2) the settled *E. coli* and the *E. coli* in poultry litter can survive much longer than airborne *E. coli*, and the mean half-life time was 5.7 ± 1.2 min for airborne *E. coli*, 9.6 ± 1.6 h for settled *E. coli*, and 15.9 ± 1.3 h for *E. coli* in poultry litter.

Author Contributions: Conceptualization, X.D.N. and Y.Z.; methodology, X.D.N., Y.Z.; formal analysis, X.D.N.; investigation, X.D.N.; resources, Y.Z., J.L., J.D.E., J.L.P.; data curation, X.D.N.; writing—original draft preparation, X.D.N.; writing—review and editing, all co-authors; visualization, X.D.N.; supervision, Y.Z.; project administration, Y.Z.; funding acquisition, Y.Z., J.D.E., J.L.P. All authors have read and agreed to the published version of the manuscript.

Funding: Financial support for the study was provided by the USDA National Program (Project No.6064-13000-013-10S).

Institutional Review Board Statement: Not applicable.

Informed Consent Statement: We have got the verbal permission from farmers to use the poultry litter for the study.

Acknowledgments: We appreciate the cooperation and assistance of the broiler grower in providing the broiler litter. This research was supported by the USDA National Institute of Food and Agriculture multistate project under accession number 1028091.

Conflicts of Interest: The authors declare no conflict of interest. The funders had no role in the design of the study; in the collection, analyses, or interpretation of data; in the writing of the manuscript, or in the decision to publish the results.

References

1. *Poultry—Production and Value 2020 Summary*; U.S. Department of Agriculture, National Agricultural Statistics Service (USDA, NASS): Washington, DC, USA, 2020.
2. Torremorell, M.; Alonso, C.; Davies, P.R.; Raynor, P.C.; Patnayak, D.; Torchetti, M.; McCluskey, B. Investigation into the airborne dissemination of H5N2 highly pathogenic avian influenza virus during the 2015 spring outbreaks in the Midwestern United States. *Avian Dis.* **2016**, *60*, 637–643. [CrossRef] [PubMed]
3. Saif, Y.; Barnes, H.; Glisson, J.; Fadly, A.; McDougald, L.; Swayne, D. *Diseases of Poultry*, 12th ed.; Blackwell Pub Professional: Ames, IA, USA, 2008; pp. 452–514.
4. Fancher, C.A.; Zhang, L.; Kiess, A.S.; Adhikari, P.A.; Dinh, T.T.; Sukumaran, A.T. Avian pathogenic Escherichia coli and Clostridium perfringens: Challenges in no antibiotics ever broiler production and potential solutions. *Microorganisms* **2020**, *8*, 1533. [CrossRef] [PubMed]
5. Zucker, B.A.; Trojan, S.; Müller, W. Airborne gram-negative bacterial flora in animal houses. *J. Vet. Med. B Infect Dis. Vet. Public Health* **2000**, *47*, 37–46. [CrossRef] [PubMed]
6. Munoz, L.; Pacheco, W.; Hauck, R.; Macklin, K. Evaluation of commercially manufactured animal feeds to determine presence of Salmonella, Escherichia coli, and Clostridium perfringens. *J. Appl. Poult. Res.* **2021**, *30*, 100142. [CrossRef]
7. Martin, S.A.; McCann, M.A.; Waltman, W.D., II. Microbiological survey of Georgia poultry litter. *J. Appl. Poult. Res.* **1998**, *7*, 90–98. [CrossRef]
8. Pitout, J.D. Extraintestinal Pathogenic *Escherichia coli*: A Combination of Virulence with Antibiotic Resistance. *Front. Microbiol.* **2012**, *3*, 9. [CrossRef]
9. Stęczny, K.; Kokoszyński, D. Effect of probiotic preparations (EM) on productive characteristics, carcass composition, and microbial contamination in a commercial broiler chicken farm. *Anim. Biotechnol.* **2021**, *32*, 758–765. [CrossRef]
10. McLeod, A.; Hovde Liland, K.; Haugen, J.E.; Sørheim, O.; Myhrer, K.S.; Holck, A.L. Chicken fillets subjected to UV-C and pulsed UV light: Reduction of pathogenic and spoilage bacteria, and changes in sensory quality. *J. Food Saf.* **2018**, *38*, e12421. [CrossRef]
11. Carpenter, G. Dust in livestock buildings—Review of some aspects. *J. Agric. Eng. Res.* **1986**, *33*, 227–241. [CrossRef]
12. Schulz, J.; Ruddat, I.; Hartung, J.; Hamscher, G.; Kemper, N.; Ewers, C. Antimicrobial-resistant *Escherichia coli* survived in dust samples for more than 20 years. *Front. Microbiol.* **2016**, *7*, 866. [CrossRef]
13. Sanz, S.; Olarte, C.; Hidalgo-Sanz, R.; Ruiz-Ripa, L.; Fernández-Fernández, R.; García-Vela, S.; Martínez-Álvarez, S.; Torres, C.J.A. Airborne Dissemination of Bacteria (Enterococci, Staphylococci and Enterobacteriaceae) in a Modern Broiler Farm and Its Environment. *Animals* **2021**, *11*, 1783. [CrossRef] [PubMed]
14. Zhao, Y.; Aarnink, A.; De Jong, M.; Groot Koerkamp, P.W.G. Airborne Microorganisms from Livestock Production Systems and Their Relation to Dust. *Crit. Rev. Environ. Sci. Technol.* **2014**, *44*, 1071–1128. [CrossRef] [PubMed]
15. Zhao, Y.; Richardson, B.; Takle, E.; Chai, L.; Schmitt, D.; Xin, H. Airborne transmission may have played a role in the spread of 2015 highly pathogenic avian influenza outbreaks in the United States. *Sci. Rep.* **2019**, *9*, 11755. [CrossRef] [PubMed]
16. Wilks, S.; Michels, H.; Keevil, C. The survival of *Escherichia coli* O157 on a range of metal surfaces. *Int. J. Food Microbiol.* **2005**, *105*, 445–454. [CrossRef]
17. Chan, W.L.; Chung, W.T.; Ng, T.W. Airborne Survival of Escherichia coli under Different Culture Conditions in Synthetic Wastewater. *Int. J. Environ. Res. Public Health* **2019**, *16*, 4745. [CrossRef]
18. Zuo, Z.; Kuehn, T.H.; Verma, H.; Kumar, S.; Goyal, S.M.; Appert, J.; Raynor, P.C.; Ge, S.; Pui, D.Y. Association of airborne virus infectivity and survivability with its carrier particle size. *Aerosol Sci. Technol.* **2013**, *47*, 373–382. [CrossRef]
19. Lighthart, B.; Shaffer, B.T. Increased airborne bacterial survival as a function of particle content and size. *Aerosol Sci. Technol.* **1997**, *27*, 439–446. [CrossRef]
20. Soliman, E.S.; Sallam, N.H.; Abouelhassan, E.M. Effectiveness of poultry litter amendments on bacterial survival and Eimeria oocyst sporulation. *Vet. World* **2018**, *11*, 1064–1073. [CrossRef]
21. Nguyen, X.D.; Zhao, Y.; Evans, J.D.; Lin, J.; Schneider, L.; Voy, B.; Hawkins, S.A.; Purswell, J.L. Evaluation of bioaerosol samplers for collecting airborne *E. coli* carried by dust particles from poultry litter. In Proceedings of the 2021 ASABE Annual International Virtual Meeting, Virtual, 12–16 July 2021; p. 1.

22. Davis, M.; Morishita, T.Y. Relative ammonia concentrations, dust concentrations, and presence of Salmonella species and Escherichia coli inside and outside commercial layer facilities. *Avian Dis.* **2005**, *49*, 30–35. [CrossRef]
23. Crane, S.; Moore, J.; Grismer, M.; Miner, J. Bacterial pollution from agricultural sources: A review. *Trans. ASAE* **1983**, *26*, 858–866. [CrossRef]
24. Zhao, Y.; Aarnink, A.J.A.; Doornenbal, P.; Huynh, T.T.T.; Koerkamp, P.W.G.G.; de Jong, M.C.M.; Landman, W.J.M. Investigation of the efficiencies of bioaerosol samplers for collecting aerosolized bacteria using a fluorescent tracer. II: Sampling efficiency and half-life time. *Aerosol Sci. Technol.* **2011**, *45*, 432–442. [CrossRef]
25. Mubiru, D.; Coyne, M.S.; Grove, J.H. *Escherichia coli* pathogen O157: H7 does not survive longer in soil than a nonpathogenic fecal coliform. *Agron. Notes* **2000**, *32*, 2.
26. Hoeksma, P.; Aarnink, A.J.A.; Ogink, N.W.M. *Effect of Temperature and Relative Humidity on the Survival of Airborne Bacteria = Effect van Temperatuur en Relatieve Luchtvochtig-Heid op de Overleving van Bacteriën in de Lucht*; 1570-8616; Wageningen UR Livestock Research: Wageningen, The Netherlands, 2015.
27. Bhattacharjee, M.K.; Sugawara, K.; Ayandeji, O.T. Microwave sterilization of growth medium alleviates inhibition of Aggregatibacter actinomycetemcomitans by Maillard reaction products. *J. Microbiol. Methods* **2009**, *78*, 227–230. [CrossRef]
28. Woo, M.-H.; Grippin, A.; Anwar, D.; Smith, T.; Wu, C.-Y.; Wander, J.D. Effects of relative humidity and spraying medium on UV decontamination of filters loaded with viral aerosols. *Appl. Environ. Microbiol.* **2012**, *78*, 5781–5787. [CrossRef] [PubMed]
29. Ketkar, G.N.; Malaiappan, S.; Muralidharan, N. Comparative Evaluation of Inherent Antimicrobial Properties and Bacterial Surface Adherence between Copper and Stainless Steel Suction Tube. *J. Pharm. Res. Int.* **2020**, *32*, 149–156. [CrossRef]
30. Terzich, M.; Pope, M.J.; Cherry, T.E.; Hollinger, J. Survey of pathogens in poultry litter in the United States. *J. Appl. Poult. Res.* **2000**, *9*, 287–291. [CrossRef]
31. Neher, D.; Cutler, A.; Weicht, T.; Sharma, M.; Millner, P. Composts of poultry litter or dairy manure differentially affect survival of enteric bacteria in fields with spinach. *J. Appl. Microbiol.* **2019**, *126*, 1910–1922. [CrossRef]
32. Zhen, H.; Han, T.; Fennell, D.E.; Mainelis, G. A systematic comparison of four bioaerosol generators: Affect on culturability and cell membrane integrity when aerosolizing Escherichia coli bacteria. *J. Aerosol Sci.* **2014**, *70*, 67–79. [CrossRef]
33. Ruiz-Gil, T.; Acuna, J.J.; Fujiyoshi, S.; Tanaka, D.; Noda, J.; Maruyama, F.; Jorquera, M.A. Airborne bacterial communities of outdoor environments and their associated influencing factors. *Environ. Int.* **2020**, *145*, 106156. [CrossRef]
34. Tuson, H.H.; Weibel, D.B. Bacteria–surface interactions. *Soft Matter* **2013**, *9*, 4368–4380. [CrossRef]
35. Fernandez, M.O.; Thomas, R.J.; Garton, N.J.; Hudson, A.; Haddrell, A.; Reid, J.P. Assessing the airborne survival of bacteria in populations of aerosol droplets with a novel technology. *J. R. Soc. Interface* **2019**, *16*, 20180779. [CrossRef] [PubMed]
36. Soupir, M.; Mostaghimi, S.; Yagow, E.; Hagedorn, C.; Vaughan, D. Transport of fecal bacteria from poultry litter and cattle manures applied to pastureland. *Water Air Soil Pollut.* **2006**, *169*, 125–136. [CrossRef]

 animals

MDPI

Article

Evaluation of Thermal Indices as the Indicators of Heat Stress in Dairy Cows in a Temperate Climate

Geqi Yan [1,2,3], Hao Li [1,2,3] and Zhengxiang Shi [1,2,3,*]

1 College of Water Resources & Civil Engineering, China Agricultural University, Beijing 100083, China; yangeqi@cau.edu.cn (G.Y.); leehcn@hotmail.com (H.L.)
2 Key Laboratory of Agricultural Engineering in Structure and Environment, Ministry of Agriculture and Rural Affairs, Beijing 100083, China
3 Beijing Engineering Research Center on Animal Healthy Environment, Beijing 100083, China
* Correspondence: shizhx@cau.edu.cn

Simple Summary: When the ambient temperature exceeds the upper limit of a certain temperature range, heat stress is triggered and then negatively affects the production, reproduction, health, and welfare of livestock animals. Due to the limitations of ambient temperature alone as a representative measure of the thermal environment, heat stress is commonly assessed by thermal indices, which contain two or more environmental parameters representing the influence of heat exchanges between the animal and its environment. To understand and utilize the thermal indices better, we evaluated several thermal indices commonly used in the heat stress assessment of dairy cows. We found that the comprehensive climate index (CCI), which includes air temperature, relative humidity, wind speed, and solar radiation, showed a better relationship with the animal-based indicators (i.e., rectal temperature, skin temperature, and eye temperature) of heat stress. According to the results of this study, the CCI has the potential to replace the temperature–humidity index in quantifying the severity of heat stress in dairy cows.

Citation: Yan, G.; Li, H.; Shi, Z. Evaluation of Thermal Indices as the Indicators of Heat Stress in Dairy Cows in a Temperate Climate. *Animals* **2021**, *11*, 2459. https://doi.org/10.3390/ani11082459

Academic Editors: Lilong Chai and Yang Zhao

Received: 29 July 2021
Accepted: 18 August 2021
Published: 21 August 2021

Abstract: Many thermal indices (TIs) have been developed to quantify the severity of heat stress in dairy cows. Systematic evaluation of the representative TIs is still lacking, which may cause potential misapplication. The objectives of this study were to evaluate the theoretical and actual performance of the TIs in a temperate climate. The data were collected in freestall barns at a commercial dairy farm. The heat transfer characteristics of the TIs were examined by equivalent air temperature change (ΔTeq). One-way ANOVA and correlation were used to test the relationships between the TIs and the animal-based indicators (i.e., rectal temperature (RT), respiration rate (RR), skin temperature (ST), and eye temperature (ET)). Results showed that the warming effect of the increased relative humidity and the chilling effect of the increased wind speed was the most reflected by the equivalent temperature index (ETI) and the comprehensive climate index (CCI), respectively. Only the equivalent temperature index for cows (ETIC) reflected that warming effect of solar radiation could obviously increase with increasing Ta. The THI and ETIC showed expected relationships with the RT and RR, whereas the CCI and ETIC showed expected relationships with the ST and ET. Moreover, CCI showed a higher correlation with RT ($r = 0.672$, $p < 0.01$), ST($r = 0.845$, $p < 0.01$), and ET ($r = 0.617$, $p < 0.01$) than other TIs ($p < 0.0001$). ETIC showed the highest correlation with RR ($r = 0.850$, $p < 0.01$). These findings demonstrated that the CCI could be the most promising thermal index to assess heat stress for housed dairy cows. Future research is still needed to develop new TIs tp precisely assess the microclimates in cow buildings.

Keywords: thermal index; dairy cattle; heat stress

1. Introduction

Heat stress, defined as the sum of external forces acting on animals that induces an increase in core body temperature and evokes a physiological response, has a negative effect on the production, reproduction, health, and welfare of livestock animals [1,2]. Dairy cows, characterized by a large quantity of metabolic heat, are vulnerable to heat stress because of their compromised cooling capacity resulting from environmental conditions [3]. China is an agricultural country, where the dairy industry plays an important role in the agricultural economy. Statistics show that from 2000 to 2018, milk production nearly doubled in China. Not surprisingly, the challenges of heat stress are the greatest in southern China, which has a subtropical climate. However, recent studies reported that dairy cows in northern China, a region with a temperate climate, also underwent extended periods of heat stress [4,5]. To reduce economic losses, dairy producers need to precisely assess the environmental risks and need to initiate cooling in a timely manner for dairy cows before heat stress occurs.

So far, there is an academic consensus that heat stress is triggered when the ambient temperature reaches the upper critical temperature (UCT) of a dairy cow's thermal neutral zone (TNZ) [3,6,7]. Nevertheless, the limitation of ambient temperature alone as a representative measure of the thermal environment is widely admitted. More often than not, the intensity of heat stress on dairy cattle is quantitatively estimated using the temperature–humidity index (THI). Since Thom first proposed the THI (originally called the discomfort index) in 1959 [8], this index has served as a de facto standard for the classification of thermal environments in animal transport and production situations and as a basis for environmental management decisions in hot seasons [9]. The THI considers the combined effect of air temperature and humidity. Because of different estimators for air humidity, there are three mainstream THI equations, which contain relative humidity, wet-bulb temperature, and dew point temperature, respectively. Previous studies have reported that different THI equations differed in their ability to detect heat stress [10,11]. Interestingly, previous researchers consistently used the equation containing relative humidity to identify the THI threshold where the physiological responses of dairy cows significantly changes [5,12,13]. Despite that, there is insufficient information on the differences among those THI equations.

In the last few decades, most efforts to develop new thermal indices (TIs) have been made along two lines: modified THI and apparent ambient temperature. The former is achieved by adding new environmental factors into the THI models or substituting old parameters, and the typical Tis that are often used are the adjusted THI (THIadj) and the black globe humidity index (BGHI) [14,15]. The TIs classified as apparent ambient temperature are achieved by converting the thermal effects of other environmental parameters into the equivalent thermal effect of air temperature, and the typical examples include the equivalent temperature index (ETI) [16], the tcomprehensive climate index (CCI) [17], and the equivalent temperature index for dairy cattle (ETIC) [18]. Previous researchers have reported the index performance under the climatic conditions they studied and have recommended some of these indices. However, systematic evaluation of the existing TIs is still lacking. Potential problems will occur if a TI is applied as an environment control strategy without a detailed examination.

A TI value has to reflect the comprehensive effect produced by the sensible and latent heat exchanges between the organism and its environment. Meanwhile, the TIs should be highly associated with the physiological responses that can indicate the thermal status of an animal. For dairy cows, effective animal-based indicators include rectal temperature, respiration rate, skin temperature, and eye temperature. Rectal temperature is a predominant indicator of core body temperature, which is a gold standard and is used in 28% of heat stress assessments [1]. Respiration rate is universally recognized as an early indicator of heat stress [19]. The skin surface is the primary site for the heat exchange process, and thus, skin temperature is highly related to thermal comfort [20]. Recent research has found that eye temperature was influenced by pain and heat stress [21,22]. Moreover, eye temperature measurements show acceptable agreement with rectal temperature measurements in dairy cows [23].

The present study aimed to examine the theoretical performance of the cow-related TIs with respect to heat transfer characteristics reflected by the parameters and to evaluate the actual relationships between the TIs and the animal-based indicators of heat stress. We restricted the current study to the temperate climate conditions in northern China. Moreover, we only evaluated the TIs mentioned earlier, which means that other TIs developed for the specified animals and environments were not included.

2. Materials and Methods

2.1. Cows, Housing, and Management

The study was conducted between July and October at a commercial dairy farm in Tianjin, China. All procedures were approved by the China Agricultural University Department of Agricultural Structure and Bioenvironmental Engineering Animal Ethics Committee (Approval ID: 20200625). This study included 826 Holstein lactating cows. Of these cows, were 161 first-lactation cows (average milk yield 27 ± 8 kg/day and average days in milk 286 ± 134 at the beginning of the study), 280 were second-lactation cows (average milk yield 30 ± 13 kg/day and average days in milk 246 ± 159), and were 384 third-lactation cows (average milk yield 29 ± 14 kg/day and average days in milk 277 ± 172). The cows were housed in free-stall barns (107.5 m × 31.0 m; double-pitched roof with a gradient of 33%), which were oriented on the east–west axis and were equipped with fans (diameter of 1.0 m; air amount of 25,430 m^3/h; spaced every 6.0 m; 2.7 m high; activated at 18 °C air temperature). The cows were milked and fed three times per day and had free access to water.

2.2. Environmental Parameters and Thermal Indices of Heat Stress

Air temperature (Ta) and relative humidity (RH) were measured every 10 min using a HOBO U23-001 thermometer (Onset Computer Corp., Bourne, MA, USA; accuracy of ± 0.2 °C from −40 to 70 °C and ±2.5% from 10% to 90%). Wind speed was measured every 3 min using a TSI 9565 anemometer (TSI Inc., Shoreview, MN, USA; accuracy of ±0.015 m/s from 0 to 50 m/s). Black globe temperature (Tbg) was measured every 10 min using a black globe thermometer (JantyTech Inc., Fengtai, Beijing, China; accuracy of ±0.6 °C from 15 to 40 °C). Solar radiation (SR) was measured using a TES-1333R solar power meter (TES Electrical Electronic Corp., Taipei, Taiwan, China; accuracy of 10 W/m^2 from 0 to 2000 W/m^2). Wet-bulb temperature (Tw) and dew point temperature (Tdp) were obtained by inputting the Ta and RH into an online calculator (www.omnicalculator.com/physics/). (accessed 20 August 2021) These environmental parameters were used to calculate the following TIs:

Temperature—humidity Index (THI):

$$\mathrm{THI1} = (1.8\mathrm{Ta} + 32) - (0.55 - 0.0055\mathrm{RH}) \times (1.8\mathrm{Ta} - 26) \quad (1)$$

$$\mathrm{THI2} = 0.72 \times (\mathrm{Ta} + \mathrm{Tw}) + 40.6 \quad (2)$$

$$\mathrm{THI3} = \mathrm{Ta} + 0.36\mathrm{Tdp} + 41.5 \quad (3)$$

Thermal environments were classified under the categories of into no stress (THI < 72) and heat stress (THI ≥ 72), according to the THI [24].

Black Globe Humidity Index (BGHI):

$$\mathrm{BGHI} = \mathrm{Tbg} + 0.36\mathrm{Tdp} + 41.5 \quad (4)$$

Thermal environments were classified into the categories of no stress (BGHI ≤ 74) and heat stress (BGHI > 74) according to the BGHI [19].

Ajusted THI (THIadj):

$$\mathrm{THIadj} = 4.51 + [0.8\mathrm{Ta} + (\mathrm{RH}/100) \times (\mathrm{Ta} - 14.4) + 46.4] - 1.992\mathrm{u} + 0.0068\mathrm{SR} \quad (5)$$

Thermal environments were classified into the categories of no stress (THIadj ≤ 74) and heat stress (THIadj > 74) according to the THIadj [15].

Equivalent Temperature Index (ETI):

$$\begin{aligned}\text{ETI} = 27.88 - 0.456\text{Ta} + 0.010754\text{Ta}^2 - 0.4905\text{RH} + 0.00088\text{RH}^2 + 1.15\text{u} - 0.12644\text{u}^2 \\ + 0.019876\text{Ta} \times \text{RH} - 0.046313\text{Ta} \times \text{u}\end{aligned} \tag{6}$$

Thermal environments were classified into the categories of no stress (ETI < 30) and heat stress (ETI ≥ 30) according to the ETI [25].

Comprehensive Climate Index (CCI):

$$\begin{aligned}\text{CCI} &= \text{Ta} + \text{Eq.(RH)} + \text{Eq.(u)} + \text{Eq.(sr)} \\ \text{Eq.(RH)} &= \exp\left(0.00182\text{RH} + 1.8 \times 10^{-5}\text{Ta} \times \text{RH}\right) \times \left(0.000054\text{Ta}^2 + 0.00192\text{Ta} - 0.0246\right) \times (\text{RH} - 30) \\ \text{Eq.(u)} &= \frac{-6.56}{\exp\left\{\frac{1}{(2.26\text{u} + 0.23)^{0.45}} \times \left[2.9 + 1.14 \times 10^{-6}\text{u}^{2.5} - \log_{0.3}(2.26\text{u} + 0.33)^{-2}\right]\right\}} - 0.00566\text{u}^2 + 3.33 \\ \text{eq.(sr)} &= 0.0076\text{sr} - 0.00002\text{sr} \times \text{Ta} + 0.00005\text{Ta}^2 \times \sqrt{\text{sr}} + 0.1\text{Ta} - 2\end{aligned} \tag{7}$$

Thermal environments were into the categories of no stress (CCI ≤ 25) and heat stress (CCI > 25) classified according to the CCI [17].

Equivalent Temperature Index for Dairy Cattle (ETIC):

$$\text{ETIC} = \text{Ta} - 0.0038\text{Ta} \times (100 - \text{RH}) - 0.1173\text{u}^{0.707} \times (39.2 - \text{Ta}) + 1.86 \times 10^{-4}\text{Ta} \times \text{sr} \tag{8}$$

Thermal environments were classified into the categories of no stress (ETIC < 18) and heat stress (ETIC ≥ 18) according to the ETIC [18].

The environmental coniditions are shown in Figure 1. The results of the environmental factors and thermal indices during this study are shown in Table 1.

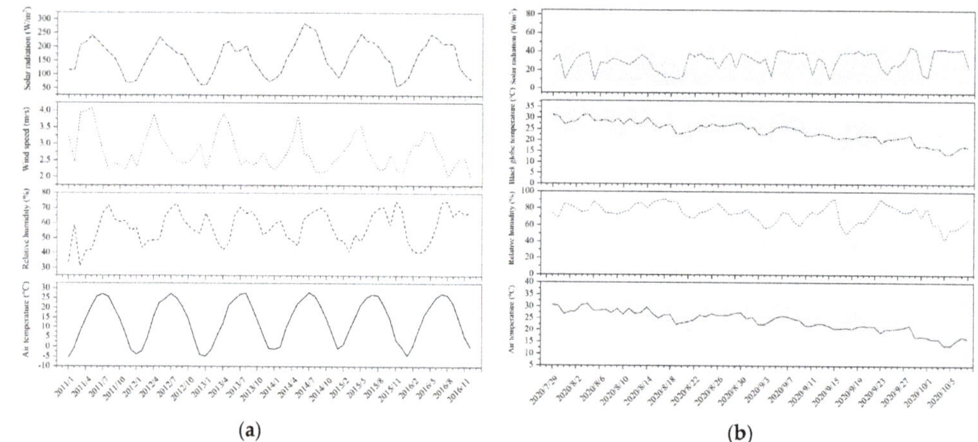

Figure 1. (**a**) Monthly variations in the average values of the outdoor environmental factors of the measured region (2011–2016) and (**b**) daily variations in the average values of the environmental factors inside the barns over the study period. The black line indicates the average value. The grey region indicates the maximum and minimum values.

Table 1. Descriptive statistics for the environmental factors and thermal indices during this study.

Item	Number	Minimum	Maximum	Mean	Standard Deviation
Air temperature (°C)	3005	13.55	36.00	26.33	4.72
Relative humidity (%)	3005	27.57	91.70	69.75	15.89
Black globe temperature (°C)	3005	13.55	36.70	26.93	4.86
Wind speed (m/s)	3005	0	4.50	2.50	0.90
Solar radiation (W/m^2)	3005	0	64.30	24.68	13.42
Temperature–humidity index (THI1) [1]	3005	56.76	87.48	75.68	6.87
Temperature–humidity index (THI2) [1]	3005	57.08	86.31	75.35	6.35
Temperature–humidity index (THI3) [1]	3005	57.02	86.44	74.99	6.28
Black globe humidity index (BGHI)	3005	57.02	87.70	75.59	6.38
Ajusted THI (THIadj)	3005	59.30	89.23	75.34	6.47
Equivalent tempeprature index (ETI)	3005	14.18	41.64	28.93	6.12
Comprehensive climate index (CCI)	3005	13.91	40.13	27.16	5.62
Equivalent temperature index for dairy cattle (ETIC)	3005	8.40	29.68	20.63	4.76

[1] THI1, THI2, and THI3 contain relative humidity, wet-bulb temperature, and dew point temperature, respectively.

2.3. Animal-Based Indicators of Heat Stress

Animal measures were conducted twice daily: in the morning (0800–1200 h) and in the afternoon (1400–1800 h). Rectal temperature (RT) was measured using a veterinary digital thermometer (ShangNong Technology Inc., Qingdao, Shandong, China; accuracy of ± 0.1 °C from 35 to 39 °C). Respiration rate (RR) was measured by counting the flank movements for 60 s. Skin temperature (ST) and eye temperature (ET) were measured using a Fotric 235 infrared thermography (FOTRIC Inc., Jingan, Shanghai, China; resolution of 336 × 252 pixels; accuracy of ±2.0 °C from −20 to 150 °C). The infrared images were analyzed using AnalyzIR software (Fortic Inc., Jingan, Shanghai, China). The methods for the measurements of ST and ET using infrared thermography agreed with those from previous studies [26,27]. The Ta and RH corresponding to the time when the image was taken were inputted into the software to adjust for these variables on the camera accuracy. The emissivity was set to 0.97. Typical examples of the regions used to obtain temperature variables are shown in Figure 2. By using the shape-drawing tool, a box was placed on the trunk to obtain the ST variable, and an oval was placed on the eye to obtain the ET variable. According to the recommendations from Hoffmann et al. [23], only the maximum temperature of the region was considered in the statistical analysis. The results of the animal-based indicators during this study are shown in Table 2.

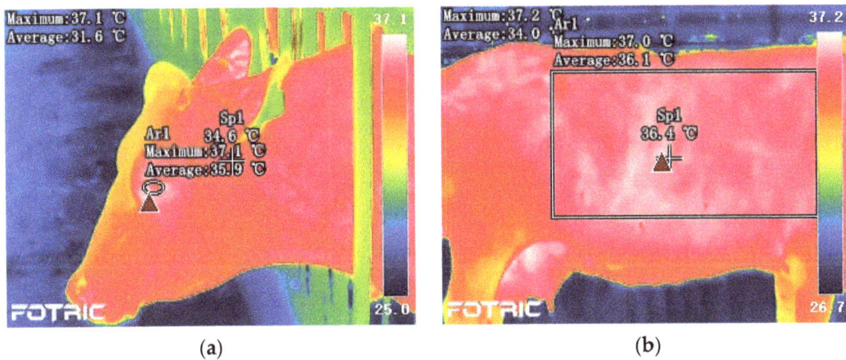

Figure 2. Illustration of the eye temperature (**a**) and skin temperature (**b**) measurements.

Table 2. Descriptive statistics for the animal-based indicators during this study.

Item	Number	Minimum	Maximum	Mean	Standard Deviation
Rectal temperature (°C)	3005	37.87	41.4	38.92	0.47
Respiration rate (bpm)	3005	19	114	62.77	19.34
Skin temperature (°C)	2996	30.5	40.7	35.23	1.67
Eye temperature (°C)	2924	33.4	39.6	36.41	0.96

2.4. Statistical Analysis

2.4.1. Equivalent Air Temperature Change

Under the assumption that variation in the TI value caused by a change in one environmental parameter can be offset by a change in air temperature, the equivalent air temperature change (ΔTeq) was defined as the difference between the newly generated and the original air temperature [28]. Equations (9) and (10) further explain the ΔTeq:

$$\begin{aligned} TI(Ta_1, RH_2, u_1, SR_1) &= TI(Ta_2, RH_1, u_1, SR_1) \text{ for a change from } RH_1 \text{ to } RH_2 \\ TI(Ta_1, RH_1, u_2, SR_1) &= TI(Ta_2, RH_1, u_1, SR_1) \text{ for a change from } u_1 \text{ to u2} \\ TI(Ta_1, RH_2, u_1, SR_2) &= TI(Ta_2, RH_1, u_1, SR_1) \text{ for a change from } SR_1 \text{ to } SR_2 \end{aligned} \quad (9)$$

$$\Delta Teq = Ta_2 - Ta_1 \quad (10)$$

A positive ΔTeq indicates a warming effect caused by the changed parameter, whereas a negative ΔTeq indicates a chilling effect. A larger absolute ΔTeq values implies a stronger warming/chilling effect in the corresponding TI.

2.4.2. Analysis of Variance

One-way analysis of variance (ANOVA) was used to test the effect of different TI levels on the animal-based indicators using the following model:

$$Y_{ij} = \mu + TI_i + \varepsilon_{ij} \quad (11)$$

where Y_{ij} indicates the jth observation of the animal-based indicator; μ indicates the overall mean; ε_{ij} indicates the random error; TI_i indicates the effect of ith TI (both THI1 and BGHI contain sixteen levels, and i is equal to 58, 60, 62, ... , 88; both THI2 and THI3 contain fifteen levels, and i is equal to 58, 60, 62, ... , 86; THIadj contains fifteen levels, and i is equal to 60, 62, 64, ... , 88; both ETI and CCI contain fifteen levels, and i is equal to 15, 17, 19, ... , 41; ETIC contains eleven levels, and i is equal to 9, 11, 13, ... , 29).

When the p-value from the ANOVA was less than 0.05, a post hoc test was conducted based on Fisher's least significant difference (LSD) criterion.

2.4.3. Correlation Analysis

Pearson correlation coefficients (r) between the TIs and the animal-based indicators were calculated and were then interpreted as follows: a coefficient that is less than 0.3 indicates a weak relationship; a coefficient that varies between 0.3 and 0.5 indicates a medium relationship; a coefficient that varies between 0.5 and 0.8 indicates a strong relationship; a coefficient that is more than 0.8 indicates a very strong relationship.

Comparison of the two overlapping dependent correlations (r.(TI1 and Y) vs. r.(TI2 and Y)) was tested by Hotelling's t statistic [29]. The null hypothesis is that the correlations are the same.

Statistical analyses were performed using SPSS statistical software (IBM Corp., Armonk, NY, USA).

3. Results

3.1. Equivalent Ambient Temperature Change

3.1.1. Relative Humidity

Figure 3 shows the changes of ΔTeq caused by an increase in RH from 40% to 60% as the Ta rises from 25 to 40 °C at a wind speed of 0.2 m/s and solar radiation of 0 W/m^2. All of the TIs treated an increase in RH as a warming effect, and the ΔTeq values increased with the Ta rising. Within a range of 25 °C to 40 °C Ta, the highest ΔTeq value was generally given by the ETI, and the smallest was given by the CCI. There were almost no differences in the value of ΔTeq and its change with the Ta between the THI1 and the THIadj. When the Ta was more than approximately 33 °C, the ETIC showed a smaller ΔTeq than the THI1.

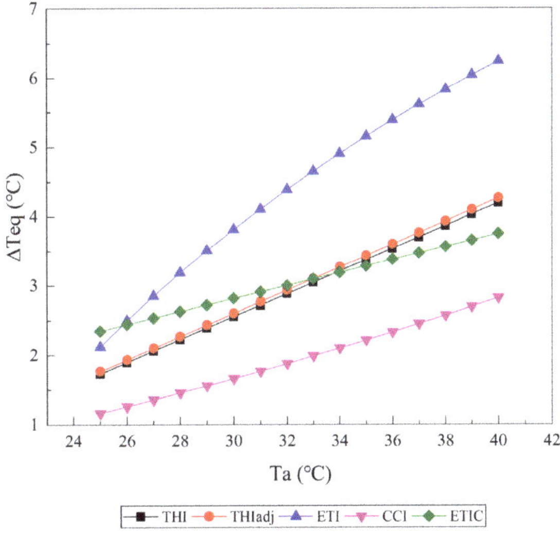

Figure 3. The changes of the equivalent ambient temperature change (ΔTeq) caused by an increase in relative humidity from 40% to 60% as the air temperature (Ta) rises from 25 to 40 °C are reported from the temperature–humidity index (THI), adjusted THI (THIadj), equivalent temperature index (ETI), comprehensive climate index (CCI), and equivalent temperature index for dairy cattle (ETIC). Wind speed is assumed to be 0.2 m/s, and solar radiation is assumed to be 0 W/m^2. THI2, THI3, and black globe humidity index (BGHI) are not included since they do not contain the parameter of relative humidity.

3.1.2. Wind Speed

Figure 4 shows the changes of the ΔTeq value caused by an increase in wind speed from 1 m/s to 2 m/s as the Ta rises from 25 to 40 °C at a relative humidity of 50% and solar radiation of 0 W/m^2. With the exception of THI1, all of the TIs treated an increase in wind speed as a chilling effect when Ta was below 39 °C. For the THIadj, the ΔTeq values were constant at −1.5 °C. Similarly, the ΔTeq values given by the CCI were almost kept constant (−2.5 °C), although they decreased when the Ta increased. The ΔTeq values given by the CCI decreased as the Ta increased, but those given by the ETIC increased as the Ta increased.

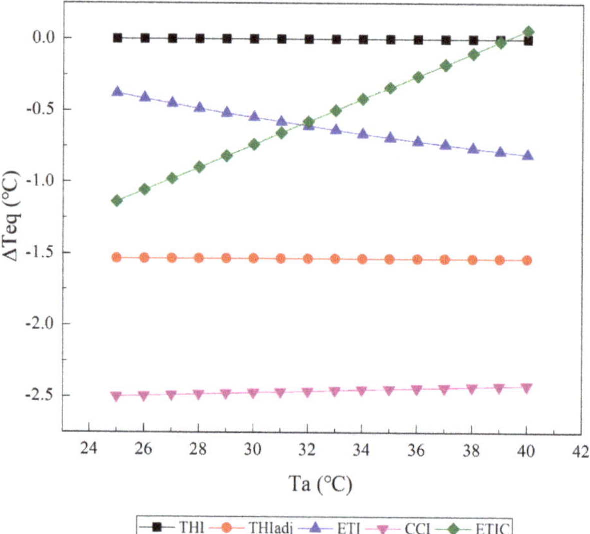

Figure 4. The changes of the equivalent ambient temperature change (ΔTeq) caused by an increase in wind speed from 1 m/s to 2 m/s as air temperature (Ta) rises from 25 to 40 °C are reported from the temperature–humidity index (THI), adjusted THI (THIadj), equivalent temperature index (ETI), comprehensive climate index (CCI), and equivalent temperature index for dairy cattle (ETIC). Relative humidity is assumed to be 50%, and solar radiation is assumed to be 0 W/m^2. THI2, THI3, and black globe humidity index (BGHI) are not included since they do not contain wind speed and relative humidity.

3.1.3. Solar Radiation

Figure 5 shows the changes of the ΔTeq values caused by an increase in solar radiation from 100 W/m^2 to 500 W/m^2 at a wind speed of 0.2 m/s and a relative humidity of 50%. THI1 and ETI did not include the parameter of solar radiation, and thus, their ΔTeq values were equal to zero. Other TIs (i.e., THIadj, CCI, and ETIC) treated an increase in solar sr as a warming effect. For the THIadj, the ΔTeq values were constant at approximately 2.1 °C. The ΔTeq values obtained from the ETIC decreased with when the Ta increased, while those obtained from the CCI decreased as the Ta increased at a comparatively slower rate.

3.2. *Effect of the TI Values on the Animal-Based Indicators*

3.2.1. Rectal Temperature

Figure 6 shows the changes in ther rectal temperature with the index value. Figure 6a shows that nine significant ($p < 0.05$) increases in RT were observed at 64, 70, 74, 76, 78, 80, 82, 84, and 86 THI1. There was a significant ($p < 0.05$) decrease in RT at 66 THI1. Figure 6b shows that ten significant ($p < 0.05$) increases in RT were observed at 60, 64, 70, 72, 74, 76, 78, 80, 82, and 84 THI2. The insignificant ($p = 0.076$) difference in RT was found between 60 and 62 THI3, although the changes of the RT with THI3 were similar to those with THI2 (Figure 6c). Figure 6d shows that nine significant ($p < 0.05$) increases in RT were found at 64, 70, 72, 74, 76, 78, 80, 82, and 84 BGHI. Likewise, nine significant ($p < 0.05$) increases in RT were reported at 19, 21, 23, 25, 27, 29, 31, 33, and 35 CCI (Figure 6g). However, only seven significant ($p < 0.05$) increases in RT were found at 68, 70, 72, 74, 78, 80, and 82 THIadj (Figure 6e), 23, 25, 27, 29, 31, 33, and 35 ETI (Figure 6f), and 9, 15, 17, 19, 21, 23, and 25 ETIC (Figure 6h), respectively.

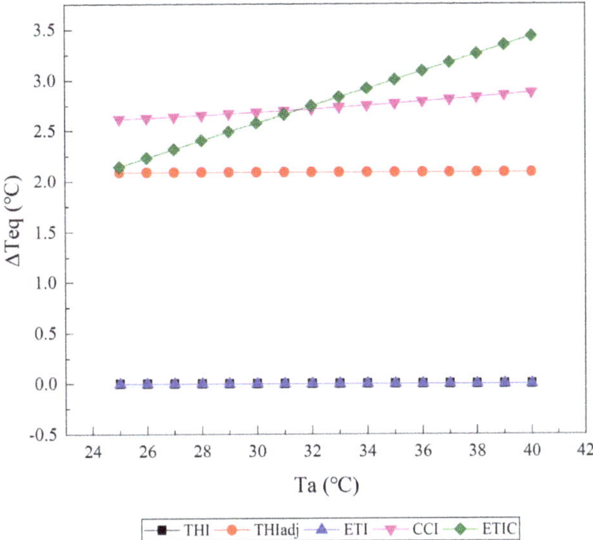

Figure 5. The changes of the equivalent ambient temperature change (ΔTeq) caused by an increase in solar radiation from 100 W/m^2 to 500 W/m^2 as air temperature (Ta) rises from 25 to 40 °C are reported from the temperature–humidity index (THI), adjusted THI (THIadj), equivalent temperature index (ETI), comprehensive climate index (CCI), and equivalent temperature index for dairy cattle (ETIC). Relative humidity is assumed to be 50%, and wind speed is assumed to be 0.2 m/s. THI2, THI3, and black globe humidity index (BGHI) are not included since they do not contain solar radiation and relative humidity.

3.2.2. Respiration Rate

Figure 7 shows the changes of the respiration rate with the index value. A total of eleven significant ($p < 0.05$) and continuous increases in RR were found from 66 to 86 THI1 (Figure 7a) and from 64 to 84 THIadj (Figure 7e). There were ten significant ($p < 0.05$) and continuous increases in RR that were found from 66 to 84 for both THI2 (Figure 7b) and THI3 (Figure 7c). Similarly, nine significant ($p < 0.05$) and continuous increases in RR were observed from 68 to 84 BGHI (Figure 7d). From Figure 7f, eight significant ($p < 0.05$) and consecutive increases in RR were observed from 21 to 35 ETI. Figure 7g shows that nine significant ($p < 0.05$) increases in RR were observed at 19, 21, 23, 25, 27, 29, 31, 33, and 37 CCI. With the exception of at 11 ETIC, nine significant ($p < 0.05$) increases in RR were found from 9 to 27 ETIC.

3.2.3. Skin Temperature

Figure 8 shows the change of the skin temperature with the index value. Figure 8a shows that eleven significant ($p < 0.05$) increases in ST were observed at 58, 64, 68, 70, 74, 76, 78, 80, 82, 84, and 86 THI1, and three significant ($p < 0.05$) decreases in ST were observed at 60, 62, and 66 THI1. For THI2, THI3, and BGHI, ST significantly (p <0.05) increased at the 58, 60, 64, 68, 70, 74, 76, 78, 80, 82, 84, and 86 THI values and decreased at the 62 and 66 THI values (Figure 8b–d). According to Figure 8e, ST at 62 THIadj were significantly ($p < 0.05$) higher than that at 60 and 64 THIadj. There were eleven significant ($p < 0.05$) and continuous increases in ST that were observed from 64 to 84 THIadj. Figure 8f shows that nine significant ($p < 0.05$) increases in ST were reported at 19, 23, 25, 27, 29, 31, 33, 35, and 39 ETI. Apart from one significant decrease observed at 17 CCI, twelve significant ($p < 0.05$) increases in ST were found from 15 to 39 CCI (Figure 8g). Likewise, there were nine significant ($p < 0.05$) increases in ST observed from 9 to 27 ETIC, apart from one decrease observed at 11 ETIC (Figure 8h).

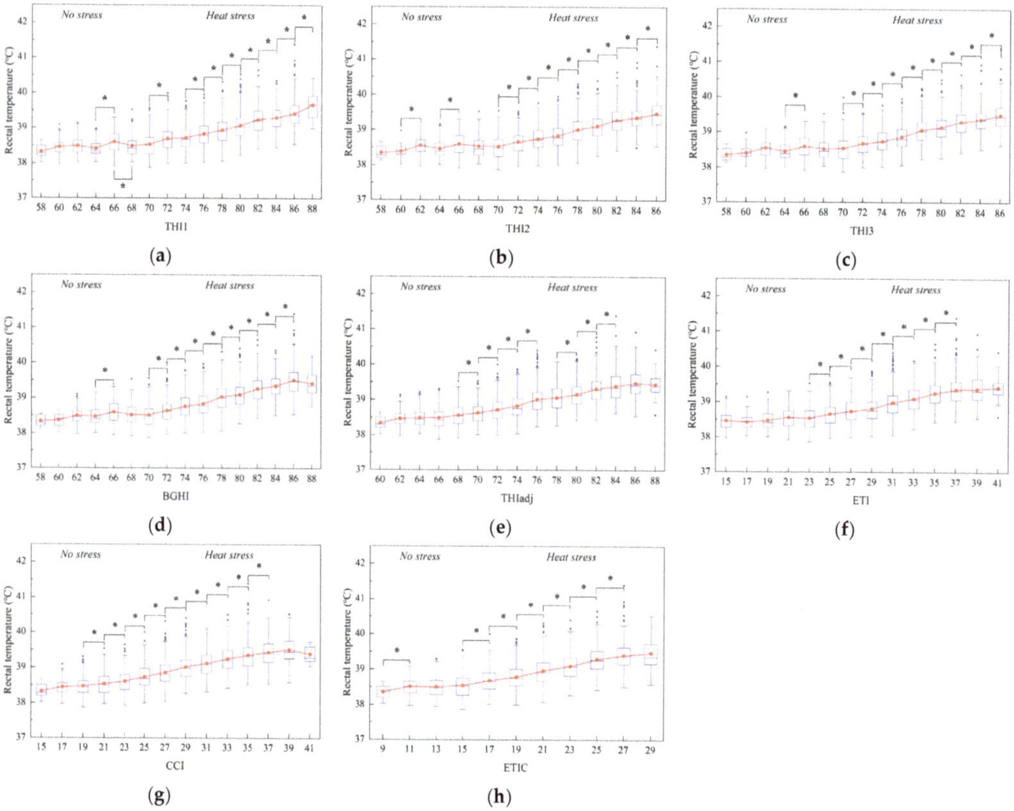

(a)

(b)

(c)

(d)

(e)

(f)

(g)

(h)

Figure 6. Effect of the thermal index values on rectal temperature. The interquartile range (IQR) is shown by a blue box, the median as a blue horizontal line, the mean as a red square, and the whiskers extend to 1.5 IQR. Outliers are indicated by blue dots. An asterisk indicates a significant ($p < 0.05$) change in the variable. The thermal indices are the temperature–humidity index (THI1 (**a**), THI2 (**b**), THI3 (**c**)), black globe humidity index (BGHI (**d**)), adjusted THI (THIadj (**e**)), equivalent temperature index (ETI (**f**)), comprehensive climate index (CCI (**g**)), and equivalent temperature index for dairy cattle (ETIC (**h**)).

3.2.4. Eye Temperature

Figure 9 shows the changes of the eye temperature with the index values. Figure 9a shows that eight significant ($p < 0.05$) increases in ET were found at 64, 70, 74, 78, 80, 82, 84, and 86 THI1, and two significant ($p < 0.05$) decreases in ET were found at 66 and 72 THI1. Figure 9b shows that nine significant ($p < 0.05$) increases in ET were observed at 60, 64, 70, 74, 76, 78, 80, 82, and 84 THI2, and one significant ($p < 0.05$) decrease in ET was observed at 66 THI2. Compared to the results from THI2, one extra significant ($p < 0.05$) decrease in ET was observed at 72 THI3 (Figure 9c). A significant ($p < 0.05$) decrease in ET occurred at 66 BGHI, and then six significant ($p < 0.05$) increases in ET occured at 70, 76, 78, 80, 82, and 84 BGHI (Figure 9d). There were six significant ($p < 0.05$) decreases in ET observed at 68, 74, 78, 80, 82, and 84 THIIadj, but one significant ($p < 0.05$) decrease in ET occurred at 86 THIadj (Figure 9e). Figure 9f shows that one significant ($p < 0.05$) decrease in ET occurred at 21 ETI, and then five significant ($p < 0.05$) increases in ET occurred at 23, 29, 31, 33, and 35 ETI. Figure 9g shows that the first significant ($p < 0.05$) increase in ET was

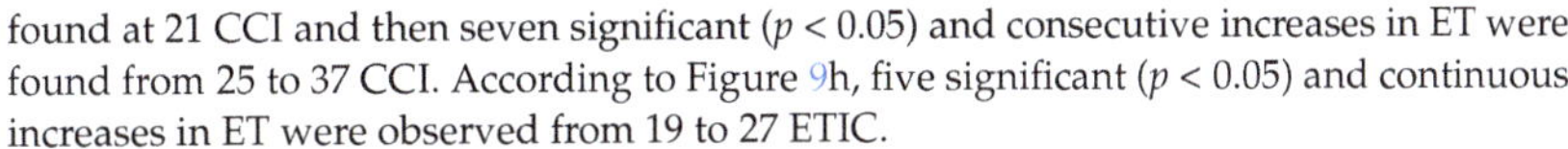
found at 21 CCI and then seven significant ($p < 0.05$) and consecutive increases in ET were found from 25 to 37 CCI. According to Figure 9h, five significant ($p < 0.05$) and continuous increases in ET were observed from 19 to 27 ETIC.

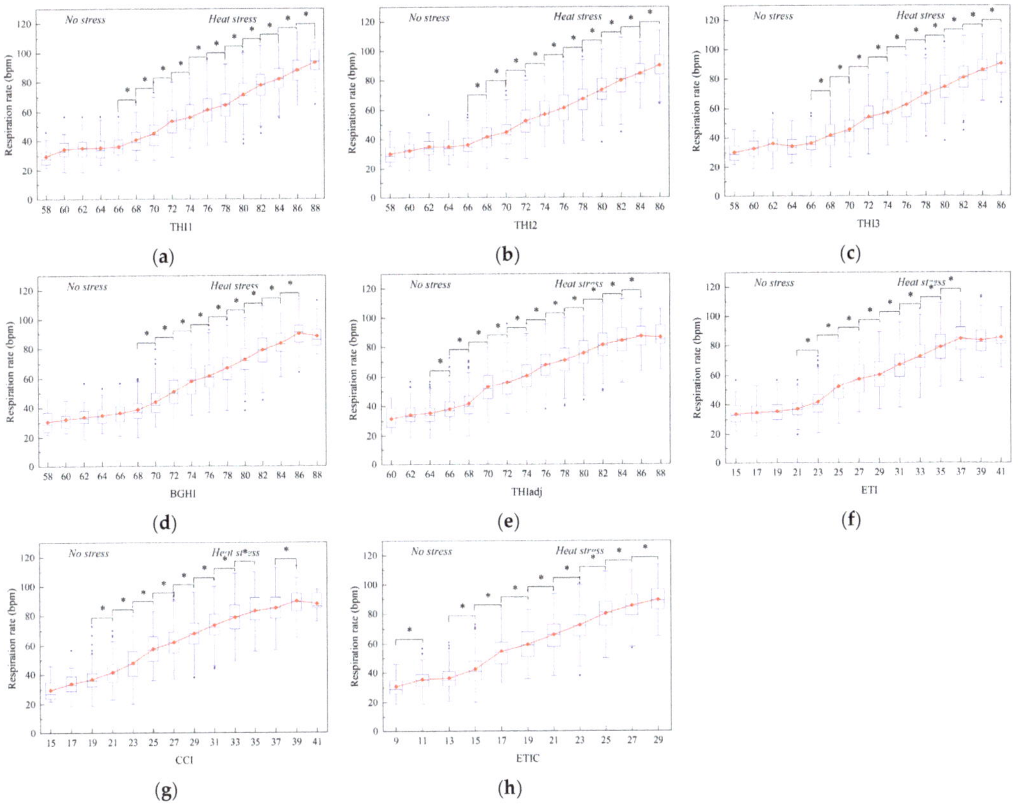

(**a**) (**b**) (**c**)

(**d**) (**e**) (**f**)

(**g**) (**h**)

Figure 7. Effect of the thermal index values on respiration rate. The interquartile range (IQR) is shown by a blue box, the median as a blue horizontal line, the mean as a red square, and the whiskers extend to 1.5 IQR. Outliers are indicated by blue dots. An asterisk indicates a significant ($p < 0.05$) change in the variable. The thermal indices are temperature–humidity index (THI1 (**a**), THI2 (**b**), THI3 (**c**)), black globe humidity index (BGHI (**d**)), adjusted THI (THIadj (**e**)), equivalent temperature index (ETI (**f**)), comprehensive climate index (CCI (**g**)), and equivalent temperature index for dairy cattle (ETIC (**h**)).

3.3. Correlations between Indices and Animal-Based Indicators

Table 3 presents that the correlations among the TIs were positive and very strong ($r \geq 0.95$), and the correlations between the TIs and the animal-based indicators were positive and high ($r \geq 0.5$). Results of the correlation comparison are listed in Table 4. The CCI showed the highest correlation with rectal temperature (r = 0.672, $p < 0.01$), followed by the THIadj (r = 0.667, $p < 0.01$; r.CCI > r.THIadj, $p < 0.0001$) and the ETIC (r = 0.662, $p < 0.01$; r.CCI > r.ETIC, $p < 0.0001$). The ETIC showed the highest correlation with respiration rate (r = 0.850, $p < 0.01$), followed by the THI3 (r = 0.847, $p < 0.01$; r.ETIC > r.THI3, $p = 0.0793$) and the BGHI (r = 0.846, $p < 0.01$; r.ETIC > r.BGHI, $p = 0.0274$). The CCI exhibited the highest correlation with skin temperature (r = 0.845, $p < 0.01$), followed by the THIadj (r = 0.827, $p < 0.01$; r.CCI > r.THIadj, $p < 0.0001$) and the ETIC (r = 0.820, $p < 0.01$; r.CCI > r.ETIC, $p < 0.001$). Additionally, the CCI presented the highest correlation with eye temperature (r = 0.617, $p < 0.01$), followed by the BGHI (r = 0.598, $p < 0.01$; r.CCI > r.BGHI, $p = 0.0001$) and the ETIC (r = 0.592, $p < 0.01$; r.CCI > r.ETIC, $p < 0.0001$).

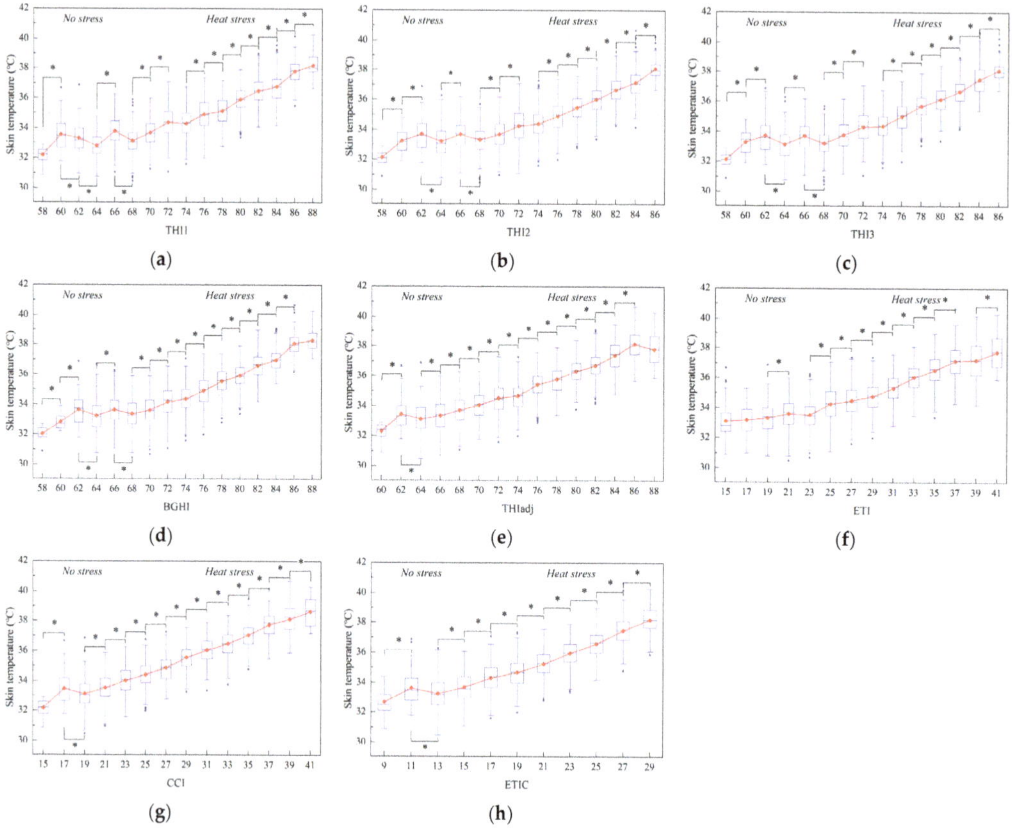

Figure 8. Effect of the thermal index values on skin temperature. The interquartile range (IQR) is shown by a blue box, the median as a blue horizontal line, the mean as a red square, and the whiskers extend to 1.5 IQR. Outliers are indicated by blue dots. An asterisk indicates a significant ($p < 0.05$) change in the variable. The thermal indices are the temperature–humidity index (THI1 (**a**), THI2 (**b**), THI3 (**c**)), black globe humidity index (BGHI (**d**)), adjusted THI (THIadj (**e**)), equivalent temperature index (ETI (**f**)), comprehensive climate index (CCI (**g**)), and equivalent temperature index for dairy cattle (ETIC (**h**)).

Table 3. Pearson correlation coefficients between thermal indices and animal-based indictors.

Variable	Statistic	THI1	THI2	THI3	BGHI	THIadj	ETI	CCI	ETIC	RT	RR	ST	ET
THI1	r	1	0.998	0.996	0.995	0.966	0.987	0.947	0.984	0.643	0.843	0.793	0.572
	p		<0.01	<0.01	<0.01	<0.01	<0.01	<0.01	<0.01	<0.01	<0.01	<0.01	<0.01
THI2	r		1	0.999	0.996	0.962	0.980	0.943	0.985	0.640	0.844	0.792	0.574
	p			<0.01	<0.01	<0.01	<0.01	<0.01	<0.01	<0.01	<0.01	<0.01	<0.01
THI3	r			1	0.999	0.961	0.974	0.946	0.984	0.645	0.847	0.801	0.586
	p				<0.01	<0.01	<0.01	<0.01	<0.01	<0.01	<0.01	<0.01	<0.01
BGHI	r				1	0.961	0.970	0.948	0.982	0.649	0.846	0.809	0.598
	p					<0.01	<0.01	<0.01	<0.01	<0.01	<0.01	<0.01	<0.01
THIadj	r					1	0.980	0.994	0.990	0.667	0.837	0.827	0.586
	p						<0.01	<0.01	<0.01	<0.01	<0.01	<0.01	<0.01
ETI	r						1	0.958	0.984	0.640	0.828	0.782	0.546
	p							<0.01	<0.01	<0.01	<0.01	<0.01	<0.01
CCI	r							1	0.980	0.676	0.833	0.845	0.617
	p								<0.01	<0.01	<0.01	<0.01	<0.01

Table 3. *Cont.*

Variable	Statistic	THI1	THI2	THI3	BGHI	THIadj	ETI	CCI	ETIC	RT	RR	ST	ET
ETIC	r								1	0.662	0.850	0.820	0.592
	p									<0.01	<0.01	<0.01	<0.01
RT	r									1	0.741	0.681	0.574
	p										<0.01	<0.01	<0.01
RR	r										1	0.775	0.598
	p											<0.01	<0.01
ST	r											1	0.715
	p												<0.01
ET	r												1

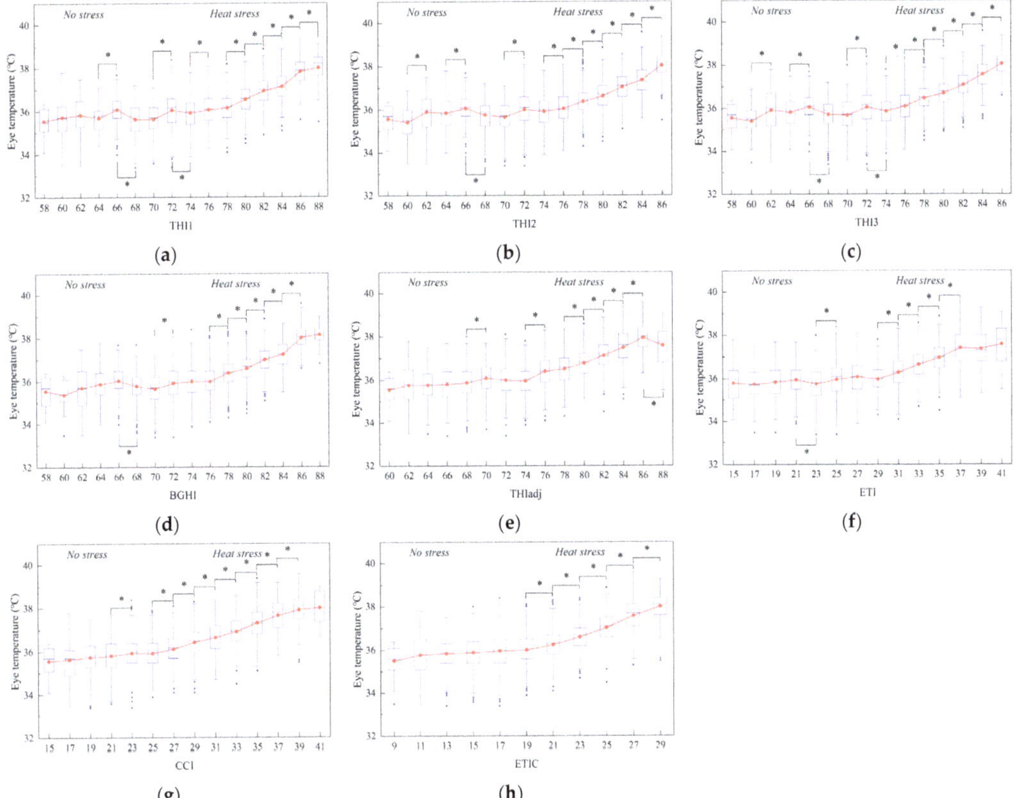

Figure 9. Effect of the thermal index values on eye temperature. The interquartile range (IQR) is shown by a blue box, the median as a blue horizontal line, the mean as a red square, and the whiskers extend to 1.5 IQR. Outliers are indicated by blue dots. An asterisk indicates a significant ($p < 0.05$) change in the variable. The thermal indices are the temperature–humidity index (THI1 (**a**), THI2 (**b**), THI3 (**c**)), black globe humidity index (BGHI (**d**)), adjusted THI (THIadj (**e**)), equivalent temperature index (ETI (**f**)), comprehensive climate index (CCI (**g**)), and equivalent temperature index for dairy cattle (ETIC (**h**)).

Table 4. Comparison of correlations with rectal temperature (RT), respiration rate (RR), skin temperature (ST), and eye temperature (ET).

Variable	RT		RR		ST		ET	
	Hotelling's t (df = 3002)	*p*	Hotelling's t (df = 3002)	*p*	Hotelling's t (df = 2993)	*p*	Hotelling's t (df = 2921)	*p*
r.THI1 vs. r.THI2	3.3956	0.0007	−1.6156	0.1063	1.4200	0.1557	−2.0880	0.0369
r.THI1 vs. r.THI3	−1.6033	0.109	−4.6097	<0.0001	−8.2067	<0.0001	−10.5800	<0.0001
r.THI1 vs. r.BGHI	−4.3239	<0.0001	−3.0837	0.0021	−15.2100	<0.0001	−18.3000	<0.0001
r.THI1 vs. r.THIadj	−6.6583	<0.0001	2.3754	0.0176	−12.6980	<0.0001	−3.5820	0.0003
r.THI1 vs. r.ETI	1.3323	0.1829	9.4858	<0.0001	6.1263	<0.0001	10.7310	<0.0001
r.THI1 vs. r.CCI	−7.5368	<0.0001	3.1935	0.0014	−16.3540	<0.0001	−9.504	<0.0001
r.THI1 vs. r.ETIC	−7.7799	<0.0001	−4.0801	<0.0001	−14.5620	<0.0001	−7.5180	<0.0001
r.THI2 vs. r.THI3	−8.0820	<0.0001	−6.9426	<0.0001	−19.3190	<0.0001	−18.8600	<0.0001
r.THI2 vs. r.BGHI	−7.2790	<0.0001	−2.2988	0.0216	−18.3300	<0.0001	−18.9800	<0.0001
r.THI2 vs. r.THIadj	−7.2026	<0.0001	2.6328	0.0085	−12.1970	<0.0001	−2.9060	0.0037
r.THI2 vs. r.ETI	0.0000	1	8.1727	<0.0001	4.4856	<0.0001	9.2882	<0.0001
r.THI2 vs. r.CCI	−7.9281	<0.0001	3.4025	0.0007	−16.0650	<0.0001	−8.7500	<0.0001
r.THI2 vs. r.ETIC	−9.3260	<0.0001	−3.6130	0.0003	−15.6510	<0.0001	−6.9840	<0.0001
r.THI3 vs. r.BGHI	−6.4730	<0.0001	2.3047	0.0212	−17.3100	<0.0001	−19.0900	<0.0001
r.THI3 vs. r.THIadj	−5.7939	<0.0001	3.7366	0.0002	−9.0665	<0.0001	0.0000	1
r.THI3 vs. r.ETI	1.5757	0.1152	8.5904	<0.0001	7.6148	<0.0001	11.8080	<0.0001
r.THI3 vs. r.CCI	−7.0155	<0.0001	4.4672	<0.0001	−13.6980	<0.0001	−6.4790	<0.0001
r.THI3 vs. r.ETIC	−6.9551	<0.0001	−1.7555	0.0793	−10.1690	<0.0001	−2.2500	0.0245
r.BGHI vs. r.THIadj	−4.7432	<0.0001	3.3562	0.0008	−6.2982	<0.0001	2.9011	0.0037
r.BGHI vs. r.ETI	2.6504	0.0081	7.5636	<0.0001	10.2609	<0.0001	14.5390	<0.0001
r.BGHI vs. r.CCI	−6.2288	<0.0001	4.2133	<0.0001	−11.4330	<0.0001	−4.0520	0.0001
r.BGHI vs. r.ETIC	−5.0088	<0.0001	−2.2070	0.0274	−5.5448	<0.0001	2.1334	0.033
r.THIadj vs. r.ETI	9.9701	<0.0001	4.5172	<0.0001	22.6397	<0.0001	13.5500	<0.0001
r.THIadj vs. r.CCI	−6.1202	<0.0001	3.6567	0.0003	−17.2370	<0.0001	−20.4900	<0.0001
r.THIadj vs. r.ETIC	2.6003	0.0094	−9.5786	<0.0001	4.8173	<0.0001	−2.84500	0.0045
r.ETI vs. r.CCI	−9.2415	<0.0001	−1.7398	0.082	−22.6040	<0.0001	−17.1700	<0.0001
r.ETI vs. r.ETIC	−9.0234	<0.0001	−12.8430	<0.0001	−20.9370	<0.0001	−17.8300	<0.0001
r.CCI vs. r.ETIC	5.2047	<0.0001	−8.8408	<0.0001	12.8251	<0.0001	8.6128	<0.0001

4. Discussion

The aim of the present study was to evaluate the performance of the TIs with respect to their heat transfer characteristics and relationships with the animal-based indicators. We calculated the equivalent ambient temperature change that resulted from the changed relative humidity, wind speed, and solar radiation. The results of this study indicate that the warming effect of the increased RH was the most reflected by the ETI, and the chilling effect of the increased wind speed was the most reflected by the CCI. These results could be explained by the environmental conditions that were used to develop the model. According to Baeta et al. [16], the ETI was developed based on variable environments in a climatic chamber with Ta ranging from 16 °C to 41 °C, RH ranging from 40% to 90%, and wind speed ranging from 0.5 m/s to 6.5m/s. However, the authors did not specifically emphasize the significance of RH. Compared to the relationship between the ETI and RH, the relationship between the CCI and wind speed seems clearer. According to Mader et al. [17], the wind chill index (WCI), which defines the relationship between wind speed and Ta, was used as the basis for modeling the CCI under cold environments (Ta < 5 °C). Although the warming effect of increased solar radiation was indicated by the THIadj, CCI, and ETIC, only the ETIC reflected that the warming effect of solar radiation could apparently increase with increasing Ta. This finding can be demonstrated by Wang et al. [18] since the authors specifically highlighted the interaction between Ta and other environmental parameters (i.e., RH, wind speed, and solar radiation) in the process of modeling the ETIC. For more complex heat transfer characteristics, Bjerg et al. [30] investigated the changes of the chilling effect of wind speed with increasing wind speed. In addition, Wang et al. [28] explored the changes in the chilling effect of wind speed with increasing RH. In this study, we only focused on the interaction between the warming or chilling effect caused by one changed environmental variable and the air temperature. It has been confirmed that evaporation is the most important mechanism that is automatically exhibited by cattle to strengthen heat loss in hot environments [31]. Evaporative heat loss consists of respiratory and cutaneous heat loss, and the latter was governed by the moisture gradient between the ambient air and the skin surface [31]. Under high Ta and RH conditions, restricted cutaneous evaporation exacerbates heat stress, which can be recognized by the TIs containing the RH parameter (i.e., THI, THIadj, ETI, CCI, and ETIC). Under high Ta and wind speed, the sweating rate is a greater driving force for cutaneous evaporation than wind speed, which means that the chilling effect of wind speed decreases with the increasing Ta. This fact was recognized by the CCI but was significantly reflected by the ETIC. It should be noted that a higher relative significance to a certain parameter or a better representation of the interaction between Ta and the other parameters is not fully equivalent to a better TI performance in actual conditions.

Further, we examined the performance of the TIs through their relationship with the physiological responses of dairy cows. As mentioned above, all of the animal-based indicators (i.e., RT, RR, ST, and ET) used in this study have been proven to be useful and effective for heat stress assessment in dairy cows by previous literature. Generally, the values of the animal-based indicators increased with the increases in heat stress magnitude. For RT, it was expected that the RT underwent a steady phase and then rose as the index value rose [5,13]. However, there were some fluctuations in RT under the no-stress conditions indicated by THI1. The expected relationship that RT remains stable under no-stress conditions and rises significantly, continuously, and linearly under heat stress conditions was better reflected by THI2, THI3, and ETIC. We noticed that the RT within the extreme heat stress level was lower (no significantly) than that within the prior heat stress level. This unexpected result was associated with the small sample size within the highest level group. The changes of the RR with the index value were generally consistent with the expectation that the RR rose slowly and then rapidly with the increase in the heat stress level. All of the TIs evaluated in this study reflected that the RR grew significantly and continuously with the increase in the index value. However, the relationship was better reflected by the THI1, THI2, and ETIC. Interestingly, the first significant increase in RR was found at 66 THI1 and THI2, which is in accordance with the critical THI threshold identified for RR from the recent studies [13]. With respect to ST, an ideal relationship could be that the ST grew approximately linearly with the increasing index value, as is to be expected for RR. Based on this, the relationships of the ST with the CCI and ETIC corresponded to the expectation. One unanticipated result was that the ST significantly decreased at 17 CCI and 11 ETIC. This can likely be attributed to the fact that the ST was measured using infrared thermography and that dirt, moisture, or other secondary factors (e.g., contact with the ground while lying) can alter the emissivity and conductivity and thus can cause inaccurately measured results [32]. Compared to ST, ET is less likely to be affected by the factors mentioned above and commonly serves as a proxy for internal temperature [26]. In the current evaluation, the ET fluctuated with the values of the THI1, THI2, THI3, BGHI, and ETI under the no-stress conditions. Additionally, the relationship between ET and THIadj was unsatisfactory because of the inconsecutive increase and unexpected decrease in ET under the heat-stress conditions indicated by the THIadj. Obviously, in comparison to other TIs, the CCI and ETIC performed better with respect to their relationship with ET.

Further, correlation, which is one of the methodologies commonly used in studies for the evaluation of thermal indices, was conducted to examine the relationships between the TIs and animal-based indicators. In the current study, we investigated and compared the correlations among the indices. We found that the RT, ST, and ET correlated the most closely with the CCI. High correlations of CCT with ST and ET agreed with another study conducted in southern China (Yancheng, Jiangsu, China), which reported that the CCI showed higher correlations ($0.443 \leq r \leq 0.849$) with the body surface temperature than other TIs (e.g., THI1, BGHI, THIadj, ETI, ETIC) [33]. Consistent with the present study, a study from Da Silva et al. [34], in which the data were collected from a tropical environment, reported that the CCI showed a higher correlation ($r = 0.374$) with RT than the BGHI and other TIs (e.g., the heat load index (HLI) and the index of thermal stress for cows (ITSC)). We also observed the highest correlation with RR shown by the ETIC. This result may be explained by the fact that RR was used for the response variable to develop the ETIC regression model. Additionally, the original authors stated that the ETIC ($r = 0.703$) performed better than the CCI ($r = 0.692$) and THIadj ($r = 0.671$) in terms of

RR [18]. With the use of different data sets the correlation results can vary. For example, a previous study also reported that ETI was significantly correlated with the RT (r = 0.293) and RR (r = 0.520) of dairy cattle in the pasture and recommended it as the best index for heat stress evaluation in tropical environments [25]. Li et al. [35] evaluated eight TIs using two data sets; they reported that the ETI showed the highest correlation with RR (r = 0.34) based on the first data set collected in three breeds of dairy cow and six breeds of feedlot heifer from five regions in the United States, and the BGHI correlated the most closely with the RR (r = 0.73) and ST (r = 0.56) based on the second data set generated from a four-day measurement in twelve Holsteins in climate-controlled chambers. Despite the good performance of some data sets, there were some doubts over the ETI since it was developed on the basis of limited animals and short treatment observation periods (3 days) [36].

To sum up the results in this study, we found that the CCI and ETIC were the two best TIs for heat stress assessment. CCI performed better with respect to its relationships with the physiological responses (i.e., the changes and correlations of RT, ST, and ET with the CCI). For precision livestock farming, the main drawback of the CCI concerning the complexity of computation can be overcome when the algorithm is inserted into the environment forewarning and controlling systems in animal housings. ETIC mainly performed well in aspects of heat transfer characteristics and correlations with RR. Moreover, ETIC includes four main factors driving heat stress, and it can be calculated relatively conveniently. Combining the findings of our previous study in southern China, we found that CCI has a satisfactory performance in assessing the heat stress of dairy cows kept in semi-confined housing systems in China. Cows are housed in free-stall barns where there is no ambient temperature regulation. The barns are typically uninsulated and are naturally ventilated with curtain sidewalls. Cooling systems such as sprinklers and panel fans are included and are only used in hot seasons to alleviate heat stress. Solar radiation has an indirect effect on animal heat stress by heating the enclosure structure of the housings. Indoor airflow is usually turbulent and is governed by outside wind, the difference between inside and outside Ta, and fans.

Caution must be applied when the findings of this study are extrapolated to other situations. The final results of index evaluation can also be influenced by other factors, including breeds (i.e., Bos taurus, Bos indicus, and water buffalo), regional climates (i.e., tropical, subtropical, and temperate climates), production systems (i.e., free-range and confined housing), and heat stress acclimation [37]. Prior studies have reported that Ta could provide similar performance in assessing heat stress compared to other TIs [11,38]. Nevertheless, there is no doubt that Ta can not represent the overall environmental stress forced upon dairy cows. Objectively, these studies reveal that some issues still exist in the current TIs. As far as we know, to date, the existing TIs are all the same type of model—an empirical model. Previous experts seemingly paid too much attention to seeking the statistical association between animal-based indicators and multi environmental factors. Future work regarding developing novel TIs should be oriented towards the essence of thermal stress—an imbalance between heat production and loss.

5. Conclusions

This study evaluated the thermal indices applied for heat stress assessment in dairy cows in a temperate climate in northern China. Compared to other investigated indices, the comprehensive climate index (CCI) performed better due to its relationships with the rectal temperature, skin temperature, and eye temperature. The equivalent temperature index for dairy cattle (ETIC) mainly performed better in regard to the heat transfer characteristics and the correlation with the respiration rate. The evaluation of the results of the thermal indices could be influenced by animal and environmental factors. Nevertheless, the current study demonstrated the findings from previous reports that the CCI could be the most promising thermal index to assess heat stress for housed dairy cows. Moreover, there is still a real need to develop new thermal indices for precision environment control of livestock buildings.

Author Contributions: Conceptualization, methodology, software, formal analysis, investigation, data curation, visualization, writing—original draft preparation, G.Y.; writing—review and editing, funding acquisition, H.L.; supervision, funding acquisition, Z.S. All authors have read and agreed to the published version of the manuscript.

Funding: This research was funded by the China Agriculture Research System of MOF and MARA and the National Key R&D program Inter-governmental/Hong Kong, Macao and Taiwan key projects (2019YFE0103800).

Institutional Review Board Statement: The study protocol was approved by the China Agricultural University Department of Agricultural Structure and Bioenvironmental Engineering Animal Ethics Committee (20200625).

Data Availability Statement: The data presented in this study are available from the corresponding author upon reasonable request.

Conflicts of Interest: The authors declare no conflict of interest.

References

1. Hoffmann, G.; Herbut, P.; Pinto, S.; Heinicke, J.; Kuhla, B.; Amon, T. Animal-related, non-invasive indicators for determining heat stress in dairy cows. *Biosyst. Eng.* **2020**, *199*, 83–96. [CrossRef]
2. Polsky, L.; von Keyserlingk, M.A. Invited review: Effects of heat stress on dairy cattle welfare. *J. Dairy. Sci.* **2017**, *100*, 8645–8657. [CrossRef]
3. West, J.W. Effects of heat-stress on production in dairy cattle. *J. Dairy. Sci.* **2003**, *86*, 2131–2144. [CrossRef]
4. Wang, T.; Zhong, R.; Zhou, D. Temporal-spatial distribution of risky sites for feeding cattle in China based on temperature/humidity index. *Agriculture* **2020**, *10*, 571. [CrossRef]
5. Yan, G.; Liu, K.; Hao, Z.; Shi, Z.; Li, H. The effects of cow-related factors on rectal temperature, respiration rate, and temperature-humidity index thresholds for lactating cows exposed to heat stress. *J. Therm. Biol.* **2021**, *100*, 103041. [CrossRef]
6. Becker, C.A.; Collier, R.J.; Stone, A.E. Invited review: Physiological and behavioral effects of heat stress in dairy cows. *J. Dairy. Sci.* **2020**, *103*, 6751–6770. [CrossRef]
7. Berman, A.; Folman, Y.; Kaim, M.; Mamen, M.; Herz, Z.; Wolfenson, D.; Arieli, A.; Graber, Y. Upper critical temperatures and forced ventilation effects for high-yielding dairy cows in a subtropical climate. *J. Dairy. Sci.* **1985**, *68*, 1488–1495. [CrossRef]
8. Thom, E.C. The discomfort index. *Weatherwise* **1959**, *12*, 57–61. [CrossRef]
9. Hahn, G.L.; Gaughan, J.B.; Mader, T.L.; Eigenberg, R.A. Chapter 5: Thermal indices and their applications for livestock environments. In *Livestock Energetics and Thermal Environmental Management*; DeShazer, J.A., Ed.; American Society of Agricultural and Biological Engineers: St. Joseph, MI, USA, 2009; pp. 113–130.
10. Bohmanova, J.; Misztal, I.; Cole, J.B. Temperature-humidity indices as indicators of milk production losses due to heat stress. *J. Dairy. Sci.* **2007**, *90*, 1947–1956. [CrossRef] [PubMed]
11. Dikmen, S.; Hansen, P.J. Is the temperature-humidity index the best indicator of heat stress in lactating dairy cows in a subtropical environment? *J. Dairy. Sci.* **2009**, *92*, 109–116. [CrossRef]
12. Heinicke, J.; Hoffmann, G.; Ammon, C.; Amon, B.; Amon, T. Effects of the daily heat load duration exceeding determined heat load thresholds on activity traits of lactating dairy cows. *J. Therm. Biol.* **2018**, *77*, 67–74. [CrossRef]
13. Pinto, S.; Hoffmann, G.; Ammon, C.; Amon, T. Critical THI thresholds based on the physiological parameters of lactating dairy cows. *J. Therm. Biol.* **2020**, *88*, 102523. [CrossRef]
14. Buffington, D.E.; Collazo-Arocho, A.; Canton, G.H.; Pitt, D.; Thatcher, W.W.; Collier, R.J. Black globe-humidity index (BGHI) as comfort equation for dairy cows. *Trans. ASABE* **1981**, *24*, 711–714. [CrossRef]
15. Mader, T.L.; Davis, M.S.; Brown-Brandl, T. Environmental factors influencing heat stress in feedlot cattle. *J. Anim. Sci.* **2006**, *84*, 712–719. [CrossRef]
16. Baeta, F.C.; Meador, N.F.; Shanklin, M.D.; Johnson, H.D. Equivalent temperature index at temperatures above the thermoneutral for lactating dairy cows. In Proceedings of the International Winter Meeting of American Society of Agricultural Engineers (ASAE), Chicago, IL, USA. Available online: https://agris.fao.org/agris-search/search.do?recordID=US8853966 (accessed on 15 February 2021).
17. Mader, T.L.; Johnson, L.J.; Gaughan, J.B. A comprehensive index for assessing environmental stress in animals. *J. Anim. Sci.* **2010**, *88*, 2153–2165. [CrossRef]
18. Wang, X.; Gao, H.; Gebremedhin, K.G.; Bjerg, B.S.; van Os, J.; Tucker, C.B.; Zhang, G. A predictive model of equivalent temperature index for dairy cattle (ETIC). *J. Therm. Biol.* **2018**, *76*, 165–170. [CrossRef] [PubMed]
19. Ji, B.; Banhazi, T.; Perano, K.; Ghahramani, A.; Bowtell, L.; Wang, C.; Li, B. A review of measuring, assessing and mitigating heat stress in dairy cattle. *Biosyst. Eng.* **2020**, *199*, 4–26. [CrossRef]

20. Romanovsky, A.A. Skin temperature: Its role in thermoregulation. *Acta. Physiol.* **2014**, *210*, 498–507. [CrossRef]
21. Godyń, D.; Herbut, P.; Angrecka, S. Measurements of peripheral and deep body temperature in cattle—A review. *J. Therm. Biol.* **2019**, *79*, 42–49. [CrossRef] [PubMed]
22. McManus, C.; Tanure, C.B.; Peripolli, V.; Seixas, L.; Fischer, V.; Gabbi, A.M.; Menegassi, S.R.O.; Stumpf, M.T.; Kolling, G.J.; Dias, E.; et al. Infrared thermography in animal production: An overview. *Comput. Electron. Agric.* **2016**, *123*, 10–16. [CrossRef]
23. Hoffmann, G.; Schmidt, M.; Ammon, C.; Rose-Meierhöfer, S.; Burfeind, O.; Heuwieser, W.; Berg, W. Monitoring the body temperature of cows and calves using video recordings from an infrared thermography camera. *Vet. Res. Commun.* **2013**, *37*, 91–99. [CrossRef] [PubMed]
24. Armstrong, D.V. Heat Stress Interaction with Shade and Cooling. *J. Dairy. Sci.* **1994**, *77*, 2044–2050. [CrossRef]
25. Da Silva, R.G.; Morais, D.A.E.F.; Guilhermino, M.M. Evaluation of thermal stress indexes for dairy cows in tropical regions. *Rev. Bras. de Zootec.* **2007**, *36*, 1192–1198. [CrossRef]
26. Macmillan, K.; Colazo, M.G.; Cook, N.J. Evaluation of infrared thermography compared to rectal temperature to identify illness in early postpartum dairy cows. *Res. Vet. Sci.* **2019**, *125*, 315–322. [CrossRef] [PubMed]
27. Montanholi, Y.R.; Lim, M.; Macdonald, A.; Smith, B.A.; Goldhawk, C.; Schwartzkopf-Genswein, K.; Miller, S.P. Technological, environmental and biological factors: Referent variance values for infrared imaging of the bovine. *J. Anim. Sci. Biotechno.* **2015**, *6*, 1–16. [CrossRef] [PubMed]
28. Wang, X.; Bjerg, B.S.; Choi, C.Y.; Zong, C.; Zhang, G. A review and quantitative assessment of cattle-related thermal indices. *J. Therm. Biol.* **2018**, *77*, 24–37. [CrossRef]
29. Diedenhofen, B.; Musch, J. cocor: A comprehensive solution for the statistical comparison of correlations. *PLoS ONE* **2015**, *10*, e0121945. [CrossRef] [PubMed]
30. Bjerg, B.; Wang, X.; Zhang, G. The effect of air velocity on heat stress at increased air temperature. In Proceedings of the CIGR-AgEng Conference, Aarhus, Denmark, 26–29 June 2016.
31. Collier, R.J.; Gebremedhin, K.G. Thermal biology of domestic animals. *Annu. Rev. Anim. Biosci.* **2015**, *3*, 513–532. [CrossRef] [PubMed]
32. Stewart, M.; Webster, J.R.; Schaefer, A.L.; Cook, N.J.; Scott, S.L. Infrared thermography as a non-invasive tool to study animal welfare. *Anim. Welf.* **2005**, *14*, 319–325.
33. Yan, G.; Li, H.; Zhao, W.; Shi, Z. Evaluation of thermal indices based on their relationships with some physiological responses of housed lactating cows under heat stress. *Int. J. Biometeorol.* **2020**, *64*, 2077–2091. [CrossRef]
34. Da Silva, R.G.; Maia, A.S.C.; de Macedo Costa, L.L. Index of thermal stress for cows (ITSC) under high solar radiation in tropical environments. *Int. J. Biometeorol.* **2015**, *59*, 551–559. [CrossRef] [PubMed]
35. Li, S.; Gebremedhin, K.G.; Lee, C.; Collier, R. Evaluation of thermal stress indices for cattle. In *Proceedings of the ASABE Annual Meeting, Reno, NV, USA, 21–24 June 2009*; American Society of Agricultural and Biological Engineers: St. Joseph, MI, USA, 2009; p. 1. [CrossRef]
36. Fournel, S.; Rousseau, A.N.; Laberge, B. Rethinking environment control strategy of confined animal housing systems through precision livestock farming. *Biosyst. Eng.* **2017**, *155*, 96–123. [CrossRef]
37. Yan, G.; Li, H.; Shi, Z.; Wang, C. Research status and existing problems in establishing cow heat stress indices. *Trans. Chin. Soc. Agric. Eng.* **2019**, *35*, 226–233.
38. Ji, B.; Banhazi, T.; Ghahramani, A.; Bowtell, L.; Wang, C.; Li, B. Modelling of heat stress in a robotic dairy farm. Part 1: Thermal comfort indices as the indicators of production loss. *Biosyst. Eng.* **2020**, *199*, 27–42. [CrossRef]

Article

Effects of Cold Exposure on Performance and Skeletal Muscle Fiber in Weaned Piglets

Jie Yu [1,2,*,†], Shuai Chen [1,†], Ziyou Zeng [2], Shuaibing Xing [1], Daiwen Chen [1], Bing Yu [1], Jun He [1], Zhiqing Huang [1], Yuheng Luo [1], Ping Zheng [1], Xiangbing Mao [1], Junqiu Luo [1] and Hui Yan [1]

[1] Key Laboratory for Animal Disease-Resistance Nutrition of Sichuan Province, Animal Nutrition Institute, Sichuan Agricultural University, Chengdu 611130, China; chenshuai@tianzow.com (S.C.); xsb19951120@163.com (S.X.); dwchen@sicau.edu.cn (D.C.); ybingtian@163.com (B.Y.); hejun8067@163.com (J.H.); zqhuang@sicau.edu.cn (Z.H.); luoluo212@126.com (Y.L.); zpind05@163.com (P.Z.); acatmxb2003@163.com (X.M.); 13910@sicau.edu.cn (J.L.); yan.hui@sicau.edu.cn (H.Y.)

[2] Sichuan Tequ Agriculture and Animal Husbandry Technology Group Co., Ltd., Chengdu 610207, China; ziyouzeng@foxmail.com

* Correspondence: yujie@sicau.edu.cn; Tel.: +86-28-82690922

† These authors contributed equally to this work.

Simple Summary: Muscle fiber is the basic unit of muscle composition. The type of skeletal muscle fiber can be transformed from fast-twitch to slow-switch or vice versa by internal and external factors. Low-temperature is one of the major environmental factors that influences the growth performance of animals. However, the influence of low-temperature on weaned piglets' skeletal muscle fiber, and whether this influence is related to mitochondrial function and antioxidant capacity, has not been reported. Our results indicated that low temperature could negatively affect growth performance and nutrient digestibility in weaned piglets. Moreover, evidence was provided to show that low-temperature induces a shift toward oxidative muscle fibers, which may occur through mitochondrial function regulation and increased antioxidative capacity.

Abstract: Low-temperature is one of the most significant risks for the animal industry. In light of this, the present study aimed to explore the effects of low-temperature on growth performance, nutrient digestibility, myofiber types and mitochondrial function in weaned piglets. A total of sixteen 21-day-old male Duroc × Landrace × Yorkshire (DLY) piglets were randomly divided into a control group (CON, 26 ± 1 °C) and a low-temperature group (LT, 15 ± 1 °C), with eight duplicate piglets in each group. The trial period lasted for 21 days. We showed that LT not only increased the ADFI ($p < 0.05$), as well as increasing the diarrhea incidence and diarrhea index of weaned piglets in the early stage of the experiment ($p < 0.01$), but it also decreased the apparent digestibility of crude protein (CP), organic matter (OM) and dry matter (DM) ($p < 0.05$). Meanwhile, in the LT group, the mRNA expression of *MyHC IIa* ($p < 0.05$) in longissimus dorsi muscle (LM) and *MyHC I* ($p < 0.01$) in psoas muscle (PM) were increased, while the mRNA expression of *MyHC IIx* in PM was decreased ($p < 0.05$). In addition, LT increased the mRNA expression of mitochondrial function-related genes citrate synthase (CS) and succinate dehydrogenase-b (SDHB) in LM, as well as increased the mRNA expression of CS ($p < 0.05$) and carnitine palmitoyl transferase-1b (CPT-1b) ($p < 0.01$) in PM. Furthermore, LT increased the T-AOC activity in serum and LM ($p < 0.01$), as well as increased the T-SOD activity in PM ($p < 0.05$). Taken together, these findings showed that low-temperature could negatively affect the growth performance and nutrient digestibility, but resulted in a shift toward oxidative muscle fibers, which may occur through mitochondrial function regulation.

Keywords: weaned piglets; cold exposure; growth performance; skeletal muscle fiber; antioxidant capacity

Citation: Yu, J.; Chen, S.; Zeng, Z.; Xing, S.; Chen, D.; Yu, B.; He, J.; Huang, Z.; Luo, Y.; Zheng, P.; et al. Effects of Cold Exposure on Performance and Skeletal Muscle Fiber in Weaned Piglets. *Animals* **2021**, *11*, 2148. https://doi.org/10.3390/ani11072148

Academic Editors: Lilong Chai and Yang Zhao

Received: 15 June 2021
Accepted: 18 July 2021
Published: 20 July 2021

1. Introduction

Temperature is one of the main environmental factors that influences the growth and development of animals. Low ambient temperature not only affects growth performance, but also reduces antioxidant capacity and immunity of young animals [1,2]. In cold environments, animals mainly rely on two ways to produce heat to maintain body temperature, largely shivering thermogenesis and non-shivering thermogenesis; shivering thermogenesis means to produce heat by the contraction of skeletal muscle, and non-shivering thermogenesis means to produce heat by nutrient metabolism [3]. Skeletal muscle is composed of a large number of muscle fibers. Different types of skeletal muscle fibers have different energy metabolisms and contraction speeds [4]. The mammalian skeletal muscle can be divided into four fiber types including type I with *MyHC I*, type II with *MyHC IIa*, *MyHC IIx* and *MyHC IIb* [5]. Previous research found that type I fibers have lower excitation thresholds and stronger oxidative metabolism capacity than type II fibers [6]. Although shivering thermogenesis of skeletal muscle is indispensable for piglets to maintain body temperature in the LT environment, there is little known about the effects of cold exposure on skeletal muscle characteristics. Therefore, in the present study, our aim was to investigate the effects of LT on growth performance, antioxidant capacity, myofiber types and mitochondrial function in weaned piglets.

2. Materials and Methods

2.1. Animals, Diets and Experimental Design

Sixteen 21-day-old male Duroc × Landrace × Yorkshire (DLY) piglets with an average body weight of 6.5 ± 0.5 kg were randomly divided into 2 groups with 8 duplicates. The animals were housed individually in pens under low temperature (LT, 15 ± 1 °C) or thermoneutral temperature (CON, 26 ± 1 °C). The thermoneutral temperature (CON, 26 ± 1 °C) in the experiment was according to the recommendations of Dividich et al. [7]. The diet was formulated according to NRC (2012). Ingredients and calculated nutrient contents of the diet were presented in Table 1. The trial period lasted for 21 days. Food and water were provided ad libitum throughout the experiment.

Table 1. Composition and calculated nutrient content of the basal diet.

Ingredients	Content (%)
Extruded corn	30.00
Corn	25.00
Soybean meal	10.50
Extruded soybean	5.50
Rice screenings	8.15
Wheat bran	1.50
Soybean protein concentrate	4.00
Spray-dried animal plasma	1.50
Fish meal	3.50
Whey powder (low protein)	3.80
Soybean oil	2.00
Sucrose	2.00
Limestone	0.88
Dicalcium phosphate	0.40
NaCl	0.30
L-Lysine HCl (78%)	0.42
DL-Methionine	0.14
L-Threonine (98.5%)	0.06
Chloride choline	0.10
Vitamin premix [a]	0.05
Mineral premix [b]	0.20

Table 1. *Cont.*

Ingredients	Content (%)
Nutrition level	
DE (Mcal/kg)	3.50
Crude protein	18.03
Calcium	0.80
Total phosphorus	0.56
Available phosphorus	0.36
Lysine	1.35
Methionine	0.44
Threonine	0.79
Tryptophan	0.24

[a] The premix provided the following per kg of diets: 8000 IU of V_A, 2000 IU of VD_3, 20 IU of V_E, 2 mg of VK_3, 1.50 mg of VB_1, 5.6 mg of VB_2, 1.5 mg of VB_6, 0.02 mg of VB_{12}, 15 mg of niacin, 10 mg of pantothenic acid, 0.60 mg of folic acid and 0.10 mg of biotin. [b] The premix provided the following per kg of diets: 100 mg of Fe, 100 mg of Cu, 100 mg of Zn, 20 mg of Mn, 0.30 mg of I and 0.30 mg of Se.

2.2. Growth Performance

Feed intake was recorded daily. The pigs were individually weighed at the start and the end of the trial to calculate average daily body weight gain (ADG), average daily feed intake (ADFI) and the ratio of feed to gain (F/G).

2.3. Diarrhea Score

The diarrhea score for each pig was monitored between 12 noon and 1 pm daily. The normal consistency of feces formed (no diarrhea) score is 0; the soft, partially formed feces (mild diarrhea) score is 1; the loose, semi-liquid feces (moderate diarrhea) score is 2; and the watery feces (severe diarrhea) score is 3. The average diarrhea incidence and diarrhea score per group was calculated daily.

2.4. Sample Collection

At the start of trial, experimental diets were sampled for nutrient digestibility analysis. In the last 4 days of trial, fresh fecal samples from total of 16 pigs were collected immediately after defecation and then placed in individual plastic bags. After collection, 10 mL of a 10% H_2SO_4 solution was added to each 100 g of wet fecal sample. All diet and fecal samples were immediately stored at −20 °C for further analysis. At the end of trial, after fasting for 12 h, 10 mL of anterior vena cava blood was taken from each piglet (an empty stomach in the morning), and placed in an inclined position at room temperature for 60 min. The blood sample was collected by centrifugation with 3000× g for 10 min at 4 °C prior to antioxidant status analysis. After blood sampling, control group and low temperature group piglets were killed in rotation (first, one pig from the control group was killed, and then one pig from the low temperature group was killed, and then repeated in this order). All piglets were given sodium pentobarbital (200 mg kg^{-1} BW), exsanguinated, dehaired, eviscerated, and split down the muscle. The samples of LM and PM were collected immediately, frozen in liquid nitrogen and then stored at −80 °C until analysis.

2.5. Chemical Analysis

The apparent nutrient digestibility was measured, using acid insoluble ash (AIA) as the indicator. For AIA determination, 5 g of finely ground feed or feces was boiled in 75 mL HCl for 15 min, then filtered through ashless filter paper, washed with boiling water until free of acid and finally ashed at 550 °C in a muffle furnace for 8 h [8]. After AIA determination, all diet and fecal samples were analyzed for crude protein, crude ash and dry matter [9]. Gross energy was measured by an automatic adiabatic oxygen bomb calorimeter (Parr Instrument Co., Moline, IL, USA). The apparent nutrient digestibility was calculated uing following formula, where A1 represents the AIA content of the diet;

A2 represents the AIA content of feces; F1 represents the nutrient content of the diet and F2 represents the nutrient content of feces:

$$Apparent\ nutrient\ digestibility\ (\%) = \left(1 - \frac{A1 * F2}{A2 * F1}\right) * 100$$

2.6. Determination of Antioxidant Parameters in Serum and Skeletal Muscle

About 0.8 g of skeletal muscle sample (LM and PM) was quickly weighed, thawed, and homogenized (1:9, wt/vol) with ice-cold physiological saline using a homogenizer. After this, the mixture of muscle and normal saline was centrifuged at 4000× *g* for 15 min at 4 °C. The supernatant was acquired and stored at −80 °C and used for antioxidant-related enzyme activity examination. Total protein concentration was determined by the BCA Protein Assay Kit (Pierce, Rockford, IL, USA). The total antioxidant capacity (T-AOC), total superoxide dismutase (T-SOD) and malondialdehyde (MDA) level in serum and muscle were determined by commercial kits (Nanjing Jiancheng Bioengineering Institute, Nanjing, China) according to the manufacturer's instructions.

2.7. Total RNA Extraction, Reverse Transcription and Quantitative Real-Time PCR

Total RNA was extracted from samples of the skeletal muscle using the TRIzol reagent (TaKaRa), according to the manufacturer's instructions. The concentration of RNA in the final preparations was calculated from the OD260. Reverse transcription was performed using the PrimeScript TM RT Reagent Kit (TaKaRa) with a 1 µg RNA sample, according to the manufacturer's instructions. Complementary DNA was used as the template for PCR. Real-time quantitative PCR was performed in an Option Monitor 3 Real-Time PCR Detection System (Bio-Rad) using the SYBR Green Supermix (TaKaRa). The gene-specific primers used are listed in Table 2. The thermal cycling conditions were 40 cycles of 95 °C for 5 s, and 60 °C for 30 s.

Table 2. The primer sequences used for real-time quantitative PCR.

Gene Name	Primer	Sequences (5′→3′)	Product Length (bp)	Accession No.
β-actin	Forward Reverse	GCGGCATCCACGAAACTAC TGATCTCCTTCTGCATCCTGTC	138	DQ-845171.1
MyHC I	Forward Reverse	AAGGGCTTGAACGAGGAGTAGA TTATTCTGCTTCCTCCAAAGGG	114	AB-053226
MyHC IIa	Forward Reverse	GCTGAGCGAGCTGAAATCC ACTGAGACACCAGAGCTTCT	137	AB-025260
MyHC IIx	Forward Reverse	AGAAGATCAACTGAGTGAACT AGAGCTGAGAAACTAACGTG	149	AB-025262
MyHC IIb	Forward Reverse	ATGAAGAGGAACCACATTA TTATTGCCTCAGTAGCTTG	166	AB-025261
CS	Forward Reverse	GGAAGTCGGCAAAGATGTGT TCATGAGGCAGGTGTTTCAG	162	NM-214276.1
CPT-1b	Forward Reverse	GCTATCTGTGTCCGCCTTCT GGCTGTATTCCTCGTCATCC	151	NM-001007191.1
SDHB	Forward Reverse	TGTGGTCCTATGGTGTTGGA TTTGTCGAGGTTGGTGTCAA	168	NM-001104953.1
Nrf-1	Forward Reverse	TCCATCAATCCGGAAGAGAC GCACCACATTCTCCAAAGGT	170	XM-021078993.1

For normalization, β-actin was chosen as the reference gene because no variation in its expression was observed between treatments. The relative mRNA abundance of the analyzed genes was calculated using the $2^{-\triangle\triangle Ct}$ method [10], and the messenger RNA (mRNA) level of each target gene for the CON group was set to 1.0.

2.8. Statistical Analysis

Data were analyzed by *t*-test using the statistical program SAS (version 9.4; SAS Inst. Inc., Cary, NC, USA). Each pig served as a statistical unit. Data are presented as the mean ± standard error, and results are reported as least square means and considered extremely significant if $p \leq 0.01$, significant if $p \leq 0.05$ and a tendency if $0.05 < p \leq 0.10$.

3. Results

3.1. Growth Performance

Compared with the CON group, LT had no effect ($p > 0.10$) on ADG, but resulted in greater ($0.05 < p \leq 0.10$) F/G, and increased ($p < 0.05$) ADFI (Table 3).

Table 3. Effects of low-temperature on growth performance in weaned piglets.

Item	Treatments		*p*-Value
	CON [a]	LT [b]	
ADG, g	269.21 ± 20.93	323.60 ± 36.41	0.214
ADFI, g	428.49 ± 24.95	557.72 ± 48.66	0.036
F/G	1.61 ± 0.05	1.77 ± 0.07	0.073

[a] CON, thermoneutral temperature; [b] LT, low-temperature.

3.2. Diarrhea Score

As shown in Table 4, compared with the CON group, diarrhea incidence and diarrhea index tended to increase ($0.05 < p \leq 0.10$) in the LT group. Moreover, there were significant increases ($p < 0.01$) in diarrhea incidence and diarrhea index during the first week, but there was no significant effect ($p > 0.05$) during the second and third week.

Table 4. Effects of low temperature on diarrhea in weaned piglets.

Item	Treatments		*p*-Value
	CON	LT	
1–7 d			
Diarrhea incidence (%)	31.42 ± 3.40	52.38 ± 3.99	0.002
Diarrhea index	0.47 ± 0.07	0.90 ± 0.09	0.003
8–14 d			
Diarrhea incidence (%)	22.86 ± 1.84	23.81 ± 1.59	0.702
Diarrhea index	0.27 ± 0.04	0.33 ± 0.03	0.276
15–21 d			
Diarrhea incidence (%)	17.14 ± 3.60	20.63 ± 3.78	0.516
Diarrhea index	0.23 ± 0.04	0.24 ± 0.04	0.880
1–21 d			
Diarrhea incidence (%)	23.81 ± 2.12	32.28 ± 3.67	0.054
Diarrhea index	0.32 ± 0.04	0.49 ± 0.07	0.051

3.3. Nutrient Digestibility

As shown in Table 5, compared with the CON group, the apparent digestibilities of CP, OM and DM were decreased ($p < 0.05$) in the LT group. In addition, compared with the CON group, LT has a tendency to decrease the apparent digestibility of GE ($0.05 < p \leq 0.10$).

Table 5. Effects of low-temperature on growth performance in weaned piglets.

Item	Treatments		*p*-Value
	CON	LT	
CP [a], %	86.13 ± 0.67	83.35 ± 0.73	0.014
DM [b], %	90.28 ± 0.24	89.16 ± 0.30	0.012
OM [c], %	91.98 ± 0.23	90.96 ± 0.27	0.013
GE [d], %	90.60 ± 0.28	89.68 ± 0.33	0.051

[a] CP, crude protein. [b] DM, dry matter. [c] OM, organic matter. [d] GE, gross energy.

3.4. Skeletal Muscle Fiber Type-Related Gene Expression

The *MyHC I*, *MyHC IIa*, *MyHC IIb* and *MyHC IIx* mRNA expressions were detected by real-time quantitative PCR. As shown in Figure 1A, LT had no effect ($p > 0.05$) on LM *MyHC I*, *MyHC IIb* and *MyHC IIx* mRNA levels, but resulted in greater ($p < 0.05$) *MyHC*

IIa mRNA levels. As shown in Figure 1B, LT had no effect ($p > 0.05$) on PM *MyHC IIa* and *MyHC IIb* mRNA levels, but resulted in greater ($p < 0.01$) *MyHC I* mRNA levels and lower ($p < 0.05$) *MyHCIIx* mRNA levels.

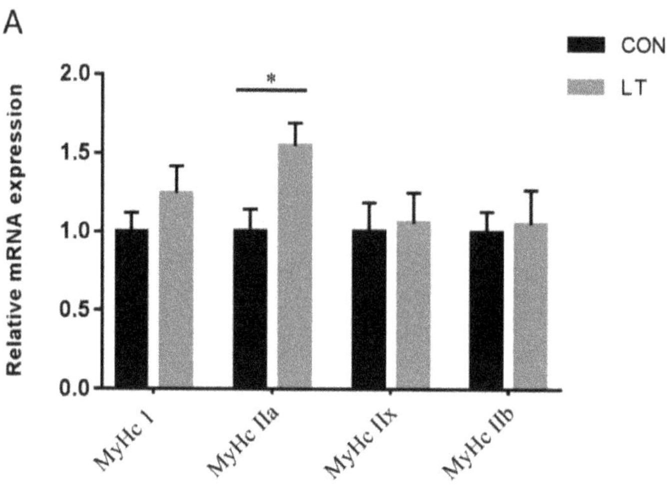

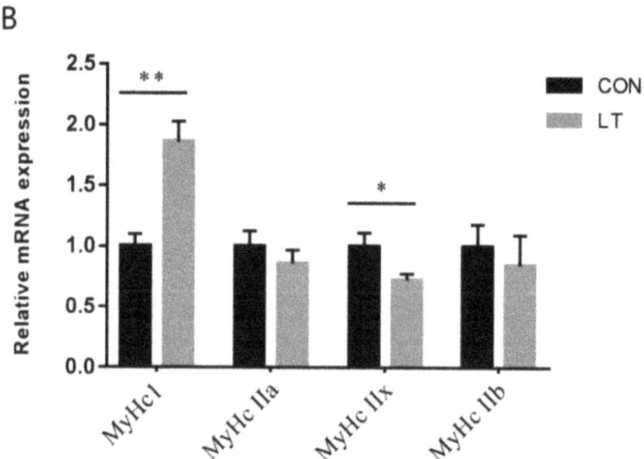

Figure 1. Effects of low-temperature on the mRNA expression level of fiber type-related genes in (**A**) LM and (**B**) PM of weaned piglets. Results were the mean and standard errors. LM, longissimus dorsi muscle; PM, psoas muscle. * Mean values were significantly different between two groups ($p < 0.05$). ** Mean values were very significantly different between two groups ($p < 0.01$).

3.5. Mitochondrial Function-Related Gene Expression

As shown in Figure 2A, compared with the CON group, LT increased ($p < 0.05$) the mRNA expression of CS and SDHB in LM, but had no effect ($p > 0.05$) on CPT-1b and Nrf-1 mRNA levels. As shown in Figure 2B, increased CS ($p < 0.05$) and CPT-1b ($p < 0.01$) mRNA levels were observed in the LT group in PM, but there were no effects ($p > 0.05$) on SDHB and Nrf-1 mRNA levels.

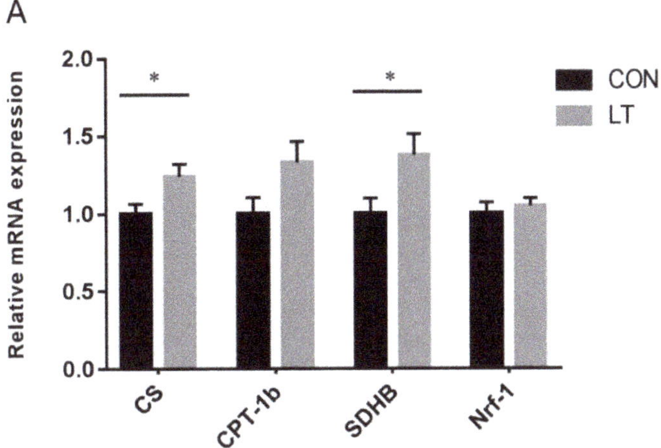

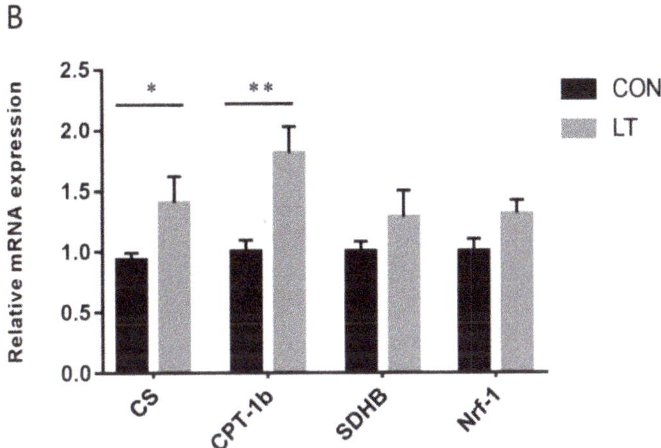

Figure 2. Effects of low-temperature on the mRNA expression level of mitochondrial function-related genes in (**A**) LM and (**B**) PM of weaned piglets. Results were the mean and standard errors. CS, citrate synthase; CPT-1b, carnitine palmitoyl transferase-1b; SDHB, succinate dehydrogenase-b; Nrf-1, nuclear respiratory factor 1. * Mean values were significantly different between two groups ($p < 0.05$). ** Mean values were very significantly different between two groups ($p < 0.01$).

3.6. Antioxidant Capacity

The data showed that LT had significantly increased ($p < 0.01$) T-AOC activity in serum (Table 6), while there was no difference ($p > 0.05$) in T-SOD activity and MDA content in serum between the LT group and the control group (Table 6). In LM, LT had no effect ($p > 0.05$) on MDA content and T-SOD activity, but resulted in greater ($p < 0.01$) T-AOC activity (Table 6). The result showed that LT significantly increased ($p < 0.05$) the T-SOD activity in PM, while there was no difference in ($p > 0.05$) T-AOC activity and MDA content in PM between the LT group and the control group (Table 6).

Table 6. Effects of low-temperature on antioxidant capacity in serum and skeletal muscles.

Item	Treatments		*p*-Value
	CON	LT	
Serum			
MDA, nmol/mL	3.86 ± 0.18	3.91 ± 0.11	0.805
T-AOC, U/mL	0.87 ± 0.05	1.25 ± 0.09	0.002
T-SOD, U/mL	145.25 ± 4.16	146.50 ± 3.56	0.825
LM			
MDA, nmol/mg prot	0.16 ± 0.01	0.14 ± 0.02	0.402
T-AOC, U/mg prot	0.16 ± 0.02	0.27 ± 0.03	0.006
T-SOD, U/mg prot	39.31 ± 1.06	42.93 ± 1.88	0.121
PM			
MDA, nmol/mg prot	0.24 ± 0.02	0.26 ± 0.02	0.466
T-AOC, U/mg prot	0.20 ± 0.02	0.24 ± 0.03	0.340
T-SOD, U/mg prot	35.11 ± 1.53	40.19 ± 1.62	0.039

4. Discussion

Previous studies have shown that livestock and poultry need to increase heat production by increasing feed intake and body energy metabolism in order to maintain body temperature in the LT environment, which may result in reduced performance and high cost. The present study showed that the F/G of weaned piglets increased in the LT environment, which was consistent with the results obtained in growing-finishing pigs [11] and weaned piglets [12]. Numerous studies have demonstrated that the diarrhea incidence of piglets was significantly increased in the LT environment [13]. The present study showed that LT increased the diarrhea incidence and diarrhea index of weaned piglets during the first week, but there were no significant effects on diarrhea of pigs during the second and third week. This indicates that weaned piglets may gradually adapt to cold exposure.

Nutrient digestibility is an important indicator for assessing the digestive capacity of animal gastrointestinal tracts. Studies have shown that LT reduces the proximal gastric diastolic function and accelerates the gastric emptying of animals, resulting in lower digestibility [14]. In the present study, LT decreased the apparent digestibility of CP, OM and DM. Thus, our data supports that cold environments would increase the diarrhea incidence of weaned piglets, leading to dysfunction of the digestive tract, which is also one of the key factors that reduces the apparent digestibility of nutrients.

The mammalian skeletal muscle can be divided into four different *MyHC* isoforms [5]. Several studies have shown that the total number of pig muscle fibers after birth basically does not change, rather, the composition of muscle fiber type changes [15,16]. The types of muscle fiber can be transformed from fast-twitch to slow-switch or slow-switch to fast-twitch by a variety of factors, including innervation, exercise, hormones and ambient temperature [17,18]. The greater abundance of fast oxidative-glycolytic *MyHC IIa* and *MyHC IIx* fibers in the psoas muscle was associated with superior meat quality traits, and the longissimus dorsi muscle are mainly composed of fast glycolytic *MyHC IIb* fibers, which could account for less favourable quality traits [19]. Our study aimed at evaluating the bare effects of cold exposure on muscle fiber types of the oxidative PM and glycolytic LM. In cold environments, pigs mainly produce heat through skeletal muscle contraction, and then maintain their body temperature [20,21]. It has been reported that skeletal muscle contractile activity might lead to muscle fiber type transformation [22]. Previous study had shown that the proportion of oxidative fibers is greater in skeletal muscle from piglets in a cold environment compared with a warm environment, under conditions of controlled food intake [23]. Bee et al. demonstrated that the LM of outdoor pigs during the winter (5 °C) had more fast oxidative-glycolytic fibers and fewer fast glycolytic fibers than muscles of indoor-housed pigs (22 °C) [24]. Similar to previous research results, we found that the proportion of *MyHC IIa* mRNA levels in LM was increased in the LT group. Moreover, the finishing pigs and early postnatal pigs in cold environments have a greater proportion of oxidative fibers in semispinalis muscle (oxidative), compared with pigs at room temperature [12,25]. Gentry et al. demonstrated that the LM muscle of pigs born outdoors (5 °C) had a higher percentage of type I than pigs born indoors (18 °C), and pigs reared outdoors had a lower percentage of IIX fibers than pigs reared indoors for the semispinalis muscle [26]. In this study, the proportion of *MyHC I* mRNA levels was increased in 15 °C compared with that under 26 °C. Therefore, our results support that low-temperature may promote muscle fiber type transformation from type II to type I due to the skeletal muscle continuous contraction caused by high frequency chills.

Mitochondria is the organelle where the main nutrients finally release energy by oxidation, which plays an important role in maintaining the balance of various physiological activities in cells [27]. CS is one of the key enzymes in the production and metabolic pathway of energy, and it has been widely used to evaluate metabolism landmarks of oxidative and respiratory capacity [28]. CPT-1b is located on the outer membrane of mitochondria, which is a key rate-limiting enzyme in fatty acid oxidation in mitochondria [29]. SDHB can be used as a marker enzyme reflecting mitochondrial function. Previous studies have found that cold exposure increased the activities of CS in early postnatal pig muscle longissimus lumborum (LL) and rhomboideus (RH) muscle [25], and SDH activity in skeletal muscle was increased in rats exposed to cold [30]. Similar results were observed in this study that LT increased the mRNA expression of CS in LM and PM. We also found that LT increased the mRNA expression of SDHB in LM muscle and CPT-1b in PM muscle. LM and PM mitochondria exhibit specific changes that are probably involved in the difference of skeletal muscle oxidative metabolism. A previous study showed that muscle fiber types are closely related to muscle mitochondrial synthesis and function [31]. Compared with type II fibers, type I fibers have higher mitochondrial content and oxidative metabolism capacity [6]. These results suggested that cold environment regulated muscle fiber types, which may occur through mitochondrial function regulation.

The antioxidant system is a complete system for scavenging free-radicals that could cause damage to membranes and tissues [32]. T-AOC is an important indicator reflecting the coordination of antioxidant systems in the body. T-SOD is the important antioxidant enzyme and has strong free-radical scavenging ability [33]. Studies have found that SOD activity was decreased in the blood of broilers exposed to the cold environment at 4 °C [34]. The antioxidant enzyme activity and free-radical scavenging ability declined when animals were under cold stress (10 days of 4 °C) [35]. Furthermore, it has been reported that chronic stress induces an increase in T-AOC activity, and the chicken's T-AOC and T-SOD decreased during acute stress [36]. We found that LT increased the activities of T-AOC in serum and LM, and increased the activities of T-SOD in PM of weaned piglets. LM, PM and serum antioxidative capacity changes are probably involved in the differences between samples characteristics. However, the result was different from previous study, presumably caused by higher temperature or chronic stress. This result indicates that chronic cold stress at 15 °C could elevate antioxidative capacity in weaning piglets. However, the antioxidative capacity would decrease when the temperature is lower than a certain threshold.

5. Conclusions

In summary, low-temperature could negatively affect the piglet growth performance and nutrient digestibility. Moreover, we also provided evidence that low-temperature induces a shift toward oxidative muscle fibers, which may occur through mitochondrial function regulation and increased antioxidative capacity.

Author Contributions: Conceptualization, J.Y. and D.C.; Data curation, S.C., Z.Z. and B.Y.; Formal analysis, J.Y. and S.X.; Funding acquisition, J.Y.; Investigation, S.C., Z.Z., P.Z., X.M. and J.L.; Methodology, S.X., J.H. and Y.L.; Project administration, J.Y. and Z.H.; Resources, J.Y. and S.C.; Software, S.X. and H.Y.; Supervision, D.C. and B.Y.; Validation, J.Y.; Writing original draft, S.C.; Writing review & editing, J.Y., S.C. and Z.Z. All authors have read and agreed to the published version of the manuscript.

Funding: This study was supported by the National Key Research and Development Program of China (2016YFD0500506) and China Agriculture Research System of MOF and MARA (CARS-35).

Institutional Review Board Statement: All animal procedures were performed according to protocols approved by the Institutional Animal Care and Use Committee of Sichuan Agricultural University (20190053).

Data Availability Statement: The data are available on request from the corresponding author.

Acknowledgments: We would like to thank Quyuan Wang and Huifen Wang for their assistance.

Conflicts of Interest: All authors declare that they have no competing interest in the present work.

References

1. Mendes, A.A.; Watkins, S.E.; England, J.A.; Saleh, E.A.; Waldroup, A.L.; Waldroup, P.W. Influence of dietary lysine levels and arginine:lysine ratios on performance of broilers exposed to heat or cold stress during the period of three to six weeks of age. *Poult. Sci.* **1997**, *76*, 472–481. [CrossRef] [PubMed]
2. Li, J.; Huang, F.; Li, X.; Su, Y.; Li, H.; Bao, J. Effects of intermittent cold stimulation on antioxidant capacity and mRNA expression in broilers. *Livest. Sci.* **2017**, *204*, 110–114. [CrossRef]
3. Janský, L. Non-shivering thermogenesis and its thermoregulatory significance. *Biol. Rev.* **2010**, *48*, 85–132. [CrossRef] [PubMed]
4. Henneman, E.; Clamann, H.P.; Gillies, J.D.; Skinner, R.D. Rank order of motoneurons within a pool: Law of combination. *J. Neurophysiol.* **1974**, *37*, 1338–1349. [CrossRef] [PubMed]
5. Lefaucheur, L.; Hoffman, R.K.; Gerrard, D.E.; Okamura, C.S.; Rubinstein, N.; Kelly, A. Evidence for three adult fast myosin heavy chain isoforms in type II skeletal muscle fibers in pigs. *J. Anim. Sci.* **1998**, *76*, 1584–1593. [CrossRef] [PubMed]
6. Berchtold, M.W.; Brinkmeier, H.; Muntener, M. Calcium ion in skeletal muscle: Its crucial role for muscle function, plasticity, and disease. *Phys. Rev.* **2000**, *80*, 1215–1265. [CrossRef]
7. Dividich, J.L.; Vermorel, M.; Noblet, J.; Bouvier, J.C.; Aumaitre, A. Effects of environmental temperature on heat production, energy retention, protein and fat gain in early weaned piglets. *Br. J. Nutr.* **1980**, *44*, 313–323. [CrossRef]
8. McCarthy, J.; Aherne, F.; Okai, D. Use of HCl insoluble ash as an index material for determining apparent digestibility with pigs. *Can. J. Anim. Sci.* **1974**, *54*, 107–109. [CrossRef]
9. AOAC International. *Official Methods of Analysis of AOAC International*, 16th ed.; Trends in Food Science & Technology; AOAC International: Rockville, MD, USA, 1995; Volume 1, p. 382.
10. Schmittgen, T.D. Real-time quantitative PCR. *Methods* **2001**, *25*, 383–385. [CrossRef]
11. Stahly, T.; Cromwell, G.; Aviotti, M. The effect of environmental temperature and dietary lysine source and level on the performance and carcass characteristics of growing swine. *J. Anim. Sci.* **1979**, *49*, 1242–1251. [CrossRef]
12. Lefaucheur, L.; Le, D.J.; Mourot, J.; Monin, G.; Ecolan, P.; Krauss, D. Influence of environmental temperature on growth, muscle and adipose tissue metabolism, and meat quality in swine. *J. Anim. Sci.* **1991**, *69*, 2844. [CrossRef] [PubMed]
13. Kelley, K.W.; Blecha, F.; Regnier, J.A. Cold exposure and absorption of colostral immunoglobulins by neonatal pigs. *J. Anim. Sci.* **1982**, *552*, 363–368. [CrossRef] [PubMed]
14. Sun, S.; Huang, X.; Hou, X.; Xie, X. Impaired accommodation of the proximal stomach to a colder meal in healthy adults. *J. Clin. Intern. Med.* **2005**, *22*, 678–681.
15. Wigmore, P.M.; Stickland, N.C. Muscle development in large and small pig fetuses. *J. Anat.* **1983**, *137*, 235–245. [PubMed]
16. Stickland, N.C.; Handel, S.E. The numbers and types of muscle fibres in large and small breeds of pigs. *J. Anat.* **1986**, *147*, 181–189. [PubMed]
17. Picard, B.; Lefaucheur, L.C.; Duclos, M.J. Muscle fibre ontogenesis in farm animal species. *Reprod. Nutr. Dev.* **2002**, *42*, 415–431. [CrossRef]
18. Chang, K.C. Key signalling factors and pathways in the molecular determination of skeletal muscle phenotype. *Animal* **2007**, *1*, 681–698. [CrossRef] [PubMed]
19. Chang, K.C.; Costa, N.D.; Blackley, R.; Southwood, O.; Evans, G.; Plastow, G.; Wood, J.D.; Richardson, R.I. Relationships of myosin heavy chain fibre types to meat quality traits in traditional and modern pigs. *Meat Sci.* **2003**, *64*, 93–103. [CrossRef]
20. Heath, M.; Ingram, D.L. Thermoregulatory heat production in cold-reared and warm-reared pigs. *Am. J. Physiol.* **1983**, *244*, 273–278. [CrossRef]
21. Dauncey, M.J.; Ingram, D.L. Acclimatization to warm or cold temperatures and the role of food intake. *J. Therm. Biol.* **1986**, *11*, 89–93. [CrossRef]
22. Bottinelli, R.; Canepari, M.; Reggiani, C.; Stienen, G.J. Myofibrillar ATPase activity during isometric contraction and isomyosin composition in rat single skinned muscle fibres. *J. Physiol.* **1994**, *481*, 663–675. [CrossRef] [PubMed]
23. Dauncey, M.J.; Ingham, D.L. Respiratory enzymes in muscle: Interaction between environmental temperature, nutrition and growth. *J. Therm. Biol.* **1990**, *15*, 325–328. [CrossRef]
24. Bee, G.; Guex, G.; Herzog, W. Free-range rearing of pigs during the winter: Adaptations in muscle fiber characteristics and effects on adipose tissue composition and meat quality traits. *J. Anim. Sci.* **2004**, *4*, 1206–1218. [CrossRef]

25. Lefaucheur, L.; Ecolan, P.; Lossec, G.; Gabillard, J.C.; Butler-Browne, G.S.; Herpin, P. Influence of early postnatal cold exposure on myofiber maturation in pig skeletal muscle. *J. Muscle Res. Cell. Motil.* **2001**, *22*, 439–452. [CrossRef] [PubMed]
26. Gentry, J.G.; Mcglone, J.J.; Miller, M.F.; Blanton, J. Environmental effects on pig performance, meat quality, and muscle characteristics. *J. Anim. Sci.* **2004**, *82*, 209–217. [CrossRef] [PubMed]
27. Sherratt, H.S.A. Mitochondria: Structure and function. *Rev. Neurol.* **1991**, *147*, 417–430.
28. Spina, R.J.; Chi, M.M.; Hopkins, M.G.; Nemeth, P.M.; Lowry, O.H.; Holloszy, J.O. Mitochondrial enzymes increase in muscle in response to 7-10 days of cycle exercise. *J. Appl. Physiol.* **1996**, *80*, 2250–2254. [CrossRef]
29. Bruce, C.R.; Hoy, A.J.; Turner, N.; Watt, M.J.; Allen, T.L.; Carpenter, K.; Cooney, G.J.; Febbraio, M.A.; Kraegen, E.W. Overexpression of carnitine palmitoyltransferase-1 in skeletal muscle is sufficient to enhance fatty acid oxidation and improve high-fat diet–induced insulin resistance. *Diabetes* **2009**, *58*, 550–558. [CrossRef]
30. Hannon, J.P. Effect of prolonged cold exposure on components of the electron transport system. *Am. J. Physiol.* **1960**, *198*, 740–744. [CrossRef]
31. Herpin, S.P. Postnatal Changes in Mitochondrial Protein Mass and Respiration in Skeletal Muscle from the Newborn Pig. *Biochem. Mol. Biol.* **1997**, *118*, 639–647.
32. Brand-Williams, W.M.; Cuvelier, M.E.; Berset, C. Use of free radical method to evaluate antioxidant activity. *LWT Food Sci. Technol.* **1995**, *28*, 25–30. [CrossRef]
33. Djordjevi, V.B. Free radicals in cell biology. *Int. Rev. Cytol.* **2004**, *237*, 57–89.
34. Ramnath, V.; Rekha, P. Brahma Rasayana enhances in vivo antioxidant status in cold-stressed chickens (Gallus gallus domesticus). *Indian J. Pharmacol.* **2009**, *41*, 115–119. [CrossRef]
35. Venditti, P.; Pamplona, R.; Ayala, V.; De Rosa, R.; Caldarone, G.; Di Meo, S. Differential effects of experimental and cold-induced hyperthyroidism on factors inducing rat liver oxidative damage. *J. Exp. Biol.* **2006**, *209*, 817–825. [CrossRef] [PubMed]
36. Zhao, F.-Q.; Zhang, Z.-W.; Qu, J.-P.; Yao, H.-D.; Li, M.; Xu, S.-W. Cold stress induces antioxidants and Hsps in chicken immune organs. *Cell Stress Chaperones* **2014**, *19*, 635–648. [CrossRef] [PubMed]

Article

A Comparison of the Behavior, Physiology, and Offspring Resilience of Gestating Sows When Raised in a Group Housing System and Individual Stalls

Xin Liu [1,†], Pengkang Song [1,2,†], Hua Yan [1], Longchao Zhang [1], Ligang Wang [1], Fuping Zhao [1], Hongmei Gao [1], Xinhua Hou [1], Lijun Shi [1], Bugao Li [2,*] and Lixian Wang [1,*]

1 Institute of Animal Science, Chinese Academy of Agricultural Sciences, Beijing 100193, China; firstliuxin@163.com (X.L.); pengkangsong2021@163.com (P.S.); zcyyh@126.com (H.Y.); zhlchias@163.com (L.Z.); ligwang@126.com (L.W.); zhaofuping@caas.cn (F.Z.); gaohongmei_123@126.com (H.G.); 7hxh73@163.com (X.H.); shilijun01@caas.cn (L.S.)
2 College of Animal Science, Shanxi Agricultural University, Jinzhong 030801, China
* Correspondence: bgli@sxau.edu.cn (B.L.); iaswlx@263.net (L.W.); Tel.: +86-010-6281-8771 (L.W.)
† These authors contributed equally to this work.

Citation: Liu, X.; Song, P.; Yan, H.; Zhang, L.; Wang, L.; Zhao, F.; Gao, H.; Hou, X.; Shi, L.; Li, B.; et al. A Comparison of the Behavior, Physiology, and Offspring Resilience of Gestating Sows When Raised in a Group Housing System and Individual Stalls. *Animals* **2021**, *11*, 2076. https://doi.org/10.3390/ani11072076

Academic Editors: Lilong Chai and Yang Zhao

Received: 9 June 2021
Accepted: 9 July 2021
Published: 12 July 2021

Publisher's Note: MDPI stays neutral with regard to jurisdictional claims in published maps and institutional affiliations.

Simple Summary: The housing patterns of gestating sows affect their health and welfare. In this study, the differences between behavior and stress hormone levels were assessed when sows were housed in a group housing system compared to individual stalls; in addition, the disease resistance and resilience of their piglets were compared. In our investigation, the group-housed sows showed more exploratory behavior, less vacuum chewing, less sitting behavior, and lower stress hormone levels throughout pregnancy. A lipopolysaccharide (LPS) injection test revealed that the offspring of group-housed sows showed better resistance and resilience to disease. Therefore, the gestating sows raised in a group housing system and their piglets are healthier and have improved welfare. Our results show that a group housing system provides higher welfare standards, with conditions that are more suitable for gestating sows in modern pig production.

Abstract: Being in a confined environment causes chronic stress in gestating sows, which is detrimental for sow health, welfare and, consequently, offspring physiology. This study assessed the health and welfare of gestating sows housed in a group housing system compared to individual gestation stalls. After pregnancy was confirmed, experimental sows were divided randomly into two groups: the group housing system (GS), with the electronic sow feeding (ESF) system; or individual stall (IS). The behavior of sows housed in the GS or IS was then compared; throughout pregnancy, GS sows displayed more exploratory behavior, less vacuum chewing, and less sitting behavior ($p < 0.05$). IS sows showed higher stress hormone levels than GS sows. In particular, at 41 days of gestation, the concentration of the adrenocorticotropic hormone (ACTH) and adrenaline (A) in IS sows was significantly higher than that of GS sows, and the A level of IS sows remained significantly higher at 71 days of gestation ($p < 0.01$). The lipopolysaccharide (LPS) test was carried out in the weaned piglets of the studied sows. Compared with the offspring of gestating sows housed in GS (PG) or IS (PS), PG experienced a shorter period of high temperature and showed a quicker return to the normal state ($p < 0.05$). Additionally, their lower levels of stress hormone ($p < 0.01$) suggest that PG did not suffer from as much stress as PS. These findings suggested that gestating sows housed in GS were more able to carry out their natural behaviors and, therefore, had lower levels of stress and improved welfare. In addition, PG also showed better disease resistance and resilience. These results will provide a research basis for the welfare and breeding of gestating sows.

Keywords: group housing system; individual stall; behavior; stress hormone; offspring; gestating sows

1. Introduction

In most parts of the world, gestating sows face stress due to space and management during gestation in intensive pig production systems. Conventional individual stall housing (IS) is commonly used for gestating sows because it makes handling easier, has a low capital cost, and reduces social stress [1]. However, the space restrictions of stalls limit the innate behaviors of gestating sows; therefore, pigs housed in IS cannot execute the behaviors needed to meet their specific needs [2,3]. These housing deficiencies cause sows to exhibit abnormal behaviors and physiology, causing chronic disease and leading to a reduction in muscle strength and bone density [4,5]. In order to improve the welfare of gestating sows, this IS practice was banned by the European Union (CD 2001/88/EC), who instead promotes group housing systems (GS) in European countries. Sows housed in GS suffer less than those housed in IS. GS with an electronic sow feeding (ESF) system provided gestating sows with a less physiologically stressful environment and greater opportunities for activity [6]. However, GS also has some disadvantages; for example, individual feeding is more difficult, and sows can be more aggressive in the early stage of mixing, leading to more injuries [7,8]. In the Chinese pig industry, gestating sows are still reared in IS in almost all pig farms. With the modernization of the pig industry and the emphasis on animal welfare, GS may be the direction of development. Therefore, it is necessary to study the effects of different housing systems on sows to provide the pig industry with more information.

Previous studies have compared the effects of reproductive performance, management, and behaviors on gestating sows housed in different housing systems. Several studies showed that sow reproductive performance was improved in GS, with others confirming that no differences were found among housing types [8–10]. Some researchers recommended that gestating sows housed in GS showed an improved welfare status, greater levels of activity, and fewer physiological abnormalities, but some studies did not find a significant difference in stress-related hormones between the two housing conditions [6,10,11]. However, previous studies have reported conflicting results, and limited data have been garnered regarding piglet resilience. Therefore, it is necessary to compare the effects on the behavior, physiology, and piglet resilience of gestating sows when housed in GS or in IS.

The objective of this study was to assess the effects of GS and IS on the health and welfare of gestating sows and their offspring, by detecting sows' behaviors, physiology, and offspring resilience. The hormonal and behavioral changes in gestating sows housed in GS or IS were observed throughout the gestation period, and the disease resistance and resilience of the piglets was detected using a lipopolysaccharide (LPS) injection model. The results of the study could provide the scientific support for improving the health and welfare of gestating sows and piglets in production.

2. Materials and Methods

2.1. Ethistall Statements

All methods and procedures in the study were carried out according to the standard guidelines on experimental animals (No. IASCAAS-AE-09), which were established by the Animal Ethical Committee of the Institute of Animal Science, Chinese Academy of Agricultural Sciences (IAS-CAAS) (Beijing, China). The experimental protocols were approved by the Science Research Department of IAS-CAAS (Beijing, China) (No. IAS2019-18).

2.2. Animals and Management

All experimental animals were Large White pigs reared in identical intensive breeding conditions (Chang Rong Nong Ke, Yuncheng, China), with the same feed and management. The nutrient requirements of sow and piglet diets refer to NRC 2012 (Nutrient Requirements of Swine of the National Research Council). A total of 60 experimental sows with the second parity were artificially inseminated; pregnancy was confirmed with an ultrasound analyzer within 28 days of insemination. Then, sows with a confirmed pregnancy were allotted to their housing groups-30 sows in IS and 30 in the GS. They were all moved to the farrowing crate three days before the expected delivery.

The offspring piglets of the test sows were used for disease resistance and resilience tests. Twenty piglets, each born from sows housed in IS or GS, were used for disease resistance. Test piglets with good physical health (remaining healthy and free of illness) and similar weaning weights were weaned at 21 days of age.

2.3. Housing Systems

In the study, the IS size was 2.40 m × 0.65 m (length × width, 1.56 m^2/head) with an individual feeder and drinker (Figure 1A). The IS was slightly larger than the size of the sow's body; there was only enough room for the sow to stand or lie down in place, with no room for the sow to turn around or move freely. The gestating sows of GS were housed in a room (10.5 m × 14.4 m, 5.04 m^2/head) with an ESF, which provided enough space for the sows (Figure 1B). Sows in the group house could move freely, which allowed them to meet some of their innate behavioral requirements. The temperature of the gestating room was approximately 20 °C, which could be controlled using a fan or by heating.

A

B

Figure 1. Types of housing facilities of sows. (**A**) Individual Stalls (IS). (**B**) Group system with ESF (GS).

2.4. Behavioral Observations

The behaviors of all experimental sows were recorded using a video surveillance system (Hikvision camera, Hangzhou, China) for data collection, which clearly recorded the movement of each experimental sow and avoided artificial observation errors. The gestating sows were continuously video recorded from 9:00 a.m. to 5:00 p.m. on each behavioral observation day (days 40, 70, and 100 of gestation). We observed and recorded the standing behavior, dog sitting behavior, lying down behavior, vacuum chewing behavior, and exploratory behavior of gestating sows; the definitions of various behaviors are listed in Table 1. The total number of instances of each behavior on the observation days was counted by recording the number of behaviors every ten minutes.

2.5. Sample Collection and Physiological Analysis

Blood samples were collected from the jugular vein of all experimental sows at days 41, 71, and 101 of gestation. The blood samples were kept at room temperature for 2 h and then the serum was separated and extracted by centrifugation at 3500 rpm for 10 min. The samples were stored at −80 °C. The samples were tested for the adrenocorticotropic hormone (ACTH), adrenaline (A) and cortisol (COR). ACTH and COR were measured through radioimmunoassay. Adrenaline was measured with the enzyme-linked immunosorbent assay (ELISA) method.

2.6. Disease Resilience Test of Piglets

Forty 21-day-old, 6 kg, healthy weaned piglets were selected for the experiment. Twenty of them were randomly selected from 10 litters of the test gestating sows housed in IS (PS), and the others were randomly selected from 10 litters of the test gestating sows housed in GS (PG). Two piglets were randomly selected from each litter and then assigned to lipopolysaccharide (LPS) and normal saline (NS) injection group, respectively. The injection dose of LPS (from *E. coli* O55:B5) or NS was 15 μg/kg BW. The ratio of male to female was half in each group. The ear temperature of piglets was measured with an animal thermometer at 1 h before injection, 1 h after injection, 2 h after injection, 3 h after injection, 4 h after injection, 5 h after injection, and 6 h after injection [14,15]. Blood was collected by jugular venipuncture 6 h after injection. Serum was extracted and frozen at −80 °C. The concentration of serum COR was measured. The determination method of COR was the same as 2.5.

Table 1. Behavior categories of pregnant sows and their definitions.

Behavior Categories	Definitions
Standing behavior	All four hooves are on the pen floor with limbs extended or the pig is walking with limbs in both extension and flexion and moving throughout the pen [12]
Dog sitting behavior	The front limbs are extended and bearing weight the rear limbs and body are in contact with the pen floor [12]
Lying down behavior	The pig's body and limbs are in contact with the pen floor [12]
Vacuum chewing behavior	Continuous chewing while no feed is present in the mouth [8]
Exploratory behavior	Actively manipulating and exploring the surrounding environment [13]

2.7. Statistical Analysis

The collected data of behavioral and physiological tests were analyzed for the homogeneity of variance and then different significance analysis was carried out. These data were tested using *t*-test in SAS (version 9.2, SAS Inst. Inc., Raleigh, NC, USA). The results of the analysis were presented as the means ± standard error. The differences and statistical significance between groups were considered at $p < 0.05$ and $p < 0.01$.

3. Results

3.1. The Behavioral Response of Gestating Sows Housed in GS or IS

The behavioral response of gestating sows was compared between GS and IS groups, as shown in Figure 2. On days 40 and 70 of gestation, the frequency of dog sitting behavior in the gestating sows housed in IS was significantly higher than that in the GS condition ($p < 0.05$). During the whole pregnancy period, the frequency of empty chewing behavior in gestating sows housed in IS was significantly higher than that of sows in GS, while the frequency of exploratory behavior was significantly lower ($p < 0.01$). The frequency of standing behavior in gestating sows housed in the GS was less than that in sows housed in IS ($P_{40\text{ day}} = 0.94$, $P_{70\text{ day}} = 0.58$, $P_{100\text{ day}} = 0.24$), while the lying down behavior increased ($P_{40\text{ day}} = 0.58$, $P_{70\text{ day}} = 0.43$, $P_{100\text{ day}} = 0.16$); however, these behavioral differences did not reach a significant level.

3.2. Effects of IS or GS Housing Systems on the Physiological Responses of Gestating Sows

The effects of the two different housing systems of gestating sows on physiological responses during gestation are presented in Figure 3. According to the data, the stress hormone (ACTH, A, COR) level of gestating sows housed in IS was higher than that of gestating sows housed in GS throughout the whole gestation period. Particularly, the concentrations of ACTH and A in gestating sows were significantly improved in IS compared to those reported in GS on day 41 of gestation; in addition, a significant increase in hormone A continued until day 71 of gestation ($p < 0.01$). The COR concentrations of sows in IS were numerically higher than the concentrations in GS sows, but this was not a significant difference ($P_{41\text{ day}} = 0.75$, $P_{71\text{ day}} = 0.35$, $P_{101\text{ day}} = 0.09$).

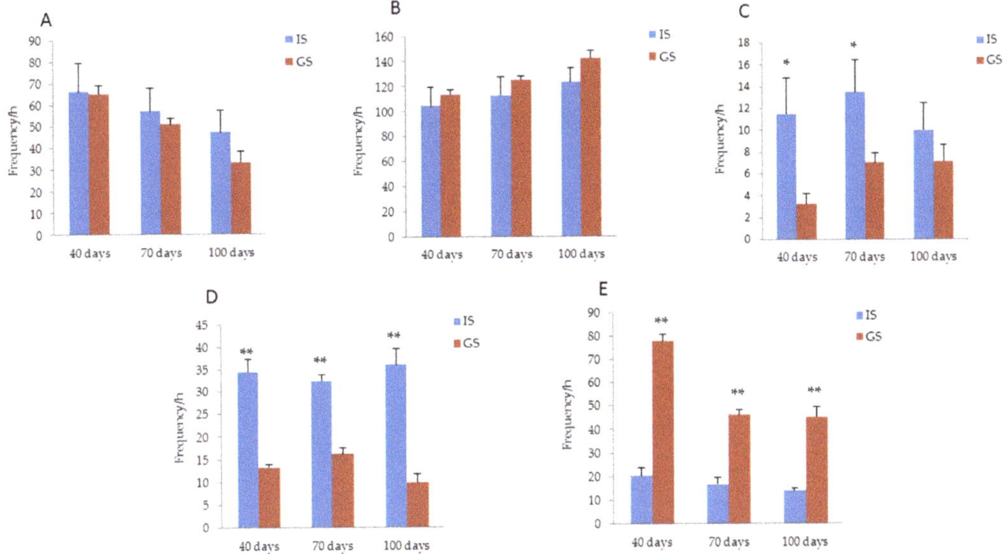

Figure 2. Comparison of the behavioral responses of gestating sows between IS and GS. (**A**) Changes in the standing behavior of gestating sows housed in IS and GS. (**B**) Changes in the lying down behavior of gestating sows housed in IS and GS. (**C**) Changes in the dog sitting behavior of gestating sows housed in IS and GS. (**D**) Changes in the vacuum chewing behavior of gestating sows housed in IS and GS. (**E**) Changes in the exploratory behavior of gestating sows housed in IS and GS. * $p < 0.05$. ** $p < 0.01$.

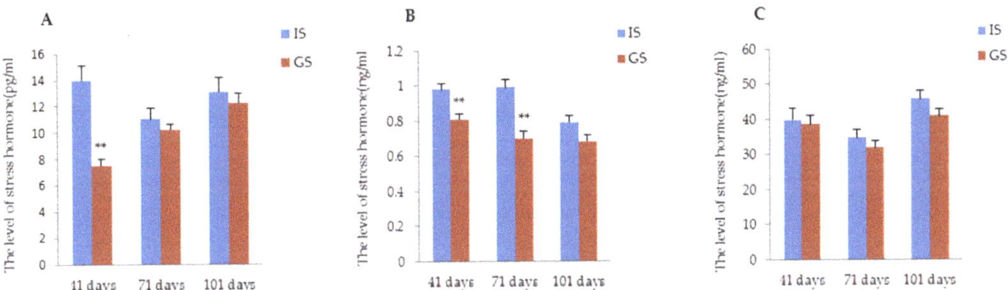

Figure 3. The influence of two different housing systems on the concentration of the physiological index in gestating sows. (**A**) Comparison of ACTH concentration of gestating sows raised in GS and IS; (**B**) Comparison of A concentration of gestating sows raised in GS and IS; (**C**) Comparison of COR concentration of gestating sows raised in GS and IS. ** $p < 0.01$.

3.3. Comparison of Resistance and Resilience of Offspring Piglets

The model of inflammatory response was constructed by injecting LPS into piglets. The ear temperature of the piglets was measured before and after injection. As shown in Figure 4A,B, with NS injection as the control group, the ear temperature of the piglets was significantly higher after the LPS injection ($p < 0.01$). After the LPS injection, the ear temperature of the offspring piglet, both PG and PS, was raised rapidly and continued to return to normal 6 h after injection. It was also found that the duration of higher ear temperature of PS was longer than that of PG, in other words, the ear temperature of PG returned to normal state significantly faster and easier ($p < 0.05$) (Figure 4C). The concentration level of hormone COR of PG was significantly lower than that of PS ($p < 0.01$) (Table 2). These results indicated that the offspring piglets of gestating sows housed in the group system had greater resistance and resilience.

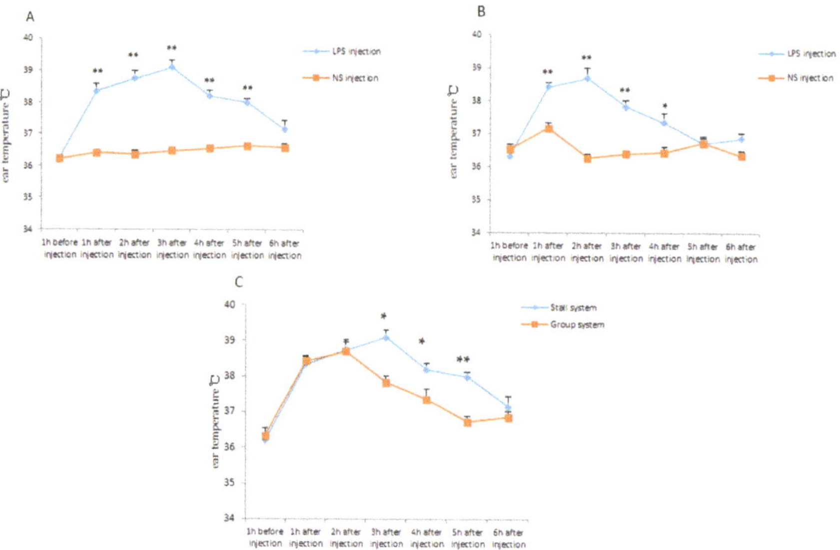

Figure 4. The ear temperature changes in the experiment of piglet resilience. (**A**), Changes of ear temperature of PS after NS or LPS injection. (**B**), Changes of ear temperature of PG after NS or LPS injection. (**C**), Changes of ear temperature of PG or PS after LPS injection. * $p < 0.05$. ** $p < 0.01$.

Table 2. The concentration of COR in piglets in LPS injection test.

Piglet	Hormone	NS Injection	LPS Injection	*p*-Value
PS	COR (ng/mL)	72.30 ± 14.27^{A}	$158.34 \pm 13.50^{a,C}$	0.0003
PG	COR (ng/mL)	32.79 ± 10.77^{B}	$112.74 \pm 21.08^{b,c}$	0.0003

^{A},^{a},^{B},^{b}: Means in the same row with different superscripts are significantly different ($p < 0.01$). ^{C},^{c}: Means in the same column with different superscripts are significantly different ($p < 0.01$).

4. Discussion

In the present study, two different housing systems (individual stall and group housing) for gestating sows were compared. Gestating sows housed in IS had limited space, and GS sows had more freedom of movement. Floor space allowance markedly affects sow welfare [16], particularly during early gestation. Accordingly, appropriate housing is important to protect embryos and to confirm pregnancy [17]. The narrow space and metal-bars of the stall restrict the behaviors of gestating sows, particularly in late pregnant period of pregnancy, when the sow's size and body weight increase [11]. IS housing is considered to be a chronic stressor for gestating sows, and has negative consequences on welfare and health [18]. Chronic stress has persistent effects on the behavior, physiology, and performance of sows and offspring [19,20]. The abnormal behavior of the sow not only reflects the response to environmental adaptability, but also the sow's own psychological welfare. If sows are not comfortable during pregnancy, they will exhibit abnormal behaviors, for example, locomotion difficulties, stereotypies, etc., resulting in physiological and psychological stress [21]. In the present study, the postural behaviours of gestating sows in the two housing systems were compared. The frequency of standing in GS gestating sows was less than IS gestating sows, while the frequency of lying down was increased (though not at a significant level). GS sows exhibited more exploratory behavior and less vacuum chewing and dog sitting behavior. This suggested gestating sows housed in GS were healthier and had better welfare. Some previous studies showed similar results. Haley's study [22] showed that sows were in a state of physical discomfort when they spent less time lying down and more time standing without eating. Confinement in stalls has been implicated in the development of oral stereotypies and repetitive, apparently functionless behaviors; the normal exploratory behavior of the sows could not be satisfied, mainly because of the extremely limited environmental stimuli [19,21]. Janssens' study also demonstrated that sows in a group-housing system showed a decrease in the frequency of sham chewing and an increase in non-agonistic social behavior [23].

Stress impacts several physiological systems and the stress hormone levels have been used for physiological measurement [24]. Animal responses to stress activate the hypothalamic-pituitary-adrenocortical (HPA) axis and cause increased plasma levels of cortisol and catecholamines [25]. With an increased confinement duration, the sows in the restraining environment became bored and showed a failure response pattern by the activation of the HPA system [26]. Under chronic stress, the activation of the HPA system increased responsiveness of the adrenal cortex to ACTH and eventually lead to increased cortisol output [23]. In our study, the concentration of stress hormones (ACTH, A, and COR) in gestating sows housed in IS was higher than that in GS. Gestating sows housed in IS produced chronic long-term stress and increased stress hormone levels. Some studies have reported study similar results. Jang [6] reported that compared with the group sows, conventional stall sows had a higher serum cortisol level at 110 days of gestation. Merlot [27] showed that the conventional system was more stressful for sows during gestation, as illustrated by the elevated cortisol levels in the saliva of gestating sows; furthermore, the conventional system moderately worsened sow health in late gestation. In Quesnel's study [28], sows raised in the conventional system had greater salivary concentrations of cortisol compared with the enriched system (larger pens and on deep straw bedding) during the gestation period. Optimizing commercial housing conditions would reduce stress levels and have positive effects on the immune status of mothers during gestation [29].

During the sow's gestation period, the environment (including housing and management systems) can generate maternal stress, which can be detrimental to sow welfare and health, and also it could influence on off-spring physiology, such as the immune function, and impairing neonatal health [30–32]. Therefore, in order to continue the study of how maternal stress caused by different housing systems during the gestation period, affected their offspring, the piglet health and resilience test was designed and implemented. LPS is a major structural part of the outer membrane of Gram-negative bacteria and can effectively stimulate the body's immune system. Therefore, LPS has been widely applied as an experimental model in vertebrate immune stress tests [33,34]. The acute phase response (APR) was induced by LPS stimulation and also caused the behavioral changes and physiological disorders in the pigs, including an elevated body temperature, increased cytokine levels, reduced feed intake, etc. [35]. In our test, we obtained the similar results. After LPS injection into piglets, APR was induced and their body temperature increased rapidly and significantly. Compared with PS in the test, PG experienced a shorter period of high temperature and the return to a normal state was faster; in addition, they suffered lower levels of stress in terms of their stress hormone levels. According to the concept of resilience, the ability of an animal to maintain performance under infection, or to rapidly return to prior performance levels after infection [36,37], PG was considered to have better resilience. Previous research reported stress piglets displayed higher levels of circulating cortisol [38], and with the same result, in the present study, PS suffered more stress in terms of their stress hormone levels. All of this suggested that the offspring of sows housed in GS during gestation had better resistance and resilience, which showed that these piglets were healthier. PS suffered with a higher level of stress and had lower resistance and resilience, which may be caused by the IS-housing-related stress experienced by their mothers during gestation.

5. Conclusions

Gestating sows were exposed to different environments and faced different challenges when they were housed in two systems (IS and GS). As a result of enjoying a more relaxed and comfortable environment, the sows housed in GS wereconductive more as in accordance with to their nature. Gestating sows housed in GS demonstrated more exploratory behavior, less vacuum chewing, and less dog sitting behavior compared with IS sows. Meanwhile, GS sows had a lower concentration of stress hormone than IS. In addition, the results of LPS injection experiment showed that PG had better resistance and resilience than PS. These findings provide a research basis for welfare breeding of gestating sows.

Author Contributions: Conceptualization, X.L., B.L. and L.W. (Lixian Wang); Formal analysis, X.L., B.L. and L.W. (Lixian Wang); Investigation, X.L., P.S., X.H., H.G., L.S. and H.Y.; Methodology, X.L., P.S., F.Z., L.Z. and L.W. (Ligang Wang). Resources: X.L. and L.W. (Lixian Wang); Supervision: X.L., B.L. and L.W. (Lixian Wang); Writing-original draft: X.L.; Writing-review & editing: X.L., P.S. and L.W. (Lixian Wang). All authors have read and agreed to the published version of the manuscript.

Funding: This work was supported by the National Key Research and Development Program (2017YFE0114400), H2020 Grant Agreement 773436—Healthy Livestock and the Agricultural Science and Technology Innovation Project (ASTIP-IAS02).

Institutional Review Board Statement: The study was conducted according to the standard guidelines on experimental animals, which were established by the Animal Ethical Committee of the Institute of Animal Science, Chinese Academy of Agricultural Sciences (IAS-CAAS) (Beijing, China). No. IASCAAS-AE-09.

Acknowledgments: We thank Bas Kemp (Wageningen University & Research) for helpful discussion and comments.

Conflicts of Interest: The authors declare no conflict of interest.

References

1. Marco-Ramell, A.; Arroyo, L.; Peña, R.; Pato, R.; Saco, Y.; Fraile, L.; Bendixen, E.; Bassols, A. Biochemical and proteomic analyses of the physiological response induced by individual housing in gilts provide new potential stress markers. *BMC Vet. Res.* **2016**, *12*, 265. [CrossRef] [PubMed]
2. Salak-Johnson, J.L.; McGlone, J.J. Making sense of apparently conflicting data: Stress and immunity in swine and cattle. *J. Anim. Sci.* **2007**, *85* (Suppl. 13), E81–E88. [CrossRef]
3. Van der Beek, E.M.; Wiegant, V.M.; Schouten, W.G.; van Eerdenburg, F.J.; Loijens, L.W.; van der Plas, C.; Benning, M.A.; de Vries, H.; de Kloet, E.R.; Lucassen, P.J. Neuronal number, volume, and apoptosis of the left dentate gyrus of chronically stressed pigs correlate negatively with basal saliva cortisol levels. *Hippocampus* **2004**, *14*, 688–700. [CrossRef] [PubMed]
4. Poole, T.B. The Nature and Evolution of Behavioural Needs in Mammals. *Anim. Welf.* **1992**, *3*, 203–220.
5. Marchant, J.N.; Broom, D.M. Factors affecting posture-changing in loose-housed and confined gestating sows. *Anim. Sci.* **1996**, *63*, 477–485. [CrossRef]
6. Jang, J.C.; Jung, S.W.; Jin, S.S.; Ohh, S.J.; Kim, J.E.; Kim, Y.Y. The Effects of Gilts Housed Either in Group with the Electronic Sow Feeding System or Conventional Stall. *Asian Australas J. Anim. Sci.* **2015**, *28*, 1512–1518. [CrossRef] [PubMed]
7. Bench, C.J.; Rioja-Lang, F.C.; Hayne, S.M.; Gonyou, H.W. Group gestation sow housing with individual feeding—II: How space allowance, group size and composition, and flooring affect sow welfare. *Livest. Sci.* **2013**, *152*, 218–227. [CrossRef]
8. Chapinal, N.; de la Torre, J.L.R.; Cerisuelo, A.; Gasa, J.; Baucells, M.D.; Coma, J.; Vidal, A.; Manteca, X. Evaluation of welfare and productivity in pregnant sows kept in stalls or in 2 different group housing systems. *J. Vet. Behav.* **2010**, *5*, 82–93. [CrossRef]
9. Min, Y.; Choi, Y.; Kim, J.; Kim, D.; Jeong, Y.; Kim, Y.; Song, M.; Jung, H. Comparison of the Productivity of Primiparous Sows Housed in Individual Stalls and Group Housing Systems. *Animal* **2020**, *10*, 1940. [CrossRef]
10. Morgan, L.; Klement, E.; Novak, S.; Eliahoo, E.; Younis, A.; Sutton, G.A.; Abu-Ahmad, W.; Raz, T. Effects of group housing on reproductive performance, lameness, injuries and saliva cortisol in gestating sows. *Prev. Vet. Med.* **2018**, *160*, 10–17. [CrossRef]
11. Zhou, Q.; Sun, Q.; Wang, G.; Zhou, B.; Lu, M.; Marchant-Forde, J.N.; Yang, X.; Zhao, R. Group housing during gestation affects the behaviour of sows and the physiological indices of offspring at weaning. *Animal* **2014**, *8*, 1162–1169. [CrossRef]
12. Colpoys, J.D.; Johnson, A.K.; Gabler, N.K. Daily feeding regimen impacts pig growth and behavior. *Physiol. Behav.* **2016**, *159*, 27–32. [CrossRef] [PubMed]
13. Greenwood, E.C.; Plush, K.J.; van Wettere, W.H.E.J.; Hughes, P.E. Group and individual sow behavior is altered in early gestation by space allowance in the days immediately following grouping. *J. Anim. Sci.* **2016**, *94*, 385–393. [CrossRef] [PubMed]
14. Schmid, S.M.; Büscher, W.; Steinhoff-Wagner, J. Suitability of Different Thermometers for Measuring Body Core and Skin Temperatures in Suckling Piglets. *Animal* **2021**, *11*, 1004. [CrossRef] [PubMed]
15. Rooney, H.B.; O'driscoll, K.; O'doherty, J.V.; Lawlor, P.G. Effect of increasing dietary energy density during late gestation and lactation on sow performance, piglet vitality, and lifetime growth of offspring. *J. Anim. Sci.* **2020**, *98*, 379. [CrossRef] [PubMed]

16. Verdon, M.; Hansen, C.F.; Rault, J.L.; Jongman, E.; Hansen, L.U.; Plush, K.; Hemsworth, P.H. Effects of group housing on sow welfare: A review. *J. Anim. Sci.* **2015**, *93*, 1999–2017. [CrossRef]
17. Koketsu, Y.; Iida, R. Sow housing associated with reproductive performance in breeding herds. *Mol. Reprod. Dev.* **2017**, *84*, 979–986. [CrossRef] [PubMed]
18. Damgaard, B.M.; Malmkvist, J.; Pedersen, L.J.; Jensen, K.H.; Thodberg, K.; Jørgensen, E.; Juul-Madsen, H.R. The effects of floor heating on body temperature, water consumption, stress response and immune competence around parturition in loose-housed sows. *Res. Vet. Sci.* **2009**, *86*, 136–145. [CrossRef]
19. Vieuille-Thomas, C.; Pape, G.L.; Signoret, J.P. Stereotypies in pregnant sows: Indications of influence of the housing system on the patterns expressed by the animals. *Appl. Anim. Behav. Sci.* **1995**, *44*, 19–27. [CrossRef]
20. Braastad, B.O. Effects of prenatal stress on behaviour of offspring of laboratory and farmed mammals. *Appl. Anim. Behav. Sci.* **1998**, *61*, 159–180. [CrossRef]
21. Zhang, M.Y.; Li, X.; Zhang, X.H.; Liu, H.G.; Li, J.H.; Bao, J. Effects of confinement duration and parity on stereotypic behavioral and physiological responses of pregnant sows. *Physiol. Behav.* **2017**, *179*, 369–376. [CrossRef]
22. Haley, D.B.; de Passillé, A.M.; Rushen, J. Assessing cow comfort: Effects of two floor types and two tie stall designs on the behaviour of lactating dairy cows. *Appl. Anim. Behav. Sci.* **2001**, *71*, 105–117. [CrossRef]
23. Janssens, C.J.; Helmond, F.A.; Wiegant, V.M. Increased cortisol response to exogenous adrenocorticotropic hormone in chronically stressed pigs: Influence of housing conditions. *J. Anim. Sci.* **1994**, *72*, 1771–1777. [CrossRef] [PubMed]
24. Mcglone, J.J.; von Borell, E.H.; Deen, J.; Johnson, A.K.; Levis, D.G.; Meunier-Salaon, M.; Morrow, J.; Reeves, D.; Salak-Johnson, J.L.; Sundberg, P.L. Compilation of the Scientific Literature Comparing Housing Systems for Gestating Sows and Gilts Using Measures of Physiology, Behavior, Performance, and Health. *Prof. Anim. Sci.* **2004**, *20*, 105–117. [CrossRef]
25. Otten, W.; Puppe, B.; Kanitz, E.; Schön, P.C.; Stabenow, B. Physiological and behavioral effects of different success during social confrontation in pigs with prior dominance experience. *Physiol. Behav.* **2002**, *75*, 127–133. [CrossRef]
26. Henry, J.P. Biological basis of the stress response. *Intgr. Physiol. Behav. Sci.* **1992**, *27*, 66–83. [CrossRef]
27. Merlot, E.; Pastorelli, H.; Prunier, A.; Père, M.C.; Louveau, I.; Lefaucheur, L.; Perruchot, M.H.; Meunier-Salaün, M.C.; Gardan-Salmon, D.; Gondret, F.; et al. Sow environment during gestation: Part I. Influence on maternal physiology and lacteal secretions in relation with neonatal survival. *Animal* **2019**, *13*, 1432–1439. [CrossRef]
28. Quesnel, H.; Père, M.C.; Louveau, I.; Lefaucheur, L.; Perruchot, M.H.; Prunier, A.; Pastorelli, H.; Meunier-Salaün, M.C.; Gardan-Salmon, D.; Merlot, E.; et al. Sow environment during gestation: Part II. Influence on piglet physiology and tissue maturity at birth. *Animal* **2019**, *13*, 1440–1447. [CrossRef]
29. Merlot, E.; Calvar, C.; Prunier, A. Influence of the housing environment during sow gestation on maternal health, and offspring immunity and survival. *Anim. Prod. Sci.* **2017**, *57*, 1751. [CrossRef]
30. Seck, J.R. Prenatal glucocorticoids and long-term programming. *Eur. J. Endocrinol.* **2004**, *151* (Suppl. 3), U49–U62. [CrossRef]
31. Couret, D.; Jamin, A.; Kuntz-Simon, G.; Prunier, A.; Merlot, E. Maternal stress during late gestation has moderate but long-lasting effects on the immune system of the piglets. *Vet. Immunol. Immunopathol.* **2009**, *131*, 17–24. [CrossRef] [PubMed]
32. Merlot, E.; Quesnel, H.; Prunier, A. Prenatal stress, immunity and neonatal health in farm animal species. *Animal* **2013**, *7*, 2016–2025. [CrossRef] [PubMed]
33. Raetz, C.R. Biochemistry of endotoxins. *Annu. Rev. Biochem.* **1990**, *59*, 129–170. [CrossRef]
34. Wyns, H.; Plessers, E.; De Backer, P.; Meyer, E.; Croubels, S. In vivo porcine lipopolysaccharide inflammation models to study immunomodulation of drugs. *Vet. Immunol. Immunopathol.* **2015**, *166*, 58–69. [CrossRef] [PubMed]
35. Nordgreen, J.; Munsterhjelm, C.; Aae, F.; Popova, A.; Boysen, P.; Ranheim, B.; Heinonen, M.; Raszplewicz, J.; Piepponen, P.; Lervik, A.; et al. The effect of lipopolysaccharide (LPS) on inflammatory markers in blood and brain and on behavior in individually-housed pigs. *Physiol. Behav.* **2018**, *195*, 98–111. [CrossRef] [PubMed]
36. Harlizius, B.; Mathur, P.; Knol, E.F. Breeding for resilience: New opportunities in a modern pig breeding program. *J. Anim. Sci.* **2020**, *98* (Suppl. 1), S150–S154. [CrossRef]
37. Mulder, H.A.; Rashidi, H. Selection on resilience improves disease resistance and tolerance to infections. *J. Anim. Sci.* **2017**, *95*, 3346–3358. [CrossRef]
38. Bacou, E.; Haurogné, K.; Mignot, G.; Allard, M.; De Beaurepaire, L.; Marchand, J.; Terenina, E.; Billon, Y.; Jacques, J.; Bach, J.M.; et al. Acute social stress-induced immunomodulation in pigs high and low responders to ACTH. *Physiol. Behav.* **2017**, *169*, 1–8. [CrossRef] [PubMed]

 animals

MDPI

Article

Modeling of Heat Stress in Sows—Part 1: Establishment of the Prediction Model for the Equivalent Temperature Index of the Sows

Mengbing Cao [1,2], Chao Zong [1,2,*], Xiaoshuai Wang [3], Guanghui Teng [1,2], Yanrong Zhuang [1,2] and Kaidong Lei [1,2]

1 College of Water Resources and Civil Engineering, China Agricultural University, Beijing 100083, China; mengbing-cao@cau.edu.cn (M.C.); futong@cau.edu.cn (G.T.); zyr123@cau.edu.cn (Y.Z.); leikaidong@cau.edu.cn (K.L.)
2 Key Laboratory of Agricultural Engineering in Structure and Environment, Ministry of Agriculture and Rural Affairs, Beijing 100083, China
3 College of Biosystems Engineering and Food Science, Zhejiang University, No. 866 Yuhangtang Road, Hangzhou 310058, China; xiaoshuai.wang@hotmail.com
* Correspondence: chaozong@cau.edu.cn

Simple Summary: Sows are susceptible to heat stress. Various indicators can be found in the literature assessing the level of heat stress in pigs, but none of them is specific to assess the sows' thermal condition. Moreover, previous thermal indices have been developed by considering only partial environment parameters, and the interaction between the index and the animal's physiological response are not always included. Therefore, this study aims to develop and assess a new thermal index specified for sows, called equivalent temperature index for sows (ETIS), with a comprehensive consideration of the influencing factors. An experiment was conducted, and the experimental data was applied for model development and validation. The equivalent temperatures have been transformed on the basis of equal effects of air velocity, relative humidity, floor heat conduction and indoor radiation on the thermal index, and used for the ETIS combination. The correlations between ETIS and sow's physiological parameters were performed. In the comparison with other thermal indices, the ETIS had the best performance (R = 0.82) using experimental data obtained from the sow house. In addition, the comfort threshold of ETIS has been classified for evaluating heat stress levels in the sow. This study concludes that the newly developed ETIS can be used to assess the degree of thermal comfort for sows.

Abstract: Heat stress affects the estrus time and conception rate of sows. Compared with other life stages of pigs, sows are more susceptible to heat stress because of their increased heat production. Various indicators can be found in the literature assessing the level of heat stress in pigs. However, none of them is specific to assess the sows' thermal condition. Moreover, thermal indices are mainly developed by considering partial environment parameters, and there is no interaction between the index and the animal's physiological response. Therefore, this study aims to develop a thermal index specified for sows, called equivalent temperature index for sows (ETIS), which includes parameters of air temperature, relative humidity and air velocity. Based on the heat transfer characteristics of sows, multiple regression analysis is used to combine air temperature, relative humidity and air velocity. Environmental data are used as independent variables, and physiological parameters are used as dependent variables. In 1029 sets of data, 70% of the data is used as the training set, and 30% of the data is used as the test set to create and develop a new thermal index. According to the correlation equation between ETIS and temperature-humidity index (THI), combined with the threshold of THI, ETIS was divided into thresholds. The results show that the ETIS heat stress threshold is classified as follows: suitable temperature ETIS < 33.1 °C, mild temperature 33.1 °C ≤ ETIS < 34.5 °C, moderate stress temperature 34.5 °C ≤ ETIS < 35.9 °C, and severe temperature ETIS ≥ 35.9 °C. The ETIS model can predict the sows' physiological response in a good manner. The correlation coefficients R of skin temperature was 0.82. Compared to early developed thermal indices, ETIS has the best predictive effect on skin temperature. This index could be a useful tool for assessing the thermal environment to ensure thermal comfort for sows.

Citation: Cao, M.; Zong, C.; Wang, X.; Teng, G.; Zhuang, Y.; Lei, K. Modeling of Heat Stress in Sows—Part 1: Establishment of the Prediction Model for the Equivalent Temperature Index of the Sows. *Animals* **2021**, *11*, 1472. https://doi.org/10.3390/ani11051472

Academic Editors: Lilong Chai and Yang Zhao

Received: 4 April 2021
Accepted: 17 May 2021
Published: 20 May 2021

Keywords: thermal index; sows; environmental parameters; heat stress threshold; skin temperature

1. Introduction

Heat stress can affect the reproductive endocrine system in sows and inhibit the ovarian function, which in turn affects estrus activity [1], causing delayed estrus, hidden estrus or even no estrus phenomenon [2]. Heat stress can reduce fertility rate [3,4], decrease piglet weight gain [5], decrease milk production [6], increase weight loss through lactation [5] and increase death rate [7]. According to reports, the annual economic losses caused by heat stress to the pig industry amounted to 299 million dollars in the US, and the number reached billions of dollars globally [8]. Sows' performance can greatly affect the profitable margin of a farm, and proper means to reduce heat stress are desperately needed. In order to alleviate heat stress of sows more effectively, it is necessary to quantify the thermal environment of the sow barn.

Over the past decades, many indices have been developed in the assessment of thermal environment, and some have been applied with pigs, such as the temperature-humidity index (THI) [9–16], the globe-humidity index (BGHI) [17,18], the effective temperature (ET) [19,20] and the enthalpy (H) [21]. However, those indices applied with pigs contain the following issues: (1) they normally include two or three environmental parameters, which are unilateral from the perspective of heat exchange [22,23]; (2) those indices mainly developed based on other animals, while directly applied in pigs; (3) they lack consideration of pigs' real-time physiological and production characteristics. In addition, there is no study focusing on heat stress in sows, so far. To overcome the limitation of those indices, a specific thermal index developed for sows is necessary.

Respiratory rate, core body temperature, rectal temperature, skin temperature, feed, water intake and other physiological responses as well as production performance (pregnancy rate, delivery times, estrus time, etc.) are affected by heat stress to varying levels [24–28]. Thus, the level of heat stress can be assessed by measuring or monitoring changes in physiological responses, animal behavior and performance. However, animal behavior, performance and most physiological indicators are either invasive or difficult to monitor continuously. In contrast, environmental parameters, such as air temperature, relative humidity and air velocity, acting as the influencing factors of heat stress, are comparatively much easier to measure. Therefore, it is necessary to research the relationship between environmental parameters and animal physiological responses, and establish a heat stress index.

The thermal index, which includes environmental parameters such as air temperature, relative humidity and air velocity, is often used to analyze heat stress in animals. The total heat dissipation of the sow will be affected by the sow's convective heat transfer, radiation heat transfer, heat conduction and respiratory heat transfer [22,23,29], and those forms of heat dissipation will be affected by environmental factors such as air temperature, relative humidity airflow velocity and so on. Heat stress of sows is mainly caused by poor heat dissipation [22,30,31]. Therefore, heat dissipation should be considered in the establishment of a thermal index.

Therefore, the objectives of this study are (1) to develop a new thermal index for sows based on environmental parameters and physiological responses; (2) to categorize the developed thermal index with heat stress levels in sows.

2. Materials and Methods

2.1. Model Development

2.1.1. Structure of the Equivalent Temperature Index for Sows (ETIS) Model

In this study, the new thermal index is expressed as the equivalent temperature index of a sow (ETIS). ETIS is composed of air temperature and equivalent temperatures adjusted from the thermal effects of air velocity, relative humidity, floor heat conduction and radiation on the heat load. Mathematically, ETIS is expressed as

$$\mathrm{ETIS} = \mathrm{T} + \mathrm{T_{rh}} + \mathrm{T_f} + \mathrm{T_u} + \mathrm{T_r} \tag{1}$$

where T is the dry-bulb temperature of the air (°C); T_{rh}, T_u, T_f and T_r are equivalent air temperatures related to relative humidity, air velocity, floor heat conduction and radiation, respectively (°C).

2.1.2. Equivalent Temperature Based on Relative Humidity

Beckett [32] provides a chart to illustrate the combined effects of air temperature and relative humidity on pig growth. Bjerg [19] compared the values from Beckett's study and found that T_{rh} can be calculated by Equation (2):

$$\mathrm{T_{rh}} = \mathrm{a}\cdot(\mathrm{RH} - 50)\cdot\mathrm{T} \tag{2}$$

where T_{rh}, is equivalent air temperature related to relative humidity (°C). a is a coefficient, and expected to be positive. RH is the relative humidity (%). T is the dry-bulb temperature of the air (°C)

It can be concluded from the equation that when the humidity is above 50%, as the humidity increases, the thermal index shows an upward trend.

2.1.3. Equivalent Temperature Based on Air Velocity

The equivalent air velocity temperature can be obtained from the equations related to convective heat transfer [22,33,34], as shown in Equations (3) and (4)

$$\mathrm{H_c} = \mathrm{h_c}\cdot\mathrm{A}\cdot(\mathrm{T_s} - \mathrm{T}) \tag{3}$$

$$\mathrm{Nu} = \frac{\mathrm{h_c}\cdot\mathrm{l}}{\mathrm{k}} = \mathrm{mRe^c} = \mathrm{m}\left(\frac{\rho \mathrm{ul}}{\mu}\right)^{\mathrm{c}} \tag{4}$$

where H_c is total heat transfer rate (W), h_c is convective heat transfer coefficient ($W{\cdot}m^{-2}{\cdot}{}^{\circ}C^{-1}$), A is surface area of the animal (m^2) and T_s is skin temperature (°C). Nu is Nusselt number, l is characteristic length (m) and k is air thermal conductivity ($W{\cdot}m^{-1}{\cdot}{}^{\circ}C^{-1}$). Re is Reynolds number. ρ is air density ($kg{\cdot}m^{-3}$). u is air velocity ($m{\cdot}s^{-1}$). μ is dynamic viscosity coefficient ($m^2{\cdot}s^{-1}$). m and c are constants determined by the relationship between Nu and Re.

According to Equations (3) and (4), it can be seen that the air velocity has an extremely important influence on the convective heat transfer coefficient, so the equivalent temperature of air velocity (T_u) is expressed as Equation (5):

$$\mathrm{T_u} = \mathrm{e}\cdot\mathrm{u^c}\cdot(\mathrm{T_s} - \mathrm{T}) \tag{5}$$

where T_u is equivalent air temperature related to air velocity (°C). e is a coefficient that represents the relationship between convective heat transfer of the sow and equivalent temperature based on air velocity. u is air velocity ($m{\cdot}s^{-1}$). c is a constant determined by the relationship between Nu and Re. T_s is skin temperature (°C). T is the dry-bulb temperature of the air (°C)

Equation (5) is consistent with Wang's equation [35], where the constant c represents the effect of air velocity changes on the convective heat transfer coefficient. This constant is usually obtained by analyzing the relationship between air velocity and object convection heat transfer coefficient [36,37].

Li [36] used computational fluid dynamic (CFD) simulation to study the convective heat transfer of a standing pig and found that it is proportional to $v^{0.66}$. However, the sow spends most of the time lying down and resting [38,39]. To know the convective heat transfer of sows, a pilot study by Cao et al. [40] was carried out, and the convective heat transfer coefficient of sows was found to be 0.6827.

2.1.4. Equivalent Temperature Based on Conductive Heat Transfer

The heat transfer between the sow body and the floor surface is driven by the temperature difference. In this study, a concrete floor instead of slatted floor was installed in the sow barn. In the summer, the surface of the floor is a source of cold. The sow heat can be transferred to the floor surface [20,22]. Then, the equivalent temperature of heat conduction at that place must be related to the temperature of the floor surface and sow body surface, and the relationship between them is indicated by D. Thus, T_f can be expressed as Equation (6):

$$T_f = D \cdot (T_s - T_d) \tag{6}$$

where T_f is equivalent air temperature related to floor heat conduction (°C). D is a coefficient related to floor heat conduction equivalent temperature. T_s is skin temperature (°C). T_d is the surface temperature for the floor (°C).

As the floor surface temperature is very close to the ambient air temperature, the air temperature is usually used as the floor surface temperature for calculations [41,42]. The equivalent temperature of floor heat conduction can be expressed as

$$T_f = D \cdot (T_s - T) \tag{7}$$

where T_f is equivalent air temperature related to floor heat conduction (°C). D is a coefficient related to floor heat conduction equivalent temperature. T_s is skin temperature (°C). T is the dry-bulb temperature of the air (°C).

2.1.5. Equivalent Temperature Based on Radiative Heat Transfer

The radiative heat transfer is driven by the difference of the fourth power of the absolute temperatures between two objects. For sows in the house, long-wave radiation was considered. To integrate the effects of the radiation between the measured object and the surrounding surfaces, the mean radiant temperature was introduced [23,43,44]. The radiative heat of a sow is related to the sow's skin temperature and mean radiant temperature, and Equation (8) is obtained:

$$T_r = R_{rad} \cdot \left((T_s + 273.15)^4 - (T_{rad} + 273.15)^4 \right) \tag{8}$$

where T_r is equivalent air temperature related to radiation. R_{rad} is a coefficient, which represents the relationship between the sow's long-wave radiation and the equivalent radiation temperature. T_s is skin temperature (°C). T_{rad} is the average radiant temperature (°C).

Since indoor climate studies usually assume that the average radiant temperature is equal to the air temperature [44–48], T_{rad} is represented here by T, and equivalent temperature of radiative heat transfer can be expressed as Equation (9).

$$T_r = R_{rad} \cdot \left((T_s + 273.15)^4 - (T + 273.15)^4 \right) \tag{9}$$

where T_r is equivalent air temperature related to radiation. R_{rad} is a coefficient, which represents the relationship between the sow's long-wave radiation and the equivalent radiation temperature. T_s is skin temperature (°C). T is the dry-bulb temperature of the air (°C).

2.1.6. Combined Equivalent Temperature Index

The ETIS is formed by integrating those equivalent temperatures into Equation (1) and shown as

$$ETIS = T + a \cdot (RH - 50) \cdot T + e \cdot u^{0.6827} \cdot (T_s - T) + D \cdot (T_s - T) + R_{rad} \cdot ((T_s + 273.15)^4 - (T + 273.15)^4) \quad (10)$$

where ETIS is the equivalent temperature index for sows. T is the dry-bulb temperature of the air (°C). a is the coefficient related to the relative humidity equivalent temperature. RH is the relative humidity (%). e is a coefficient that represents the relationship between convective heat transfer of the sow and equivalent temperature based on air velocity. u is air velocity ($m \cdot s^{-1}$). T_s is skin temperature (°C). D is a coefficient related to floor heat conduction equivalent temperature. R_{rad} is a coefficient, which represents the relationship between the sow's long-wave radiation and the equivalent radiation temperature.

2.2. Experimental Set Up

2.2.1. Animal and Housing

The study was conducted in a sow barn at the National Feed Research Center of China Agricultural University from June to August 2018. Sows were crossbreeds between Large White and Landrace.

The barn housed 30 non-pregnant multiparous sows. Sows were fed in this facility before being moved for breeding. The rectal temperature, respiration rate and skin temperature of the sows were measured. Data collection was done during non-pregnancy. Data were collected on each batch of 10 sows and on a total of 4 batches. The sows were kept in crates with concrete solid floors, as shown in Figure 1a,b. The length, width and height of the crate was 2.2 m × 0.64 m × 1 m. Each crate was equipped with one feeder and one drinker, and the sow was raised with ad libitum feeding and drinking. The slurry was removed regularly by workers, and the urine ran into the drain pipe beneath the floor. A tunnel ventilation system with one exhaust fan and 2 air inlets was used in the house. The fan (YH900, Yinghe Company, Shenzhen, China) had a capacity of 28,500 $m^3 \cdot h^{-1}$, and the ventilation rate was controlled based on indoor air temperature.

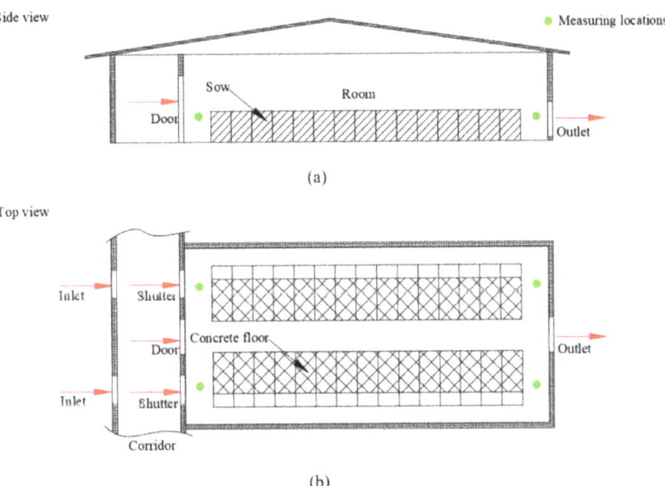

Figure 1. Barn schematic: (**a**) Side view; (**b**) Top view. Barn can house 30 sows, usually 20–24 non-pregnant sows per batch. Sow barn is 11.7 m × 7.7 m × 2.5 m. The length, width and height of the crate is 2.2 m × 0.64 m × 1 m. The sow barn has four measuring points, and each measuring point measures the air temperature, relative humidity and air velocity. This research assumed that the difference between different measuring points in environmental parameters is linear (for example, the green dots in figure (**a**) were set to be 30 °C and 31.5 °C, the environmental parameters for the 15 sows were 30.1, 30.2, 30.3, 30.4, ... , 31.5 °C), and the environmental parameters were determined at different sow locations based on this assumption. Physiological parameters of the sows in the air outlet area, the air inlet area, and the middle area of the room were measured.

2.2.2. Measurements

Environment

Air velocity, air temperature and relative humidity were measured by wireless hot wire anemometers (testo405i, Schwarzwald, Germany) and a wireless air temperature and relative humidity measuring instrument (testo605i, Schwarzwald, Germany) at 4 locations (Figure 1b). The measuring points were 0.6 m above the floor. During the experiment, a set of data was collected every 2 h from 8:00 to 18:00 per day. The air temperature, relative humidity and air velocity at the measuring point in Figure 1 were measured. The research assumed that the difference between different measuring points in environmental parameters is linear, and the environmental parameters were determined at different locations based on this assumption. Finally, the environmental conditions of sows were determined.

Physiological Parameters

The rectal temperature, respiratory rate and skin temperature of the sows in the air outlet area, the air inlet area and the middle area of the room were measured. Physiological parameters were measured on each batch of 10 sows in each of 4 batches during the non-pregnant period. The rectal temperature probe (Huaxu, Xinxiang, China) was used to measure the rectal temperatures of sows. The rate of respiration was calculated by counting the number of rises and falls of the sow's chest within one minute. The skin temperature was measured by a handheld infrared thermometer (Raynger ST, Raytek, Santa Cruz, California, CA, USA) with an accuracy of $\pm 1\%$ of the measured temperature. The physiological parameters were measured every 2 h from 8:00 to 18:00 every day. The testing of physiological data was performed in parallel with the testing of environmental data.

2.3. *Data Analysis*

2.3.1. Correlation Analysis between Temperature and Humidity Index and Physiological Parameters

Correlation analysis was used to determine the correlations between THI [10,49] and core temperature, between THI [10,49] and respiration rate, and the correlations between THI [10,49] and skin temperature. According to the correlation analysis between THI and three physiological parameters, the physiological parameter that had a strong connection with the thermal index was determined, and this physiological parameter was used as the dependent variable for creating the ETIS.

2.3.2. Linear Regression Model

In the regression model, instead of using three environmental parameters (T, RH, and u), four new corresponding terms—T, RH, $u^{0.6827} \cdot (T_S - T)$ and $(T + 273.15)^4$—were used as explanatory variables in the model fitting and analysis.

Skin temperature was selected as the only response variable because skin temperature is influenced by both the thermal environment and the sow body, acting as the window for transfer heat.

The linear regression model associated with Equation (10) is now expressed as follows:

$$y = b_0 + b_1 \cdot T + b_2 \cdot (50) \cdot T + b_3 \cdot (RH) \cdot T + b_4 \cdot u^{0.6827} \cdot (T_s - T) + b_5 \cdot T_s + b_6 \cdot T + b_7 \cdot (T_s + 273.15)^4 + b_8 \cdot (T + 273.15)^4 \quad (11)$$

where y is the response variable (skin temperature); b_0 is the intercept; b_1, b_2, b_3, b_4, b_5, b_6, b_7 and b_8 are the regression coefficients. T is the dry-bulb temperature of the air (°C). RH is the relative humidity (%). u is air velocity ($m \cdot s^{-1}$). T_s is skin temperature (°C).

Since the skin temperature is used as the response variable, T_s in Equation (11) is fixed at 38 °C to ensure that physiological parameters are not included in the dependent variable. Equation (11) eventually becomes

$$y = b_0 + b_1 \cdot T + b_2 \cdot (50) \cdot T + b_3 \cdot (RH) \cdot T + b_4 \cdot u^{0.6827} \cdot (38 - T) + b_5 \cdot 38 + b_6 \cdot T + b_7 \cdot (38 + 273.15)^4 + b_8 \cdot (T + 273.15)^4 \quad (12)$$

where y is the response variable (skin temperature); b_0 is the intercept; b_1, b_2, b_3, b_4, b_5, b_6, b_7 and b_8 are the regression coefficients. T is the dry-bulb temperature of the air (°C). RH is the relative humidity (%). u is air velocity ($m \cdot s^{-1}$).

2.3.3. Regression Analysis

The experimental data were randomly divided into two data sets, of which 70% were used to create the model and the remaining 30% were used to test the model. The relationship between skin temperature and multiple environmental parameters was determined by linear regression analysis in Matlab (2019a, MathWorks, Natick, MA, USA). The determination coefficient (R^2) was used to evaluate the predictive performance of the model. The model was validated using the skin temperature data in the test set.

The statistical significance of the explanatory variable to the response variable was determined in the form of p-value. A small p-value indicated that the corresponding explanatory variable had a high statistical significance. When $p < 0.001$, the investigated parameter was considered to be highly significant.

2.3.4. Stress Categories (Thresholds)

The threshold was the category of heat stress level that causes loss of animal production. The stress threshold categories are mild, moderate, severe and urgent. Mellado et al. [49] used THI [10] to analyze the pregnancy rate of sows. In their study, when THI < 74, the pregnancy rate was 93%; when 74 < THI < 78, the pregnancy rate was 91.8%; when 78 < THI < 82, the pregnancy rate was 91.4%; when THI > 82, the sow pregnancy rate was 89.8%. Following the categorization from the study of Mellado et al. [49], heat stress level can be classified as follows: THI < 74 indicates an appropriate environmental level, $74 \leq$ THI < 78 indicates mild thermal stress, $78 \leq$ THI < 82 indicates moderate thermal stress and THI ≥ 82 indicates severe thermal stress. A relationship between THI and ETIS has been established, and the ETIS defines heat stress thresholds according to the fitting equation.

2.3.5. Comparative Analysis of Various Thermal Indices

The ETIS thermal index model was compared with models such as THI [9–16], BGHI [17,18], ET [19] and H [33]. Pearson correlation coefficients between the thermal index and the selected physiological parameters were calculated. The larger the correlation coefficient, the better the prediction of the thermal index [21].

3. Results

3.1. Experimental Data Reliability Verification

3.1.1. Experimental Data

The summarized dataset of air temperature, humidity, air velocity, skin temperature, respiration rate and core temperature are given in Table 1.

Table 1. Statistics of the integrated dataset. N represents the total amount of data. SD represents the standard deviation of the data.

Item	N	Mean	SD	Maximum	Minimum
Air temperature (T), °C	1029	28.7	2.6	34.0	21.9
Relative humidity (RH), %	1029	65.8	10.0	89.8	40.4
Air velocity (u), $m \cdot s^{-1}$	1029	0.07	0.07	0.29	0.00
Skin temperature (T_s), °C	1029	34.9	1.4	37.8	28.6
Respiration rate (RR), breaths·min^{-1}	1029	49	28	168	12
Core temperature (T_c), °C	1029	38.44	0.46	41.27	37.21

3.1.2. The Relationship between Temperature and Humidity Index and Core Temperature

Figure 2 shows the correlations between THI and core temperature. The relationship between THI and core temperature can be described by a linear regression equation: y = 0.0454x + 34.873 (where y stands for core temperature (°C) and x stands for THI) ($R^2 = 0.0972$, $p < 0.0001$).

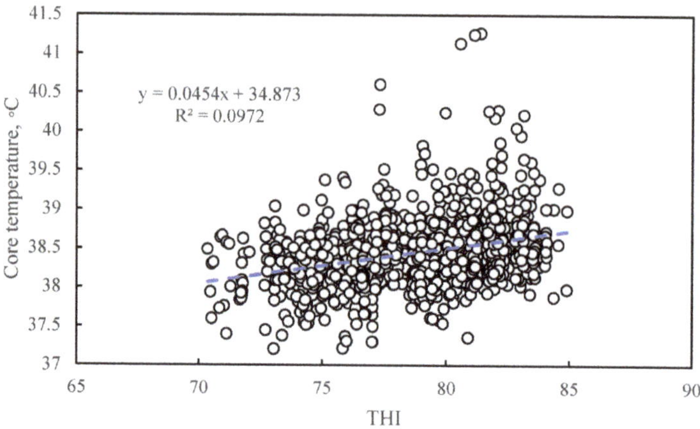

Figure 2. The correlations between temperature and humidity index and core temperature. The THI value was calculated based on the experimental data of 1029 sets of air temperature and relative humidity. The dots represent the scatter plot of the core temperature of the sow corresponding to the THI. The dotted line represents the model estimate based on the linear regression of sow core temperature with THI. The linear regression equation here is: y = 0.0454x + 34.873 (where y stands for core temperature (°C) and x stands for THI) ($R^2 = 0.0972$, $p < 0.0001$).

3.1.3. The Relationship between Temperature and Humidity Index and Respiration Rate

Figure 3 shows the correlation between THI and respiration rate. The relationship between THI and respiration rate can be described by a linear regression equation: y = 2.2137x − 135.98 (where y stands for respiration rate (breaths·min^{-1}) and x stands for THI) ($R^2 = 0.1386$, $p < 0.0001$).

3.1.4. The Relationship between Temperature and Humidity Index and Skin Temperature

Figure 4 shows the correlation between THI and sow skin temperature. The relationship between THI and skin temperature can be described by a linear regression equation: y = 0.3495x +7.3646 (where y stands for skin temperature (°C) and x stands for THI) ($R^2 = 0.6165$, $p < 0.0001$).

3.2. Development of the Equivalent Temperature Index for Sows Model

Equation (10) was used to perform multiple linear regression in Matlab. Then, the estimated ratios of T, 50·T, RH·T, $u^{0.6827}\cdot(T_S - T)$, 38, T, $(38 + 273.15)^4$ and $(T + 273.15)^4$ were obtained; these ratios were 0, 0.1152, 0.0006, −0.3132, 0, 0, 2.9370×10^{-8} and -4.8957×10^{-8}, respectively. To be as consistent as possible with the definition of ETIS, the final ETIS can be determined as Equation (13). Figure 5 shows the correlation between ETIS and sow skin temperature.

$$\begin{aligned}\mathrm{ETIS} = \mathrm{T} + 0.0006\cdot(\mathrm{RH} - 50)\cdot\mathrm{T} - 0.3132\cdot u^{0.6827}\cdot(38 - \mathrm{T}) - 4.79\cdot(1.0086\cdot 38 - \mathrm{T}) + 4.8957\cdot 10^{-8} \\ \cdot\left((38 + 273.15)^4 - (\mathrm{T} + 273.15)^4\right)\end{aligned} \quad (13)$$

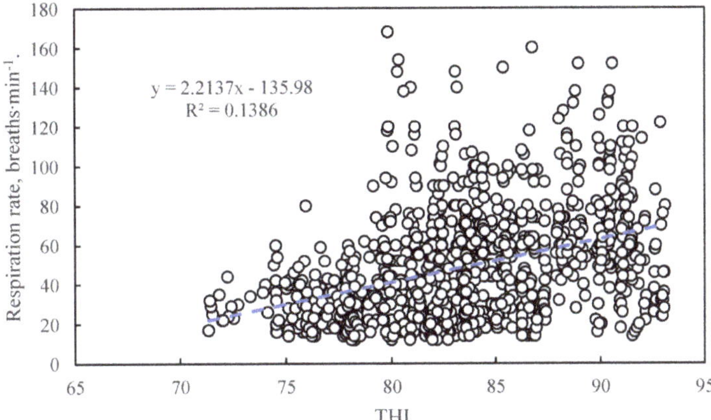

Figure 3. The correlation between temperature and humidity index and respiration rate. The THI value was calculated based on the experimental data from 1029 sets measuring air temperature and relative humidity. The dots represent the scatter plot of the respiration rate of the sow corresponding to the THI. The dotted line represents model estimate based on the linear regression of sow respiration rate with THI. The linear regression equation here is: y = 2.2137x − 135.98 (where y stands for respiration rate (breaths·min^{-1}) and x stands for THI) (R^2 = 0.1386, $p < 0.0001$).

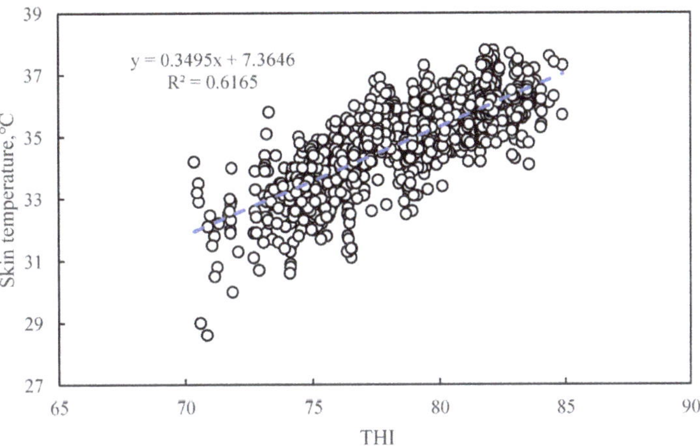

Figure 4. The correlation between temperature and humidity index and sow skin temperature. The THI value was calculated based on the experimental data from 1029 sets measuring air temperature and relative humidity. The dots represent the scatter plot of the skin temperature of the sow corresponding to the THI. The dotted line represents the model estimate based on the linear regression of sow skin temperature with THI. The linear regression equation here is: y = 0.3495x + 7.3646 (where y stands for skin temperature (°C) and x stands for THI) (R^2 = 0.6165, $p < 0.0001$).

3.3. Validation of the Equivalent Temperature Index for Sows Model

Figure 6 shows a scatter plot of the test data set of ETIS versus skin temperature. The measured coefficient of the skin temperature (R^2) was 0.7317. This shows that the model derived from the training set can well estimate the skin temperature of the test data.

3.4. Classification of Heat Stress Threshold Based on Equivalent Temperature Index for Sows

Figure 7 shows a scatterplot of THI versus ETIS, and inductive regression equation is described: y = 0.3533x − 6.9249 (where y stands for ETIS (°C) and x stands for THI) ($R^2 = 0.9026$, $p < 0.0001$).

Based on the linear regression model in Figure 7, heat stress threshold for ETIS can be developed based on thresholds developed for THI. The categories are as follows: ETIS < 33.1 °C is considered to be suitable, 33.1 °C ≤ ETIS < 34.5 °C is considered to be mild, 34.5 °C ≤ ETIS < 35.9 °C is considered to be moderate, and ETIS ≥ 35.9 °C is considered to be severe. As shown in Table 2, from the heat stress zone of ETIS, it can be concluded that the sow Farm of China Feed Research Center, located in North China, falls mostly in the mild and moderate heat stress range during the summer period.

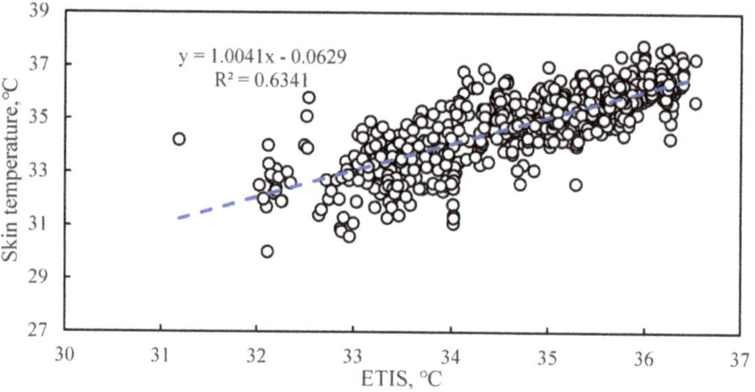

Figure 5. Regression of equivalent temperature index for sows to skin temperature based on training data. The dots represent the scatter plot of the skin temperature of the sow corresponding to the ETIS. The dotted line represents the model estimate based on the linear regression of sow skin temperature with 720 ETIS training data. The linear regression equation here is: y = 1.0041x − 0.0629 (where y stands for skin temperature (°C) and x stands for ETIS (°C)) ($R^2 = 0.6341$, $p < 0.0001$).

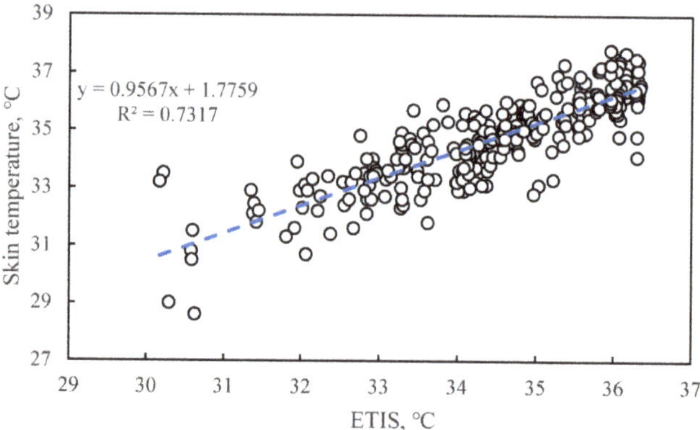

Figure 6. Regression of equivalent temperature index for sows to skin temperature based on test data. The dots represent the scatter plot of the skin temperature of the sow corresponding to the ETIS. The dotted line represents the model estimate based on the linear regression of sow skin temperature with 309 ETIS testing data. The linear regression equation here is: y = 0.9567x + 1.7759 (where y stands for skin temperature (°C) and x stands for ETIS (°C)) ($R^2 = 0.7317$, $p < 0.0001$).

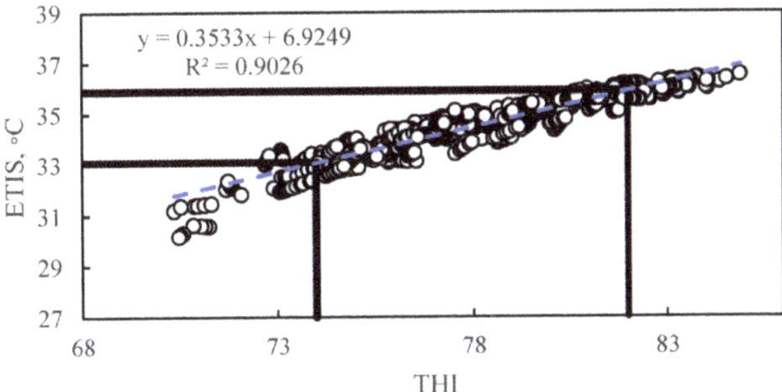

Figure 7. Regression on temperature and humidity index versus equivalent temperature index for sows. The dots represent the scatter plot of the ETIS corresponding to the THI. The dotted line represents model estimate based on linear regression of ETIS with THI. The linear regression equation here is: y = 0.3533x + 6.9249 (where y stands for ETIS (°C) and x stands for THI) ($R^2 = 0.9026$, $p < 0.0001$).

Table 2. ETIS heat stress threshold. Threshold distribution of ETIS was determined according to the threshold distribution of THI and the correlation equation between THI and ETIS.

Category	THI	ETIS, °C
Suitable	THI < 74	ETIS < 33.1
Mild	74 ≤ THI < 78	33.1 ≤ ETIS < 34.5
Moderate	78 ≤ THI < 82	34.5≤ ETIS < 35.9
Severe	82 ≤ THI	35.9 ≤ ETIS

Note: THI is temperature and humidity index. ETIS is the equivalent temperature index for sows.

3.5. Comparison of Equivalent Temperature Index for Sows with Other Indices

ETIS was compared with other environmental thermal indices [9–19,21], and Pearson correlation coefficients were calculated to determine the relationship between selected physiological responses (skin temperature) and various thermal indices. As shown in Table 3, the thermal index positively correlated with the physiological response.

Table 3. Pearson correlation coefficient between the thermal index and physiological responses (skin temperature) (all *p*-values < 0.0001). The thermal index and the skin temperature are positively correlated. When the Pearson correlation coefficient R is closer to 1, the correlation is stronger.

Thermal Index	Equation	Skin Temperature Correlation (R)
ETIS	$ETIS = T + 0.0006\cdot(RH - 50)\cdot T - 0.3132\cdot u^{0.6827}\cdot(38 - T) - 4.79 \cdot(1.0086\cdot 38 - T) + 4.8957\cdot 10^{-8}\cdot\left((38 + 273.15)^4 - (T + 273.15)^4\right)$	0.82
THI [9]	$THI = T + 0.36\cdot T_{wb} + 41.5$	0.78
THI [10]	$THI = 0.8\cdot T + \left(\frac{RH\cdot(T-14.4)}{100}\right) + 46.4$	0.79
THI [11]	$THI = 0.65\cdot T + 0.35\cdot T_{wb}$	0.79
THI [12]	$THI = T^{\circ} - (0.55 - 0.0055\cdot RH)\cdot\left(T^{\circ} - 58\right)$	0.79
THI [13]	$THI = 0.72\cdot T + 0.72\cdot T_{wb} + 40.6$	0.78
THI [16]	$THI = (1.8\cdot T + 32) - \left[0.55\cdot\left(\frac{RH}{100}\right)\right]\cdot[(1.8\cdot T + 32) - 58]$	0.69
THI [14]	$THI = T - (0.55 - 0.0055\cdot RH)\cdot(T - 14.5)$	0.79
THI [15]	$THI = 0.27\cdot T + 1.35\cdot T_{wb} + 34.07$	0.73
BGHI [17]	$BGHI = T_g + 0.36\cdot T_{dp} + 41.5$	0.71

Table 3. *Cont.*

Thermal Index	Equation	Skin Temperature Correlation (R)
ET [19]	$ET = T + 0.0015 \cdot (RH - 50) \cdot T + \left[-1.0 \cdot 42 - T \cdot \left(v^{0.66} - 0.2^{0.66}\right)\right]$	0.75
H [21]	$H = 1.006 \cdot T + \frac{RH}{Pm} \cdot 10^{\left(\frac{7.5 \cdot T}{273.3 + T}\right)} \cdot (71.28 + 0.052 \cdot T)$	0.68

Note: ETIS is the equivalent temperature index for sows. THI is temperature and humidity index. BGHI is the globe-humidity index. ET is the effective temperature. H is the enthalpy ($kJ \cdot kg^{-1}$). T is the dry bulb temperature (°C). T° is the dry bulb temperature(°F). T_g is the black globe temperature (°C). T_{wb} is the wet bulb temperature (°C); T_{dp} is the dew point temperature (°C). RH is the relative humidity (%). P_m is high mercury of barometric pressure (mmHg).

4. Discussion

4.1. Experimental Data

It can be seen from the data that the highest temperature in the summer environment (34 °C) exceeded the upper limit of comfort for sows [7], but the heat dissipation process of the sow is not only related to the ambient temperature [22,23,29], and more factors should be considered overall.

4.2. The Relationship between THI and Physiological Parameters

4.2.1. The Relationship between THI and Core Temperature

The coefficient of determination (R^2) between THI and core temperature was 0.0972, which means that 9.7% of the core temperature change can be explained by the variation in THI. Sows are warm-blooded animals, and their rectal temperatures are relatively stable under normal conditions. Previous studies have shown that when the air temperature is above 25 °C, every 1 °C increase in air temperature will increase the rectal temperature by 0.099 °C [50]. However, conventional rectal thermometers are not precise enough to measure 0.099 °C, and minor changes cannot be accurately monitored [51]. Previous studies on the effect of air temperature on rectal temperature vary widely [50]. It is reasonable that the correlation between THI and core temperature is low.

4.2.2. The Relationship between THI and Respiration Rate

The coefficient of determination (R^2) between THI and respiration rate was 0.1386, which means that 13.9% of the respiration rate change can be explained by the variation in THI. As sows have fewer sweat glands, they dissipate differently from other animals. Meanwhile, sows are restricted by crates, and they recline or lie most of the time. When the sow reclines, it often alternates between being awake and asleep, which affects the sow's breathing status [52]. The breathing status includes deep breathing and non-deep breathing. Under the same thermal conditions, if the sow breathes the same amount of air, the frequency between deep breathing and non-deep breathing will be different [53]. Individual differences and the lying position may also have an impact on a sow's breathing. In previous studies, the monitored frequency of respiration rate was low [5], or the number of samples was small [51], so that the influences of the above factors were ignored. As an environmental thermal index, THI only includes air temperature and relative humidity. Relief of heat stress is essentially a process of heat dissipation, and air temperature and relative humidity are only part of the parameters of the heat dissipation process. As a result, it can be predicted that the correlation between THI and respiration rate is weak. The core temperature change is usually small, and it is also not suitable to be a crucial variable of the thermal index.

4.2.3. The Relationship between THI and Skin Temperature

The coefficient of determination (R^2) between THI and sow skin temperature was 0.6165, which means that 61.7% of the skin temperature change can be explained by the variation in THI. Since the sow's skin is directly exposed to the air, skin temperature is greatly affected by the environment. Apart from the influence of external conditions, skin temperature is also affected by the sow's internal heat production. The skin of the sow is the main channel for heat exchange. Skin temperature is the result of heat production and environmental factors [29]. Compared with other physiological parameters, skin temperature has the highest correlation with THI. The respiration rate changes after receiving the thermal sensory signal provided by the brain of the sow. The respiration rate is also affected by the state of sleep [52]. This process can inevitably have a delay compared to the skin. Since sows are warm-blooded animals, whose core temperature tends to be stable, the correlation between core temperature and thermal index is weak. Therefore, skin temperature is more suitable as the physiological response variable in the ETIS model.

4.3. Development and Verification of the ETIS Model

Each coefficient in ETIS represents the contribution or weight of each equivalent temperature to heat stress. The coefficient of determination (R^2) between ETIS and sow skin temperature was 0.6341. The ETIS equation predicts the skin temperature of 63% of sows using the training data set. Skin temperature is not only affected by the environment but also by the sow's internal factors, and there are individual differences among different sows. The ETIS model can predict 73% of skin temperature changes using the test data set. Therefore, the ETIS model construction can be considered reasonable.

4.4. Classification of Heat Stress Threshold Based on ETIS

In addition to the influence of air velocity, ETIS also includes the influence of air temperature and relative humidity. Therefore, the correlation between ETIS and THI is relatively high. According to the threshold partition of THI and the correlation equation between ETIS and THI, the threshold of ETIS can be determined. However, the balance between heat dissipation and heat production is not only affected by environmental parameters, but also by other factors related to the sow itself and management, such as sow genotype, hair thickness, health status, productivity level, activity level, etc. These animal or management-related factors change the range of threshold and heat stress categories.

4.5. Comparison of ETIS with Other Indices

The prediction of skin temperature was better with ETIS compared to other indices. In a hot environment, sows will not only be affected by air temperature and relative humidity, but also by air velocity. When the air temperature is lower than the skin temperature of the sow, increasing the air velocity is beneficial to alleviate heat stress. The main reason is that increasing the air velocity can increase the convective heat transfer coefficient of the sow, which is beneficial to increase the convective heat transfer of the sow. Compared with THI, BGHI and H, ETIS has an air velocity term. ET simply combines air temperature, relative humidity and air velocity, and it uses a small amount of test data based on sensible heat dissipation of 3.4–70 kg pig [19]. Sensible heat is an indirect measurement, so errors will inevitably occur. Moreover, large pigs tend to avoid heat, and little pigs tend to avoid cold. The characteristic lengths of big pigs and small pigs are different, so the heat dissipation characteristics will be slightly different. Therefore, ETIS is significantly better than ET for predicting the skin temperature of sows.

The skin is a window for heat exchange between the sow and the external environment. When the thermal environment is severe, the sow skin is stimulated at first. ETIS can reasonably predict the influence of the environment on the sow skin temperature, so it can be considered that the ETIS model can be used to evaluate heat stress level of sows. However, it should be noted that the index is created in a fully enclosed sow barn, so the applicable conditions should be similar. Sows have individual differences, and sows in different regions have different environmental adaptations. These will affect the use of ETIS.

4.6. Summary of the Study and Research Perspectives

In this study, by considering the influencing factors of air temperature, relative humidity and air velocity, according to the law of heat transfer between the sow and the surrounding environment, the sows equivalent temperature model was integrated. The equivalent temperature and the sow skin temperature were used in the multiple linear regression analysis to determine the unknown coefficients of the sow equivalent model species, and to finally determine the ETIS equation. The ETIS equation has a good correlation with THI. THI is used to partition the THI thermal threshold. The classification is as follows: suitable temperature ETIS < 33.1 °C, mild temperature 33.1 °C ≤ ETIS < 34.5 °C, moderate stress temperature 34.5 °C ≤ ETIS < 35.9 °C, and severe temperature ETIS ≥ 35.9 °C. ETIS was also compared with other various indices. Finally, ETIS was concluded to have the highest correlation with skin temperature.

In this study, the correlation between ETIS and THI was used to divide the ETIS index. However, under actual conditions, the living environment of sows is different, and the characteristics of heat discomfort should be inconsistent. The thermal threshold in different regions should be slightly different. Thus, we must proceed with future studies. The actual production data will be used to verify the ETIS thermal stress threshold or make some corrections to the ETIS thermal stress threshold.

5. Conclusions

The following conclusions can be drawn from this study:

(1) A thermal index model called ETIS was developed. This model was used to predict the level of heat stress in sows. The thermal index takes into account the heat transfer characteristics of the sows. The correlation between the ETIS index and THI index (R^2) was 0.90, and the correlation with sow skin temperature (R^2) was 0.67.

(2) The ETIS heat stress threshold was classified according to the threshold defined by THI. The classification was as follows: suitable temperature ETIS < 33.1 °C, mild temperature 33.1 °C ≤ ETIS < 34.5 °C, moderate stress temperature 34.5 °C ≤ ETIS < 35.9 °C, and severe temperature ETIS ≥ 35.9 °C.

(3) Compared with other thermal indexes, the ETIS model has the best prediction of skin temperature (R = 0.82).

Author Contributions: Conceptualization, M.C. and C.Z.; methodology, M.C., C.Z. and X.W.; software, M.C.; validation, M.C., C.Z. and X.W.; formal analysis, M.C.; investigation, M.C.; resources, M.C. and Y.Z.; data curation, M.C., C.Z. and Y.Z.; writing—original draft preparation, M.C.; writing—review and editing, M.C. and C.Z.; visualization, K.L.; supervision, C.Z. and G.T.; project administration, C.Z. and G.T.; funding acquisition, G.T. All authors have read and agreed to the published version of the manuscript.

Funding: This research was funded by National Key Research and Development Plan of China (2016YFD0700204).

Institutional Review Board Statement: Not applicable.

Informed Consent Statement: Not applicable.

Data Availability Statement: All data sets during the current study are available from the corresponding author on fair request.

Conflicts of Interest: The authors declare no conflict of interest. The funders had no role in the design of the study; in the collection, analyses, or interpretation of data; in the writing of the manuscript, or in the decision to publish the results.

References

1. Barb, C.R.; Estienne, M.J.; Kraeling, R.R.; Marple, D.N.; Rampacek, G.B.; Rahe, C.H.; Sartin, J.L. Endocrine changes in sows exposed to elevated ambient temperature during lactation. *Domest. Anim. Endocrinol.* **1991**, *8*, 117–127. [CrossRef]
2. Almond, P.K.; Bilkei, G. Seasonal infertility in large pig production units in an Eastern-European climate. *Aust. Vet. J.* **2005**, *83*, 344–346. [CrossRef]
3. Bloemhof, S.; Mathur, P.K.; Knol, E.F.; van der Waaij, E.H. Effect of daily environmental temperature on farrowing rate and total born in dam line sows. *J. Anim. Sci.* **2013**, *91*, 2667–2679. [CrossRef] [PubMed]
4. Tummaruk, P.; Tantasuparuk, W.; Techakumphu, M.; Kunavongkrit, A. Effect of Season and Outdoor Climate on Litter Size at Birth in Purebred Landrace and Yorkshire Sows in Thailand. *J. Vet. Med Sci.* **2004**, *66*, 477–482. [CrossRef] [PubMed]

5. Quiniou, N.; Noblet, J. Influence of high ambient temperatures on performance of multiparous lactating sows. *J. Anim. Sci.* **1999**, *77*, 2124. [CrossRef]
6. Black, J.L.; Mullan, B.P.; Lorschy, M.L.; Giles, L.R. Lactation in the sow during heat stress. *Livest. Prod. Sci.* **1993**, *35*, 153–170. [CrossRef]
7. D'Allaire, S.; Drolet, R.; Brodeur, D. Sow mortality associated with high ambient temperatures. *Can. Vet. J.* **1996**, *37*, 237–239.
8. St-pierre, N.R.; Cobanov, B.; Schnitkey, G. Economic Losses from Heat Stress by US Livestock Industries1. *J. Dairy Sci.* **2003**, *86*, 52–77. [CrossRef]
9. Thom, E.C. Cooling degrees-days air conditioning, heating, and ventilating. *Trans. ASAE* **1958**, *55*, 65–72.
10. Thom, E.C. The discomfort index. *Weatherwise* **1959**, *12*, 57–61. [CrossRef]
11. Ingram, D.L. The effect of humidity on temperature regulation and cutaneous water loss in the young pig. *Res. Vet. Sci.* **1965**, *6*, 9. [CrossRef]
12. Bond, C.F.K.T.E. Bioclimatic Factors and Their Measurements. In *A Guide to Environmental Research in Animals*; National Academy of Science Press: Washington, DC, USA, 1971.
13. Maust, L.E.; Mcdowell, R.E.; Hooven, N.W. Effect of Summer Weather on Performance of Holstein Cows in Three Stages of Lactation. *J. Dairy Sci.* **1972**, *55*, 1133–1139. [CrossRef]
14. NOAA (National Oceanic and Atmospheric Administration). *Livestock Hot Weather Stress*; US Department of Commerce; National Weather Service Central Region: Kansas City, MO, USA, 1976.
15. Fehr, R.L.; Priddy, K.T.; Mcneill, S.G.; Overhults, D.G. Limiting Swine Stress with Evaporative Cooling in the Southwest. *Trans. ASAE* **1983**, *26*, 0542–0545. [CrossRef]
16. Wu, Z.; Chen, Z.; Zang, J.; Wang, M.; Yang, H.; Ren, F.; Liu, J.; Feng, G. Cooling performance of wet curtain fan-fabric duct ventilation system in house of pregnant sows. *Trans. Chin. Soc. Agric. Eng.* **2018**, *34*, 268–276.
17. Buffington, D.E.; Collazo-Arocho, A.; Canton, G.H.; Pitt, D.; Thatcher, W.W.; Collier, R.J. Black globe-humidity index (bghi) as comfort equation for dairy cows. *Trans. ASAE* **1981**, *24*, 0711–0714. [CrossRef]
18. Júnior, G.M.d.O.; Ferreira, A.S.; Oliveira, R.F.M.; Silva, B.A.N.; de Figueiredo, E.M.; Santos, M. Behaviour and performance of lactating sows housed in different types of farrowing rooms during summer. *Livestock Sci.* **2011**, *141*, 194–201. [CrossRef]
19. Bjerg, B.; Rong, L.; Zhang, G. Computational prediction of the effective temperature in the lying area of pig pens. *Comput. Electron. Agric.* **2018**, *149*, 71–79. [CrossRef]
20. Bjerg, B.S.; Kai, P. CFD Prediction of Heat Transfer in Heated or Cooled Concrete Floors in Laying Areas for Pig. In Proceedings of the 2019 ASABE Annual International Meeting, Boston, MA, USA, 7–10 July 2019.
21. Rodrigues, V.C.; da Silva, I.J.O.; Vieira, F.M.C.; Nascimento, S.T. A correct enthalpy relationship as thermal comfort index for livestock. *Int. J. Biometeorol.* **2011**, *55*, 455–459. [CrossRef] [PubMed]
22. Yunus, A.C.D.; Afshin, J.G. *Heat and Mass Transfer: Fundamentals and Applications*; McGraw-Hill Education: New York, NY, USA, 2014.
23. Modest, M.F. *Radiative Heat Transfer*; Academic Press: Cambridge, MA, USA, 2013.
24. Marple, D.N.; Jones, D.J.; Alliston, C.W.; Forrest, J.C. Physiological and endocrinological changes in response to terminal heat stress in swine. *J. Anim. Sci.* **1974**, *39*, 79–82. [CrossRef]
25. Patience, J.; Chaplin, R. Physiological and nutritional effects of heat-stress in the pig. *FASEB J.* **1991**, *5*, A768.
26. Patience, J.F.; Umboh, J.F.; Chaplin, R.K.; Nyachoti, C.M. Nutritional and physiological responses of growing pigs exposed to a diurnal pattern of heat stress. *Livest. Prod. Sci.* **2005**, *96*, 205–214. [CrossRef]
27. Ross, J.W.; Hale, B.J.; Gabler, N.K.; Rhoads, R.P.; Keating, A.F.; Baumgard, L.H. Physiological consequences of heat stress in pigs. *Anim. Prod. Sci.* **2015**, *55*, 1381–1390. [CrossRef]
28. Ross, J.W.; Hale, B.J.; Seibert, J.T.; Romoser, M.R.; Adur, M.K.; Keating, A.F.; Baumgard, L.H. Physiological mechanisms through which heat stress compromises reproduction in pigs. *Mol. Reprod. Dev.* **2017**, *84*, 934–945. [CrossRef] [PubMed]
29. Bond, T.E.; Kelly, C.F.; Heitman, A.H., Jr. Hog house air conditioning and ventilation data. *Trans. ASAE* **1959**, *2*, 0001–0004. [CrossRef]
30. Bianca, W. The signifiance of meterology in animal production. *Int. J. Biometeorol.* **1976**, *20*, 139–156. [CrossRef]
31. Parsons, K. *Human Thermal Environments: The Effects of Hot, Moderate, and Cold Environments on Human Health, Comfort, and Performance*; CRC Press, Inc.: Boca Raton, FL, USA, 2014.
32. Beckett, F.E. Effective temperature for evaluating or designing hog environments. *Trans. ASAE* **1965**, *8*, 0163–0166. [CrossRef]
33. Bergman, T.L.; Lavine, A.S.; Incropera, F.P.; DeWitt, D.P. *Fundamentals of Heat and Mass Transfer*; John Wiley & Sons: Hoboken, NJ, USA, 2011.
34. Kreith, F.; Bohn, M. *Principles of Heat Transfer*; Cengage Learning: Boston, MA, USA, 2010.
35. Wang, X.S.; Gao, H.D.; Gebremedhin, K.G.; Bjerg, B.S.; Van Os, J.; Tucker, C.B.; Zhang, G.Q. A predictive model of equivalent temperature index for dairy cattle (ETIC). *J. Therm. Biol.* **2018**, *76*, 165–170. [CrossRef]
36. Li, H.; Rong, L.; Zhang, G. Study on convective heat transfer from pig models by CFD in a virtual wind tunnel. *Comput. Electron. Agric.* **2016**, *123*, 203–210. [CrossRef]
37. Wang, X.S.; Zhang, G.Q.; Choi, C.Y. Effect of airflow speed and direction on convective heat transfer of standing and reclining cows. *Biosyst. Eng.* **2018**, *167*, 87–89. [CrossRef]
38. Gravas, L. The exercise needs for tied and free-moving dry sows. *Appl. Anim. Ethol.* **1981**, *7*, 389–390. [CrossRef]

39. Massabie, P.; Granier, R. Effect of Air Movement and Ambient Temperature on the Zootechnical Performance and Behavior of Growing-Finishing Pigs. In Proceedings of the the 94th ASAE Annual International Meeting, Sacramento, CA, USA, 29 July–1 August 2001; p. DF-1.5.
40. Cao, M.B.; Zong, C.; Wang, X.S.; Teng, G.H.; Zhuang, Y.R.; Lei, K.D. Numerical simulations of airflow and convective heat transfer of a sow. *Biosyst. Eng.* **2020**, *200*, 23–39. [CrossRef]
41. Bruce, J.M.; Clark, J.J. Models of heat production and critical temperature for growing pigs. *Anim. Prod. Sci.* **1979**, *28*, 353–369. [CrossRef]
42. Collier, R.J.; Gebremedhin, K.G. Thermal Biology of Domestic Animals. *Annu. Rev. Anim. Biosci.* **2014**, *3*, 513–532. [CrossRef] [PubMed]
43. Hwang, R.L.; Lin, T.P.; Matzarakis, A. Seasonal effects of urban street shading on long-term outdoor thermal comfort. *Build. Environ.* **2011**, *46*, 863–870. [CrossRef]
44. Kantor, N.; Unger, J. The most problematic variable in the course of human-biometeorological comfort assessment—The mean radiant temperature. *Cent. Eur. J. Geosci.* **2011**, *3*, 90–100. [CrossRef]
45. Kaynakli, O.; Kilic, M. Investigation of indoor thermal comfort under transient conditions. *Build. Environ.* **2005**, *40*, 165–174. [CrossRef]
46. Langner, M.; Scherber, K.; Endlicher, W.R. Indoor heat stress: An assessment of human bioclimate using the UTCI in different buildings in Berlin. *ERDE* **2013**, *144*, 260–273.
47. Matzarakis, A.; Amelung, B. Physiological Equivalent Temperature as Indicator for Impacts of Climate Change on Thermal Comfort of Humans. In *Seasonal Forecasts, Climatic Change and Human Health*; Springer: Berlin/Heidelberg, Germany, 2008.
48. Olesen, B.W.; Parsons, K.C. Introduction to thermal comfort standards and to the proposed new version of EN ISO 7730. *Energy Build.* **2002**, *34*, 537–548. [CrossRef]
49. Mellado, M.; Gaytán, L.; Macías-Cruz, U.; Avendaño, L.; Meza-Herrera, C.; Lozano, E.A.; Rodríguez, Á.; Mellado, J. Effect of climate and insemination technique on reproductive performance of gilts and sows in a subtropical zone of Mexico. *Austral. J. Vet. Sci.* **2018**, *50*, 27–34. [CrossRef]
50. Bjerg, B.; Brandt, P.; Pedersen, P.; Zhang, G. Sows' responses to increased heat load—A review. *J. Therm. Biol.* **2020**, *94*, 102758. [CrossRef]
51. Cabezón, F.; Schinckel, A.P.; Marchant-Forde, J.N.; Johnson, J.S.; Stwalley, R.M. Effect of floor cooling on late lactation sows under acute heat stress. *Livestock Sci.* **2017**, *206*, 113–120. [CrossRef]
52. Li, W.; Yang, X.D.; Dai, A.N.; Chen, K. Sleep and Wake Classification Based on Heart Rate and Respiration Rate. *IOP Conf. Ser. Mater. Sci. Eng.* **2018**, *428*, 012017. [CrossRef]
53. Cheng, K.S.; Lee, P.F. A Physiological/Model Study on the Effects of Deep Breathing on the Respiration Rate, Oxygen Saturation, and Cerebral Oxygen Delivery in Humans. *Neurophysiology* **2018**, *50*, 351–356. [CrossRef]

Article

Modeling of Heat Stress in Sows Part 2: Comparison of Various Thermal Comfort Indices

Mengbing Cao [1,2], Chao Zong [1,2,*], Yanrong Zhuang [1,2], Guanghui Teng [1,2,*], Shengnan Zhou [1,2] and Ting Yang [1,2]

1 College of Water Resources and Civil Engineering, China Agricultural University, Beijing 100083, China; mengbing-cao@cau.edu.cn (M.C.); zyr123@cau.edu.cn (Y.Z.); zhoushengnan@cau.edu.cn (S.Z.); sy20193091754@cau.edu.cn (T.Y.)
2 Key Laboratory of Agricultural Engineering in Structure and Environment, Ministry of Agriculture and Rural Affairs, Beijing 100083, China
* Correspondence: chaozong@cau.edu.cn (C.Z.); futong@cau.edu.cn (G.T.)

Simple Summary: Various thermal indices have been developed to evaluate the heat stress in animals. In this study, the temperature and humidity index (THI), effective temperature (ET), enthalpy (H), and the equivalent temperature index of sows (ETIS) from the part 1 of this paper series have been reviewed and analyzed in the process of sow production. Four approaches have been proposed to analyze these commonly applied thermal indices: (1) equivalent temperature change method; (2) the method based on the change of thermal index under different wind speeds; (3) the psychrometric chart method; and (4) CFD simulation method. In the analysis among those thermal indices, the ETIS performed best in evaluating the sow's thermal environment, followed by THI2, THI4 and THI7. This research provides a theoretical basis for selecting an appropriate thermal index for thermal environment evaluations in the sow production.

Abstract: Heat stress has an adverse effect on the production performance of sows, and causes a large economic loss every year. The thermal environment index is an important indicator for evaluating the level of heat stress in animals. Many thermal indices have been used to analyze the environment of the pig house, including temperature and humidity index (THI), effective temperature (ET), equivalent temperature index of sows (ETIS), and enthalpy (H), among others. Different heat indices have different characteristics, and it is necessary to analyze and compare the characteristics of heat indices to select a relatively suitable heat index for specific application. This article reviews the thermal environment indices used in the process of sow breeding, and compares various heat indices in four ways: (1) Holding the value of the thermal index constant and analyzing the equivalent temperature changes caused by the relative humidity. (2) Analyzing the variations of ET and ETIS caused by changes in air velocity. (3) Conducting a comparative analysis of a variety of isothermal lines fitted to the psychrometric chart. (4) Analyzing the distributions of various heat index values inside the sow barn and the correlation between various heat indices and sow heat dissipation with the use of computational fluid dynamics (CFD) technology. The results show that the ETIS performs better than other thermal indices in the analysis of sows' thermal environment, followed by THI2, THI4, and THI7. Different pigs have different heat transfer characteristics and different adaptability to the environment. Therefore, based on the above results, the following suggestions have been given: The thermal index thresholds need to be divided based on the adaptability of pigs to the environment at different growth stages and the different climates in different regions. An appropriate threshold for a thermal index can provide a theoretical basis for the environmental control of the pig house.

Keywords: temperature and humidity index; black globe-humidity index; effective temperature; equivalent temperature index for sows; enthalpy

Citation: Cao, M.; Zong, C.; Zhuang, Y.; Teng, G.; Zhou, S.; Yang, T. Modeling of Heat Stress in Sows Part 2: Comparison of Various Thermal Comfort Indices. *Animals* **2021**, *11*, 1498. https://doi.org/10.3390/ani11061498

Academic Editors: Lilong Chai and Yang Zhao

Received: 4 April 2021
Accepted: 20 May 2021
Published: 21 May 2021

1. Introduction

High temperatures reduce a sow's estrus [1] and pregnancy rate [2,3], decrease milk production [4], cause weight loss during lactation [5], and increase mortality [6]. As the core driving force in pig farm production, sow production directly affects the number and quality of piglets, and further influences the overall performance of a farm. According to reports, heat stress can cause hundreds of millions of dollars in losses to pig farms [7]. With the advent of global warming, the heat stress issue on pig farms is receiving more attention [8].

Reasonable quantification of heat stress can provide guidance for effectively regulating the thermal environment of the pig house. The commonly used thermal indices are temperature and humidity index (THI) [9–16], the globe-humidity index (BGHI) [17,18], effective temperature (ET) [19], equivalent temperature index of sows (ETIS) [20], and enthalpy (H) [21]. Among them, THI has many varieties derived from the original, and eight types of THI index have commonly been used:

(1) THI1 was developed by Thorn [9] and used as a human comfort index by the US Meteorological Administration [17]. Sales et al. [22] used the THI1 environmental index to analyze the impact of thermal environment on the reproductive performance of sows. The thresholds of THI1 were divided as follows: heat comfort zone of $61 < \text{THI1} \leq 65$, mild heat stress of $65 < \text{THI1} \leq 69$, and heat stress zone of $69 < \text{THI1} \leq 73$.

(2) The thermal index THI2, developed by Thom [10], was also for human comfort investigations. Later, the index was also used to determine the degree of heat stress in cattle and pigs. Mader et al. [23,24] used THI2 to analyze the heat stress in cattle, while Vashi et al. [25] used it to analyze the changes of different hormones in pigs during different seasons to determine the adaptability of pigs to the environmental changes of different seasons. Godyn et al. [26] used THI2 to study the effect of the atomization system on the microclimate of the farrowing room and the effect this had on sow welfare. The respiratory rates and rectal and skin surface temperatures of sows in different environments were analyzed. Yosuke et al. [27] also studied the effects of THI2 and maximum temperature on the farrowing rate of sows. Mellado et al. [28] used THI2 to analyze the relationship between THI2 and the reproductive performance of sows. When $\text{THI2} < 74$, the pregnancy rate was 93%, and as THI2 further decreased, the pregnancy rate continued to increase. When $74 \leq \text{THI2} < 78$, the pregnancy rate was 91.8%, and the pregnancy rate in this interval was relatively stable. When $78 \leq \text{THI2} < 82$, the pregnancy rate was 91.4%, which was lower than before, and the pregnancy rate was relatively stable. When $\text{THI2} > 82$, the pregnancy rate of sows was 89.8%. When THI2 was lower than 82, the pregnancy rate continued to decline. Therefore, according to the research of Mellado et al., THI2 can be classified into certain heat stress thresholds: $\text{THI2} < 74$ indicates comfortable environment, $74 \leq \text{THI2} < 78$ indicates mild heat stress, $78 \leq \text{THI2} < 82$ indicates moderate heat stress, and $\text{THI2} \geq 82$ indicates severe heat stress.

(3) Ingram [11] determined the weight of wet bulb temperature using THI3 according to the degree of influence of relative humidity in the air on pigs. Wang [29] analyzed the correlation between the THI3 index and the behavioral physiological response of pigs. Studies have shown that when THI3 is greater than 28, heat stress responses, such as increased body temperature, will occur.

(4) Kelly et al. [12] proposed THI4 in their literature. Tummaruk et al. [30] used high temperature, high relative humidity and high THI4 to analyze the impact of the thermal environment on a sow's litter size. Pu [31] used this index to analyze the effects of thermal environment on pig feeding behavior and pig physiological indicators. However, the thermal threshold division of this index is still unclear.

(5) Maust et al. [13] used the index THI5 to analyze the effect of the comprehensive thermal environment on the performance of dairy cows during lactation. Lucas et al. [32] used THI5 to analyze the impact of the thermal environment on pigs and proposed that an evaporative cooling system could be a feasible and economical solution to the

heat stress in pigs. When the index value reached 75, it indicated a heat stress level, and a value of 83 was a dangerously high level [33].

(6) The National Weather Service Central Region (NWSCR) [16] of the United States proposed THI6, and a large number of scholars have used this index to analyze the impact of climate on the reproductive performance of sows. The threshold zone of the index [34] is: THI6 $\leq$ 74 means suitable environmental level, 74 < THI6 $\leq$ 78 means mild heat stress level, 78 < THI $\leq$ 84 means moderate heat stress level, THI > 84 means severe heat stress level.

(7) The National Oceanic and Atmospheric Administration (NOAA, 1976) of the United States [14] proposed THI7. Iida et al. [35] used this index to analyze the reproductive performance of sows and determine the impact of climate on sow production, but the index did not have a threshold division.

(8) Fehr et al. [15] used THI8 to analyze the effect of evaporative cooling on pigs, and used evaporative cooling to reduce the heat stress level in pigs, but the index did not have a good threshold division.

Aside from THI indices, the black globe-humidity index (BGHI) is another type of thermal index commonly applied in animal production, which is also a variant of THI1. Buffington et al. [17] used the THI1 index for analysis of the environment of cattle, and the dry bulb temperature in THI1 was replaced by the black globe temperature to reveal the effect of radiative heat transfer. Junior et al. [18] used BGHI to study the effect of high temperature environments on sow lactation rate in a conventional environment, a floor cooling environment, and a semi-outdoor environment.

The effective temperature (ET) was used by Beckett [36] to evaluate environmental conditions, and humidity and air velocity were integrated into this index to study the impact of the environment on animals. Beckett provided a chart to illustrate the combined effect of air temperature and humidity on pig growth. Bjarne et al. [19] used the research of Beckett [36], Ingram [11], and Roller et al. [37] to determine the relative humidity item of ET. It was supposed that when the air temperature got close to the temperature of the pig body, the cooling effect achieved by the ventilation would be gradually weakened. It was also assumed that the cooling effect was proportional to the power function of air velocity, and the power function of the cooling effect and air velocity was determined according to the relationship between air velocity and pig convective heat transfer coefficient studied by Li et al. [38]. Finally, ET was expressed through the terms of ambient temperature, relative humidity, and air velocity. Bjarne et al. analyzed the correlation between ET and pig heat flux, and the ET performed well in pigs weighing between 3.4 kg and 70 kg.

The equivalent temperature index of sows (ETIS) was established in the part 1 of this paper series [20], which looked at the combined effect of the heat transfer characteristics of sows, integrated air temperature, relative humidity, and air velocity on physiological characteristics of sows. Based on the correlation between ETIS and THI2, ETIS was divided into several thresholds. The ETIS was mainly used to assess the heat stress in sows.

Enthalpy is an important state parameter of the energy of the material system [39,40], and enthalpy has often been used together with the psychrometric chart when describing environmental conditions. Enthalpy includes the influence of temperature and relative humidity in the air, so it can also be used as an index to evaluate the environment. Rodrigues et al. [21] extended the enthalpy equation to a certain extent, and finally determined the enthalpy value as the equation H, which was used as the comfort index of livestock and poultry.

The above mentioned heat stress indices have all been used to assess the heat stress in sows, but most heat indices are not developed based on sows [41,42], of which ET is used to analyze pigs from 3.4 kg to 70 kg, and only ETIS was developed for sows. The heat stress thresholds of different heat indices are also different. Using different methods for establishing a thermal index will also lead to different evaluation effects [41,42]. Therefore, for different environments, the applied thermal index should be evaluated for better performance. In order to avoid improper use of the heat index, a comprehensive understanding and comparison of the heat indices should be carried out [41].

The heat index usually mainly includes temperature, relative humidity and air velocity. The air temperature is the dominant factor of the overall thermal environment. Relative humidity is an important factor affecting the thermal environment. Whether the change of relative humidity has a reasonable influence on the thermal index is usually analyzed by the equivalent temperature method [41,43]. Air velocity is also an important factor affecting the thermal environment, as the air velocity directly affects the convective heat transfer of pigs. The influence of air velocity on the thermal index requires further analysis. The psychrometric chart has often been used to compare the variation trend of different heat indices. Meanwhile, the heat indices have also been combined with CFD simulations [19,44] to analyze the environmental distributions in the pig house. Therefore, the correlation between the heat index around the sow at different locations and the heat transfer of the sow can be used as a method in assessing the applicability of the heat indices.

Therefore, this study uses the following methods to analyze various thermal indices: (1) The equivalent temperature change method; (2) variation trend of heat indices under different air velocities; (3) psychrometric charts; and (4) CFD simulation. This study provides a reference for the selection of a heat index to analyze a sow's environment. A reasonable heat index is a powerful basis for regulating and optimizing the environment.

2. Materials and Methods

2.1. Various Thermal Indices

When the air temperature is kept constant, different relative humidity will make animals have different heat stress responses, mainly because high humidity will inhibit latent heat dissipation. Different air velocity will make animals have different heat stress responses, mainly because the air velocity directly affects the convective heat transfer of animals. The expression of environmental conditions requires comprehensive consideration of multiple environmental factors, so the temperature and humidity index (THI) [9–16], black globe-humidity index (BGHI) [17,18], effective temperature (ET) [19], equivalent temperature index of sows (ETIS) [20], and enthalpy (H) [21] are often used to analyze the environment of the sow house. The equations for different thermal indices are shown in Table 1, and only some of the thermal indices have thresholds divisions.

Table 1. Various thermal indices. Because there are many types of THI, a suffix of 1–8 is added for distinction. Among the various thermal indices, only some have threshold divisions.

Index	Calculation Equation	Threshold under Different Heat Stress Levels	Year
THI1	$THI1 = T + 0.36 \cdot T_{wb} + 41.5$	Thermal comfort: 61 < THI1 ≤ 65, Intermediate: 65 < THI1 ≤ 69, Thermal stress: 69 < THI1 ≤ 73	1958
THI2	$THI2 = 0.8 \cdot T + \left(\frac{RH \cdot (T - 14.4)}{100}\right) + 46.4$	Suitable: THI2 < 74, Mild: 74 ≤ THI2 < 78, Moderate: 78 ≤ THI2 < 82, Severe: THI2 ≥ 82	1959
THI3	$THI3 = 0.65 \cdot T + 0.35 \cdot T_{wb}$	Heat stress: THI2 ≥ 28	1965
THI4	$THI4 = T^{\circ} - (0.55 - 0.0055 \cdot RH) \cdot (T^{\circ} - 58)$	-	1971
THI5	$THI5 = 0.72 \cdot T + 0.72 \cdot T_{wb} + 40.6$	Moderate: 75 ≤ THI5 < 78. Severe: THI5 ≥ 83	1972
THI6	$THI6 = (1.8 \cdot T + 32) - \left(0.55 \cdot \left(\frac{RH}{100}\right)\right) \cdot ((1.8 \cdot T + 32) - 58)$	Suitable: THI6 ≤ 74, Mild: 74 < THI6 ≤ 78, Moderate: 78 < THI6 ≤ 84, Severe: THI6 > 84	1976
THI7	$THI7 = T - (0.55 - 0.0055 \cdot RH) \cdot (T - 14.5)$	-	1976
THI8	$THI8 = 0.27 \cdot T + 1.35 \cdot T_{wb} + 34.07$	-	1983
BGHI	$BGHI = T_g + 0.36 \cdot T_{dp} + 41.5$	-	1981

Table 1. *Cont.*

Index	Calculation Equation	Threshold under Different Heat Stress Levels	Year
ET	$ET = T + 0.0015 \cdot (RH - 50) \cdot T + (-1.0 \cdot 42 - T \cdot (v^{0.66} - 0.2^{0.66}))$	-	2018
ETIS	$ETIS = T + 0.0006 \cdot (RH - 50) \cdot T - 0.3132 \cdot u^{0.6827} \cdot (38 - T) - 4.79 \cdot (1.0086 \cdot 38 - T) + 4.8957 \cdot 10^{-8} \cdot \left((38 + 273.15)^4 - (T + 273.15)^4\right)$	Suitable: ETIS < 33.1, Mild: 33.1 ≤ ETIS < 34.5, Moderate: 34.5 ≤ ETIS < 35.9, Severe: 35.9 ≤ ETIS	2021
H	$H = 1.006 \cdot T + \frac{RH}{Pm} \cdot 10^{\left(\frac{7.5 \cdot T}{273.3 + T}\right)} \cdot (71.28 + 0.052 \cdot T)$	-	2011

Note: T is the dry bulb temperature (°C); T° is the dry bulb temperature (°F); T_g is the black globe temperature (°C); T_{wb} is the wet bulb temperature (°C); T_{dp} is the dew point temperature (°C); RH is the relative humidity (%); H is the enthalpy ($kJ \cdot kg^{-1}$); and P_m is high mercury of barometric pressure (mmHg). THI is the temperature and humidity index. The numbers 1-8 after THI represent different forms of THI. BGHI is black globe-humidity index. ET is effective temperature, ETIS is equivalent temperature index of sows. H is enthalpy.

2.2. Definition of Equivalent Air Temperature Change

In order to analyze the performance of each thermal index in evaluating the environmental thermal effect, this study uses a comparison method called the equivalent temperature change to analyze the change of the thermal index due to humidity changes. To calculate the equivalent air temperature (T_{equ}) change of the other parameters, e.g., relative humidity changes, the heat index is held constant in the calculating process. Taking the change of relative humidity from 50% to 60% as an example, as shown in Equation (1), when the temperature T_1 is 25 °C, the relative humidity RH_1 is 50%, and 60% is RH_2. In order to ensure that THI remains constant before and after environmental changes, T_2 has been adjusted. The equivalent temperature change corresponding to this process is shown in Equation (2). The air temperatures selected are 25 °C, 30 °C, 35 °C, and 40 °C, and the relative humidity changes from 50% to 60%, and the corresponding equivalent temperature changes of the thermal index in different environments have been calculated [41]. The ET and ETIS indices contain the parameter of air velocity, so when comparing with other temperature and humidity indices, the air velocity in ET and ETIS is determined to be 1 $m \cdot s^{-1}$. The dew point temperature contained in other heat indices can be transformed by Equation (3) [45], the wet bulb temperature can be transformed by Equation (4) [46], and the black bulb temperature can be transformed by Equation (5) [47] to the air temperatures. The air temperature using degrees Fahrenheit involved in the heat index is transformed by Equation (6) to degrees Celsius.

$$\mathrm{THI}(T_1, RH_1) = \mathrm{THI}(T_2, RH_2) \tag{1}$$

$$T_{equ} = T_1 - T_2 \tag{2}$$

$$T_{dp} = (0.198 + 0.0017 \cdot T) \cdot RH + 0.84 \cdot T - 19.2 \tag{3}$$

$$T_{wb} = \frac{-5.86154 + 0.58174 \cdot T + 0.1485 \cdot RH - 0.00191 \cdot RH^2 + 1.01768 \cdot 10^{-5} \cdot RH^3}{1 + 0.0036 \cdot T - 9.79822 \cdot 10^{-5} \cdot T^2 + 9.26824 \cdot 10^{-7} \cdot T^3 - 0.00899 \cdot RH + 4.38111 \cdot 10^{-5} \cdot RH^2} \tag{4}$$

$$T_g = 0.567 \cdot T + 0.393 \cdot \left[\frac{RH}{100} \cdot 6.105 \cdot \exp\left(\frac{17.27 \cdot T}{237.7 + T}\right)\right] + 3.94 \tag{5}$$

$$T^{\circ} = 1.8 \cdot T + 32 \tag{6}$$

where T_1 is air temperature of 25 °C, 30 °C, 35 °C, or 40 °C. RH_1 is relative humidity of 50%. RH_2 is relative humidity of 60%. T_2 is the air temperature (°C) calculated by Equation (1). T_{equ} is the air temperature (°C) difference between T_1 and T_2. T is the dry bulb temperature (°C). T° is the dry bulb temperature in Fahrenheit degree (°F). T_g is the black globe temperature (°C). T_{wb} is the wet bulb temperature (°C). T_{dp} is the dew point temperature (°C). RH is the relative humidity (%).

2.3. *The Effect of Air Velocity on Effective Temperature and Equivalent Temperature Index of Sows*

In order to determine the changes of ET and ETIS under different air velocities, the air temperature is set to 25 °C, 30 °C, 35 °C, and 40 °C, the relative humidity to 60%, and the air velocity is changed from 0 $m{\cdot}s^{-1}$ to 4 $m{\cdot}s^{-1}$ [43].

When the thermal environment is winter, assuming a temperature of 10 °C, a relative humidity of 60%, and an air velocity of 0 $m{\cdot}s^{-1}$, ET and ETIS are 21.2 °C (ET_{winter}) and 18.6 °C ($ETIS_{winter}$), respectively. With the increasing airflow velocity, the intersection points of the ET and ETIS curves with ET_{winter} and $ETIS_{winter}$ are analyzed.

2.4. *Psychrometric Chart*

A psychrometric chart has often been used to determine the temperature and humidity conditions in livestock and poultry houses under different environmental conditions. The heat index contains the influence of temperature and humidity in the psychrometric chart, so each heat index can be looked up in the psychrometric chart to make comparisons. Taking the temperature of 22 °C and the relative humidity of 70% as the standard value for all the heat indices, keeping each heat index constant, when the temperature and relative humidity change, the constant-index line in the psychrometric chart can be drawn. The airflow velocity of ET and ETIS was set to 0 $m{\cdot}s^{-1}$, as the applied minimum ventilation and the blockage between sows make the air speed near the sow quite low.

2.5. *Computational Fluid Dynamics Analysis*

The CFD model in this investigation is simplified by having a pig house with an air inlet at one end and a mechanical exhaust on the other. There are a total of 10 sows in the pig house. It is assumed that the inlet temperature of 30 °C, the relative humidity of 60%, the sow body temperature of 38 °C, and the exhaust port velocity of 1 $m{\cdot}s^{-1}$ are used as boundary conditions, and the distributions of heat indices in the pig house are simulated by the CFD method. The maximum values of all contours using different thermal indices are a temperature of 34 °C, a relative humidity of 60%, and an air speed of 0 $m{\cdot}s^{-1}$, and the minimum values are a temperature of 30 °C, a relative humidity of 60%, and an air speed of 1 $m{\cdot}s^{-1}$. The average heat index value is calculated from the values at four points around the sow (up, down, left, and right, 0.1 m away from the sow). The correlation between the average heat index around sows at different locations and the convective heat transfer of the corresponding sows are analyzed.

2.5.1. Geometric Model

The geometric model is a simplified pig house. The length, width, and height of the pig house are 7.15 m × 3 m × 3 m, respectively. 10 cylinders with a diameter-to-length ratio of of 1:4 [38] are used to represent those sows. Assuming the sow is 200 kg, the area of the sow is 2.775 m^2 [48,49]. To ensure that the environment inside the pig house is distinguishable, and to analyze the influence of different environments on the heat transfer of sows, the position of the air inlet is set higher than that of the sows, and the position of the air outlet is set lower than that of the sows. The details are shown in Figure 1.

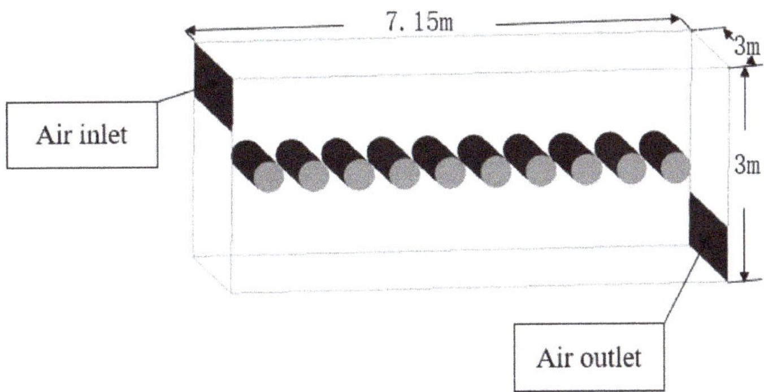

Figure 1. Computational fluid dynamics geometric model. The distance between the sows is 0.65 m. The length and width of the air inlet and air outlet are both 3 × 0.75 m.

2.5.2. Grid and Grid Independence Test

In this study, unstructured grids were used around the sow body, and structured grids were used in other areas. In order to ensure the boundary layer of the sow body [38,50], grid Y+ is kept less than 1. In this study, the first layer of grid is set to 0.07 mm. In order to minimize the influence of grid thickness, it is necessary to check the grid independence. Three resolutions of grids are built, namely fine grid (5, 139, 749), medium grid (4, 302, 437), and coarse grid (3, 016, 796). During simulation, the air velocity at the exhaust outlet is 1 m·s^{-1}, the inlet temperature is 30 °C, and the sow body temperature is 38 °C. The convective heat transfer of sows in the three grid resolutions have been calculated, and the values are 1576.63 W, 1583.02 W, and 1567.69 W for fine, medium, and coarse grid, respectively. The relative difference of the convective heat transfer coefficient between the cases of fine and medium grids is 0.4%, which indicates that the medium grid is sufficient to achieve grid convergence.

2.5.3. Boundary Condition Setting and Solution

The settings of boundary conditions are shown in Table 2. All simulations use the standard k-ε turbulence model [51]. The second-order precision SIMPLE algorithm has been chosen. Continuity, velocity, energy, turbulent kinetic energy (k), turbulent energy dissipation rate (ε), absolute residuals of heat transfer on the surface of sows, and absolute residuals of net flow at the inlet and outlet in the computational domain are monitored. When the heat flux on the body surface of the sow is less than 0.01% in 50 iterations and the net mass flow between the inlet and outlet is less than 10^{-4} kg·s^{-1}, the iteration is considered to be convergent.

Table 2. Boundary conditions of each surface in Computational fluid dynamics simulation.

Boundary	Boundary Condition
Outlet	Velocity outlet, air velocity is 1 m·s^{-1}
Inlet	Pressure outlet, the temperature is 30 °C relative humidity is 60%
Sow	No slip wall, temperature is 38 °C
Other (walls)	No slip wall, heat flux = 0

3. Results

3.1. Comparison Using Equivalent Temperature Change Method

The equivalent temperature change method is used to compare the 12 environmental thermal indices. Figure 2 shows the equivalent temperature change of the relative humidity rising from 50% to 60% at temperatures of 25 °C, 30 °C, 35 °C, and 40 °C. It can be seen from Figure 2 that, except for THI6, the equivalent temperatures of other thermal indices are all positive. At different air temperatures, the equivalent temperature changes of THI1, THI3, THI5, and ET caused by changes in relative humidity are different, but the changing trends are nearly parallel to each other. The change trends of THI2, THI4, and THI7 are similar. For H, THI8, and BGHI at an air temperature of 25 °C, the change of 10% relative humidity causes an equivalent temperature of more than 1.5 °C. For ETIS, the equivalent temperature change becomes larger as the air temperature rises from 25 °C to 40 °C.

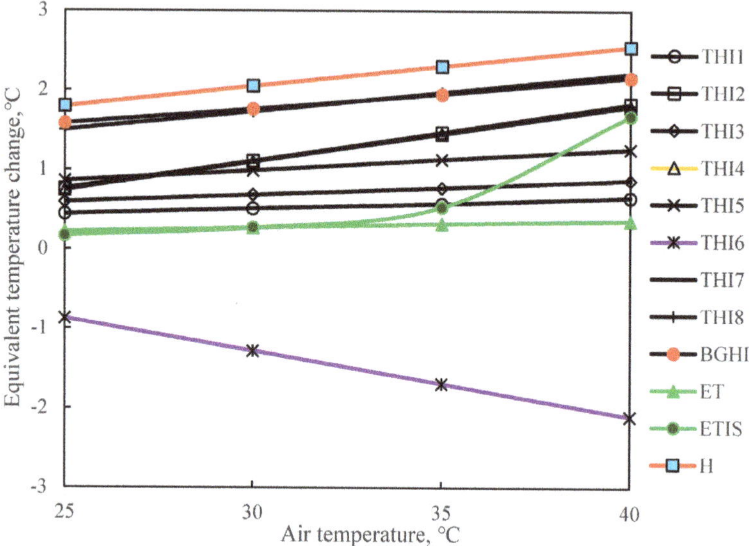

Figure 2. Equivalent temperature change. To calculate the equivalent air temperature (T_{equ}) change of the other parameters, e.g., relative humidity changes, the heat index is held constant in the calculating process. In this comparison, the air velocity of effective temperature and equivalent temperature index of sows is set as 1 m·s^{-1}. THI is the temperature and humidity index. The numbers 1–8 after THI represent different forms of THI, which are shown in Table 1. BGHI is black globe-humidity index. Et is effective temperature, ETIS is equivalent temperature index of sows. H is enthalpy.

3.2. *The Effect of Air Velocity*

Figure 3 shows the change of ET/ETIS with the change of air velocity. When the relative humidity is 60% and the air temperature is 20 °C, 30 °C, 35 °C, and 40 °C, the air velocity increases from 0 m·s^{-1} to 4 m·s^{-1} to analyze the changes in ET and ETIS. As the air velocity increases, ET curves show downward trends. When the ambient temperature is 25 °C and the air velocity increases from 0.4 m·s^{-1}–0.5 m·s^{-1}, the $ET_{25\,°C}$ line intersects the ET_{winter} line. When the ambient temperature is 30 °C, the intersects between the $ET_{30\,°C}$ line and ET_{winter} line occurs at a velocity of 1.1 m·s^{-1}–1.2 m·s^{-1}. For the ambient temperatures of 20 °C, 30 °C, and 35 °C, as the air velocity increases, ETIS curves also show downward trends, but those curves do not intersect with the $ETIS_{winter}$ line. For the ambient temperature of 40 °C, as the air velocity increases, ETIS shows an upward trend.

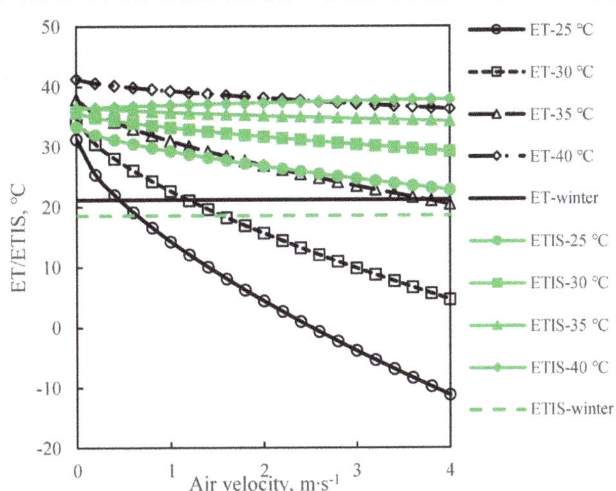

Figure 3. The change of effective temperature or equivalent temperature index of sows with the change of air velocity. When the relative air humidity is 60% and the air temperature is 20 °C, 30 °C, 35 °C, and 40 °C, the air velocity increases from 0 m·s^{-1} to 4 m·s^{-1} to analyze the changes in ET and ETIS. ET_{winter} and $ETIS_{winter}$ represent a winter climate with an air temperature of 10 °C, a relative humidity of 60%, and an air velocity of 0 m·s^{-1}(ET_{winter} = 21.2 °C, $ETIS_{winter}$ = 18.6 °C). With the increase of airflow velocity, a chilling effect will occur, and both ET and ETIS will decrease with the increasing air velocity. When the ET has been applied, there will be interlaces between ET_{winter} and ET of 25 °C, 30 °C, and 35 °C, which does not match the real conditions. In this situation, the ET performs inaccurately predicts the thermal conditions in the sow barn.

3.3. Comparison Using the Psychrometric Chart

Figure 4 shows the thermal indices fitted on the psychrometric chart. The heat index value of THI6 at an air temperature of 25.4 °C and a relative humidity of 100% is consistent with the heat index value of an air temperature of 22 °C and a relative humidity of 70%. When the air temperatures are 46 °C, 37.41 °C, and 31.6 °C for H, THI8, and BGHI, respectively, under a relative humidity of 0%, the calculated H, THI8, and BGHI are the same as those of H, THI8, and BGHI with an air temperature of 22 °C and a relative humidity of 70%. The results of THI2, THI4, and THI7 are similar. For the relative humidity changes from 0% to 70%, assuming the THI2 constant, the air temperature will change from 22 °C to 28.4 °C. In THI5, the temperature varies greatly under different humidities. There is little difference in the performance between THI1, THI3, and ET. The heat index value of ETIS at an air temperature of 23.34 °C and a relative humidity of 100% is the same as the value under an air temperature of 22 °C and a relative humidity of 70%. It is important to note that the ETIS curve and the constant-temperature line in the psychrometric chart are different.

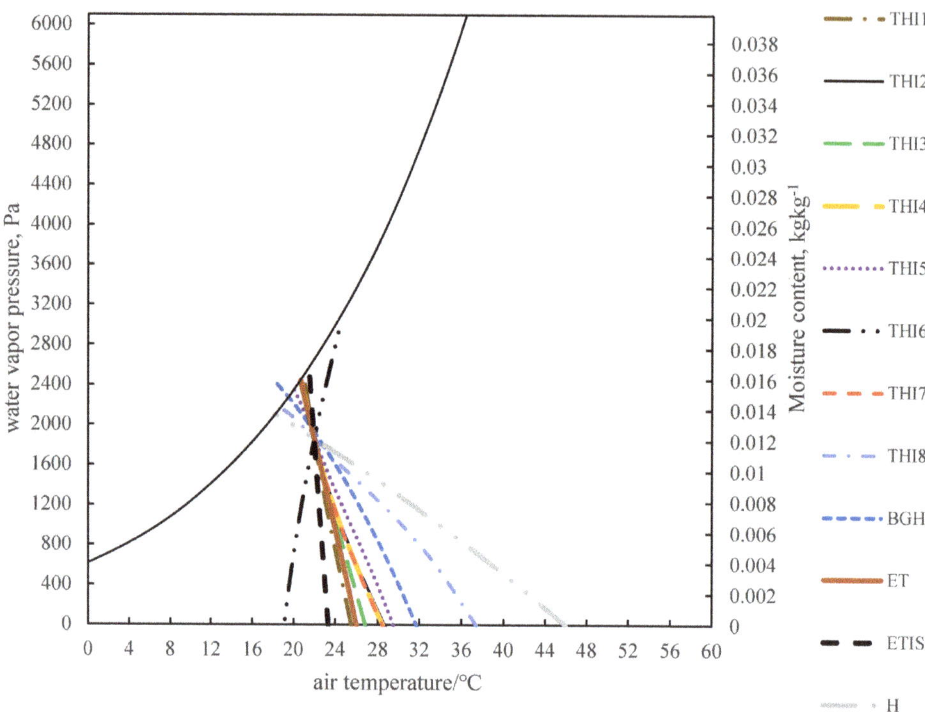

Figure 4. The patterns of the thermal indices in the psychrometric chart. All heat indices are consistent with the heat index value when the air temperature is 22 °C and the relative humidity is 70%. In this way, the iso-index lines of various heat indices are drawn. When compared with other thermal indices, the air velocity of effective temperature and equivalent temperature index of sows is taken as 0 $m \cdot s^{-1}$. THI is the temperature and humidity index. The numbers 1–8 after THI represent different forms of THI, which are shown in Table 1. BGHI is black globe-humidity index. ET is effective temperature, ETIS is equivalent temperature index of sows. H is enthalpy.

3.4. Comparison Using Computational Fluid Dynamics Methods

Figure 5 shows the distributions of the thermal index values. When using THI2, THI4, and THI7 to analyze the thermal conditions at different zones in the pig barn, it has been found that the patterns of all the three indices are the same, with a lower index value at the air inlet area and a higher value near the animal occupied zone, which indicates that there is a difference of thermal distribution in the pig house. The distribution patterns of THI1, THI3, and THI5 are similar. The index values of THI1, THI3, and THI5 in the air inlet area are lower than the THI2 value. The difference between THI8 and BGHI distributions is not clear. Both animal occupied zone and animal non-occupied zone have a high THI6 index. For ET and ETIS, as they include the influence of air velocity, they have better performances than other indices. The maximum values of all contours using different thermal indices are a temperature of 34 °C, a relative humidity of 60%, and an air speed of 0 $m \cdot s^{-1}$, and the minimum values are a temperature of 30 °C, a relative humidity of 60%, and an air speed of 1 $m \cdot s^{-1}$, while the sow is at 38 °C. In the contours of H, the color around the sow is lighter, indicating that the H index cannot conspicuously display the thermal conditions around the sow.

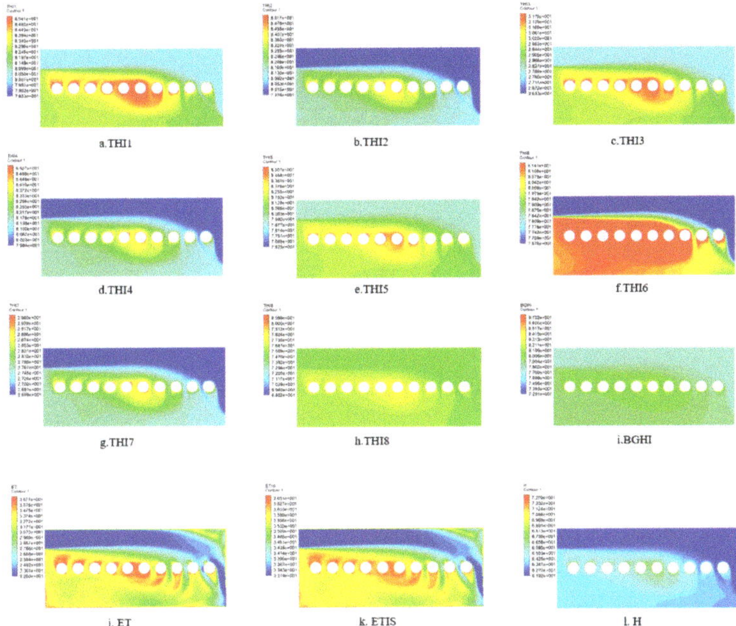

Figure 5. The distribution of various heat indices in the cross section of the pig barn. THI is the temperature and humidity index. The numbers 1–8 after THI represent different forms of THI, which are shown in Table 1. BGHI is black globe-humidity index. ET is effective temperature. ETIS is equivalent temperature index of sows. H is enthalpy.

Table 3 shows the correlation between the heat index and the heat dissipation of sows. ET and ETIS have the strongest correlation with the heat dissipation of sow among those analyzed indices. The THI2, THI4, and THI7 performs better than other THI indices, and these three indices have coefficients of determination higher than 0.8349. The THI1, THI3, and THI5 are significantly correlated with the heat transfer of the sow, and R^2 is at least 0.826. THI1, THI3, and THI5 perform relatively poorly in predicting the heat exchange of a sow compared with THI2, THI4, and THI7. The correlation between H and THI6 and the heat transfer of sows are 0.8235 and 0.8189, respectively. However, THI8 and BGHI performed the worst in predicting heat transfer of sows.

Table 3. The Coefficient of Determination (R^2) between the sow's heat transfer and heat index.

Heat Index	The Coefficient of Determination (R^2)
THI1	0.8261
THI2	0.8351
THI3	0.8264
THI4	0.8349
THI5	0.8267
THI6	0.8189
THI7	0.835
THI8	0.7575
BGHI	0.8005
ET	0.9883
ETIS	0.9863
H	0.8235

Note: THI is the temperature and humidity index. The numbers 1-8 after THI represent different forms of THI, which are shown in Table 1. BGHI is black globe-humidity index. Et is effective temperature, ETIS is equivalent temperature index of sows. H is enthalpy.

4. Discussion

4.1. Comparison Using Equivalent Temperature Change Method

In a hot environment, an increase in air temperature will cause the sow's respiration rate to increase, which removes a large amount of latent heat. The increase in humidity negatively impacts the latent heat dissipation, and the increase in relative humidity will increase the heat stress in sows. However, THI6 shows a downward trend as the relative humidity increases, which is inconsistent with reality. Previous studies have shown that at an air temperature of 30 °C, a 40% increase in relative humidity is equivalent to an increase of 1.9 °C in air temperature [52]. When H, THI8, and BGHI are at an air temperature of 30 °C, the change of 10% relative humidity can cause an increase in temperature of more than 1.73 °C, which is inconsistent with previous research. Previous studies have shown that when the temperature is higher than 32 °C, the possibility of cardiac failure in sows increases [6,53]. Studies have also shown that 32 °C is the upper critical ambient temperature for sows [54]. When ETIS is between 30 °C and 35 °C, the equivalent temperature changes drastically. When the air temperature is between 35 °C and 40 °C, as the relative humidity increases, the equivalent temperature of ETIS changes the most. The equivalent temperature performance of ETIS is more in line with previous studies [6,53]. Other indices did not show this effect. Therefore, compared with other heat indices, ETIS is more appropriate in predicting heat stress in sows.

4.2. The Effect of Air Velocity

When the air velocity is 0.4 $m{\cdot}s^{-1}$–0.5 $m{\cdot}s^{-1}$, the $ET_{25\,°C}$ curve and ET_{winter} curve intersect, which means the values of $ET_{25\,°C}$ and ET_{winter} are the same, but the skin surface feelings under 25 °C and in winter are different. This also applies to the $ET_{30\,°C}$ curve, which intersects with the ET_{winter} curve at an air velocity of 1.1 $m{\cdot}s^{-1}$–1.2 $m{\cdot}s^{-1}$. In summer, the ETIS won't decrease as much as the ET does when the air velocity increases. When the air temperature is 40 °C, ET decreases as the air velocity increases. An air temperature of 40 °C is already higher than the core temperature of a sow. Based on the theory of heat transfer [55,56], when the air temperature is higher than the object temperature, the object is in a heated state, and increasing the airflow speed will cause more convective heat transfer from ambient air to the pig body. When the air temperature is 40 °C, ETIS increases as the air velocity increases. When taking the effect of air velocity into account, the ETIS is more appropriate than ET for predicting heat stress in sows.

4.3. Comparison Using Psychrometric Chart

In the psychrometric chart, the iso-line of THI6 under high temperature and high humidity is consistent with low temperature and low humidity, which obviously does not comport with reality. The main reason is that the relative humidity term in THI6 is a decreasing function, and the air temperature term is an increasing function. In order to keep the value of THI6 unchanged, it is necessary to increase the air temperature while increasing the relative humidity. For H, THI8, and BGHI, in the case of low relative humidity, higher air temperature can ensure that the heat index value is consistent with the heat index value of the air temperature of 22 °C and the relative humidity of 70%. This is also inconsistent with reality, because the temperature difference between 31.6 °C and 22 °C will be felt by animals. Assuming that the THI2 is constant, when the relative humidity changes from 0% to 70%, the air temperature will change from 22 °C to 28.4 °C. The change in air temperature is 6.4 °C, which can be felt by the sow. The temperature difference of the iso-ETIS line under different relative humidities is small, so ETIS neither completely depends on the air temperature, nor does it amplify the influence of relative humidity.

4.4. Comparison Using Computational Fluid Dynamics Method

ET and ETIS have the best predictions of environmental distribution and convective heat transfer of sows, mainly because ET and ETIS include the parameter of air velocity. As CFD mainly analyzes the impact of airflow on sows, the air velocity has a greater impact on the sow's convective heat transfer coefficient than other parameters. H and THI include the influence of air temperature and relative humidity, but ignore the influence of air velocity. Therefore, the predictive effect of THI and H is lower than that of ET and ETIS. Due to different ratios and combinations of related parameters of air temperature and relative humidity, different THIs have slightly different prediction effects on environmental distribution. In addition, BGHI performs poorly in prediction, mainly because BGHI is the ratio of various parameters and considers radiation, but the CFD analysis does not consider the influence of radiation heat transfer.

4.5. Analysis of Existing Problems and Future Prospects

The ETIS performed better in the above-mentioned comparisons than other indices. This result is consistent with the result of Part 1 of the paper series [20]. It indicates that in the process of establishing the heat index, it is extremely important to select the heat transfer characteristics of sows as the main reference. The best performing THIs were THI2, THI4, and THI7, but this type of index does not include the impact of important factors such as air velocity. Although ET also performed well in the CFD analysis, ET showed unreasonable changes with the air velocity changing.

Although ETIS performs well in the comparison using various methods, it still has the following problems: (1) The heat transfer characteristics of pigs at different stages are different. Large pigs are afraid of heat and small pigs are afraid of cold. Different sizes and shapes will have different characteristic lengths [20,56,57]. Therefore, the ETIS threshold should be adjusted according to the growth stage of the pig. (2) Animals have different adaptability to different environments. In different climate zones, different species of animals have different comfort requirements for the thermal environment [57]. Different regions on the earth have different climatic characteristics. Therefore, ETIS thresholds in different regions will be different.

5. Conclusions

The following conclusions have been drawn:

1. It is better to use ETIS to analyze the heat stress in sows than using other thermal indices.
2. When the sow house only has temperature and humidity sensors and lacks air speed data, THI2, THI4 and THI7 can also be used to evaluate the sow's thermal comfort index.
3. The threshold division of each index still needs to be improved, and the thermal stress threshold is an important parameter for evaluating the degree of heat stress. However, the heat dissipation characteristics of sows at different stages are different, and in different climate zones, different breeds of animal have different comfort requirements for the thermal environment. Therefore, threshold zones should be established according to the characteristics of sows at different stages and different climatic conditions.

Author Contributions: Conceptualization, M.C. and C.Z.; methodology, Y.Z.; software, M.C.; validation, M.C., C.Z., and Y.Z.; formal analysis, M.C.; investigation, M.C.; resources, S.Z.; data curation, S.Z.; writing—original draft preparation, M.C.; writing—review and editing, C.Z.; visualization, M.C.; supervision, T.Y.; project administration, C.Z.; funding acquisition, G.T. All authors have read and agreed to the published version of the manuscript.

Funding: This research was funded by the National Key Research and Development Plan of China (2016YFD0700204).

Institutional Review Board Statement: Not applicable.

Informed Consent Statement: Not applicable.

Data Availability Statement: All data sets during the current study are available from the corresponding author on fair request.

Conflicts of Interest: The authors declare no conflict of interest. The funders had no role in the design of the study; in the collection, analyses, or interpretation of data; in the writing of the manuscript, or in the decision to publish the results.

References

1. Almond, P.K.; Bilkei, G. Seasonal infertility in large pig production units in an Eastern-European climate. *Aust. Vet. J.* **2005**, *83*, 344–346. [CrossRef]
2. Bloemhof, S.; Mathur, P.K.; Knol, E.F.; van der Waaij, E.H. Effect of daily environmental temperature on farrowing rate and total born in dam line sows. *J. Anim. Sci.* **2013**, *91*, 2667–2679. [CrossRef] [PubMed]
3. Tummaruk, P.; Tantasuparuk, W.; Techakumphu, M.; Kunavongkrit, A. Effect of Season and Outdoor Climate on Litter Size at Birth in Purebred Landrace and Yorkshire Sows in Thailand. *J. Vet. Med. Sci.* **2004**, *66*, 477–482. [CrossRef]
4. Black, J.L.; Mullan, B.P.; Lorschy, M.L.; Giles, L.R. Lactation in the sow during heat stress. *Livest. Prod. Sci.* **1993**, *35*, 153–170. [CrossRef]
5. Quiniou, N.; Noblet, J. Influence of high ambient temperatures on performance of multiparous lactating sows. *J. Anim. Sci.* **1999**, *77*, 2124. [CrossRef] [PubMed]
6. D'Allaire, S.; Drolet, R.; Brodeur, D. Sow mortality associated with high ambient temperatures. *Can. Vet. J. Rev. Vet. Can.* **1996**, *37*, 237–239.
7. St-pierre, N.R.; Cobanov, B.; Schnitkey, G. Economic Losses from Heat Stress by US Livestock Industries1. *J. Dairy Sci.* **2003**, *86*, 52–77. [CrossRef]
8. Mikovits, C.; Zollitsch, W.; Hörtenhuber, S.J.; Baumgartner, J.; Niebuhr, K.; Piringer, M.; Anders, I.; Andre, K.; Hennig-Pauka, I.; Schönhart, M. Impacts of global warming on confined livestock systems for growing-fattening pigs: Simulation of heat stress for 1981 to 2017 in Central Europe. *Int. J. Biometeorol.* **2019**, *63*, 1–10. [CrossRef]
9. Thom, E.C. Cooling degrees-days air conditioning, heating, and ventilating. *Trans. ASAE* **1958**, *55*, 65–72.
10. Thom, E.C. The discomfort index. *Weatherwise* **1959**, *12*, 57–61. [CrossRef]
11. Ingram, D.L. The effect of humidity on temperature regulation and cutaneous water loss in the young pig. *Res. Vet. Sci.* **1965**, *6*, 9. [CrossRef]
12. Bond, C.F.K.T.E. Bioclimatic factors and their measurements. In *A Guide to Environmental Research in Animals*; National Academy of Science Press: Washington, DC, USA, 1971.
13. Maust, L.E.; Mcdowell, R.E.; Hooven, N.W. Effect of Summer Weather on Performance of Holstein Cows in Three Stages of Lactation. *J. Dairy Sci.* **1972**, *55*, 1133–1139. [CrossRef]
14. NOAA (National Oceanic and Atmospheric Administration). *Livestock Hot Weather Stress*; US Department Commerce, National Weather Service Central Region: Orlando, FL, USA, 1976.
15. Fehr, R.L.; Priddy, K.T.; Mcneill, S.G.; Overhults, D.G. Limiting Swine Stress with Evaporative Cooling in the Southwest. *Trans. ASAE* **1983**, *26*, 0542–0545. [CrossRef]
16. NWSCR (National Weather Service Central Region). *Livestock Hot Weather Stress: Regional Operations Manual Letter*; NWSCR (National Weather Service Central Region): Kansas City, MO, USA, 1976.
17. Buffington, D.E.; Collazo-Arocho, A.; Canton, G.H.; Pitt, D.; Thatcher, W.W.; Collier, R.J. Black globe-humidity index (bghi) as comfort equation for dairy cows. *Trans. ASAE* **1981**, *24*, 0711–0714. [CrossRef]
18. Júnior, G.M.d.O.; Ferreira, A.S.; Oliveira, R.F.M.; Silva, B.A.N.; Figueiredo, E.M.d.; Santos, M. Behaviour and performance of lactating sows housed in different types of farrowing rooms during summer. *Livest. Sci.* **2011**, *141*, 194–201. [CrossRef]
19. Bjerg, B.; Rong, L.; Zhang, G. Computational prediction of the effective temperature in the lying area of pig pens. *Comput. Electron. Agric.* **2018**, *149*, 71–79. [CrossRef]
20. Cao, M.; Zong, C.; Wang, X.; Teng, G.; Zhuang, Y.; Lei, K. Modeling of Heat Stress in Sows. Part 1: Establishment of the Prediction Model for the Equivalent Temperature Index of the Sows. *Animals* **2021**, *11*, 1472. [CrossRef]
21. Rodrigues, V.C.; da Silva, I.J.O.; Vieira, F.M.C.; Nascimento, S.T. A correct enthalpy relationship as thermal comfort index for livestock. *Int. J. Biometeorol.* **2011**, *55*, 455–459. [CrossRef]
22. Sales, G.T.; Fialho, E.T.; Junior, T.Y.; Freitas, R.T.F.D.; Day, G.B. Thermal Environment Influence on Swine Reproductive Performance. In Proceedings of the Livestock Environment VIII, Iguassu Falls, Brazil, 31 August–4 September 2008.
23. Mader, T.L.; Kreikemeier, W.M. Effects of growth-promoting agents and season on blood metabolites and body temperature in heifers. *J. Anim. Sci.* **2006**, *84*, 1030. [CrossRef]
24. Mader, T.L.; Davis, M.S.; Brown-Brandl, T. Environmental factors influencing heat stress in feedlot cattle. *J. Anim. Sci.* **2006**, *84*, 712–719. [CrossRef]
25. Vashi, Y.; Naskar, S.; Chutia, T.; Banik, S.; Singh, A.K.; Goswami, J.; Sejian, V. Comparative assessment of native, crossbred and exotic pigs during different seasons (winter, spring and summer) based on rhythmic changes in the levels of serum cortisol, lactate dehydrogenase levels and PBMC HSP70 mRNA expression pattern. *Biol. Rhythm Res.* **2018**, *49*, 725–734. [CrossRef]
26. Godyn, D.; Herbut, P.; Angrecka, S. Impact of Fogging System on Thermal Comfort of Lactating Sows. *Trans. ASABE* **2018**, *61*, 1933–1938. [CrossRef]
27. Yosuke, S.; Madoka, F.; Shingo, N.; Tadahiro, K. Quantitative assessment of the effects of outside temperature on farrowing rate in gilts and sows by using a multivariate logistic regression model. *Anim. Sci. J.* **2018**, *89*, 1187–1193.

28. Mellado, M.; Gaytán, L.; Macías-Cruz, U.; Avendaño, L.; Meza-Herrera, C.; Lozano, E.A.; Rodríguez, Á.; Mellado, J. Effect of climate and insemination technique on reproductive performance of gilts and sows in a subtropical zone of Mexico. *Austral J. Vet. Sci.* **2018**, *50*, 27–34. [CrossRef]
29. Wang, K. *Study on the Relationship between the Humid and Hot Pig Housing Environment and the Thermoregulatory Behavior of Growing Swine and Sows in Southeast China*; Zhejiang University: Hangzhou, China, 2001.
30. Tummaruk, P.; Tantasuparuk, W.; Techakumphu, M.; Kunavongkrit, A. Seasonal influences on the litter size at birth of pigs are more pronounced in the gilt than sow litters. *J. Agric. Sci.* **2010**, *148*, 421–432. [CrossRef]
31. Pu, H. *Effect of Hot-Humid Environment on Feeding Behavior, Physiological and Biochemical Parameters of Swine*; Sichuan Agricultural University: Ya'an, China, 2014.
32. Lucas, E.M.; Randall, J.M.; Meneses, J.F. Potential for Evaporative Cooling during Heat Stress Periods in Pig Production in Portugal (Alentejo). *J. Agric. Eng. Res.* **2000**, *76*, 363–371. [CrossRef]
33. Vitt, R.; Weber, L.; Zollitsch, W.; Hrtenhuber, S.J.; Baumgartner, J.; Niebuhr, K.; Piringer, M.; Anders, I.; Andre, K.; Hennig-Pauka, I. Modelled performance of energy saving air treatment devices to mitigate heat stress for confined livestock buildings in Central Europe. *Biosyst. Eng.* **2017**, *164*, 85–97. [CrossRef]
34. Wu, Z.; Chen, Z.; Zang, J.; Wang, M.; Yang, H.; Ren, F.; Liu, J.; Feng, G. Cooling performance of wet curtain fan-fabric duct ventilation system in house of pregnant sows. *Trans. Chin. Soc. Agric. Eng.* **2018**, *34*, 268–276.
35. Iida, R.; Koketsu, Y. Quantitative associations between outdoor climate data and weaning-to-first-mating interval or adjusted 21-day litter weights during summer in Japanese swine breeding herds. *Livest. Sci.* **2013**, *152*, 253–260. [CrossRef]
36. Beckett, F.E. Effective temperature for evaluating or designing hog environments. *Trans. ASAE* **1965**, *8*, 0163–0166. [CrossRef]
37. Roller, W.L.; Goldman, R.F. Response of swine to acute heat exposure. *Trans. ASAE* **1969**, *12*, 0164–0169.
38. Li, H.; Rong, L.; Zhang, G.Q. Study on convective heat transfer from pig models by CFD in a virtual wind tunnel. *Comput. Electron. Agric.* **2016**, *123*, 203–210. [CrossRef]
39. William, C.; Reynolds, P.C. *Thermodynamics: Fundamentals and Engineering Applications*; Cambridge University Press: Cambridge, UK, 2018.
40. Jorgensen, S.E. *Thermodynamics and Ecological Modelling (Environmental & Ecological (Math) Modeling)*; CRC Press: Boca Raton, FL, USA, 2018.
41. Wang, X.S.; Bjerg, B.S.; Choi, C.Y.; Zong, C.; Zhang, G.Q. A review and quantitative assessment of cattle-related thermal indices. *J. Therm. Biol.* **2018**, *77*, 24–37. [CrossRef] [PubMed]
42. Yan, G.; Li, H.; Shi, Z.; Wang, C. Research status and existing problems in establishing cow heat stress indices. *Trans. Chin. Soc. Agric. Eng.* **2019**, *35*, 226–233.
43. Mader, T.L.; Johnson, L.J.; Gaughan, J.B. A comprehensive index for assessing environmental stress in animals. *J. Anim. Sci.* **2010**, *88*, 2153–2165. [CrossRef]
44. Zhou, B.; Wang, X.S.; Mondaca, M.R.; Rong, L.; Choi, C.Y. Assessment of optimal airflow baffle locations and angles in mechanically-ventilated dairy houses using computational fluid dynamics. *Comput. Electron. Agric.* **2019**, *165*, 104930. [CrossRef]
45. Xu, X.; Yu, J.; Li, H.; Yang, L. Comparative study on calculation methods of dew-point temperature. *J. Meteorol. Environ.* **2016**, *32*, 107–111.
46. Xu, Z.; Zhang, Q.; Xu, Y. Study of the calculation method of wet bulb temperature. *Hebei J. Ind. Sci. Technol.* **2018**, *35*, 123–127.
47. Shao, L.; Meng, Q.; Lu, S.; Zhong, K. Simplified calculation method of outdoor wet bulb globe temperature for areas with low latitude and high latitude. *J. Chongqing Univ.* **2020**, *43*, 112–120.
48. Hoving, L.L.; Soede, N.M.; van der Peet-Schwering, C.M.; Graat, E.A.; Feitsma, H.; Kemp, B. An increased feed intake during early pregnancy improves sow body weight recovery and increases litter size in young sows. *J. Anim. Sci.* **2011**, *89*, 3542–3550. [CrossRef]
49. Brody, S.; Comfort, E.; Matthews, J.S. Further investigations on surface area, with special reference to its significance in energy metabolism. *MO Agric. Exp. Stn. Res. Bull.* **1928**, *115*, 8–60.
50. Wang, X.S.; Zhang, G.Q.; Choi, C.Y. Effect of airflow speed and direction on convective heat transfer of standing and reclining cows. *Biosyst. Eng.* **2018**, *167*, 87–89. [CrossRef]
51. Li, H.; Rong, L.; Zhang, G. Numerical study on the convective heat transfer of fattening pig in groups in a mechanical ventilated pig house. In Proceedings of the Cigr/agegen/cfd Symposium, Aarhus, Denmark, 26–29 June 2016.
52. Bjerg, B.; Brandt, P.; Pedersen, P.; Zhang, G. Sows' responses to increased heat load—A review. *J. Therm. Biol.* **2020**, *94*, 102758. [CrossRef] [PubMed]
53. Drolet, R.; D'Allaire, S.; Chagnon, M. Some observations on cardiac failure in sows. *Can. Vet. J. La Rev. Vet. Can.* **1992**, *33*, 325.
54. Wathes, C.M.; Charles, D.R. *Livestock Housing*; CAB International: Oxfordshire, UK, 1994.
55. Kreith, F.; Bohn, M. *Principles of Heat Transfer*; Cengage Learning: San Francisco, CA, USA, 2010.
56. Bergman, T.L.; Lavine, A.S.; Incropera, F.P.; DeWitt, D.P. *Fundamentals of Heat and Mass Transfer*; John Wiley & Sons: Hoboken, NJ, USA, 2011.
57. Yasmeen, S.; Liu, H. Evaluation of thermal comfort and heat stress indices in different countries and regions—A Review. *IOP Conf. Ser. Mater. Sci. Eng.* **2019**, *609*, 052037. [CrossRef]

Article

Evaluation of TiO_2 Based Photocatalytic Treatment of Odor and Gaseous Emissions from Swine Manure with UV-A and UV-C

Myeongseong Lee [1], Jacek A. Koziel [1,*], Wyatt Murphy [1], William S. Jenks [2], Baitong Chen [1], Peiyang Li [1] and Chumki Banik [1]

[1] Department of Agricultural and Biosystems Engineering, Iowa State University, Ames, IA 50011, USA; leefame@iastate.edu (M.L.); wyatt.murphy@jetinc.net (W.M.); baitongc@iastate.edu (B.C.); peiyangl@iastate.edu (P.L.); cbanik@iastate.edu (C.B.)
[2] Department of Chemistry, Iowa State University, Ames, IA 50011, USA; wsjenks@iastate.edu
* Correspondence: koziel@iastate.edu; Tel.: +1-515-294-4206

Simple Summary: Poor indoor air quality and gaseous emissions are undesirable side effects of livestock and poultry production. Gaseous emissions of odor, odorous volatile organic compounds (VOCs), ammonia (NH_3), hydrogen sulfide (H_2S), and greenhouse gases (GHGs) have detrimental effects on the quality of life in rural communities, the environment, and climate. Proven mitigation technologies are needed to increase the sustainability of animal agriculture. This study's objective was to evaluate the ultraviolet (UV) light treatment of odor and common air pollutant emissions from stored swine manure on a pilot-scale. To our knowledge, this is the first study of this scope that was needed for scaling up technologies treating gaseous emissions of odor, odorous VOCs, NH_3, H_2S, ozone, and GHGs. The study bridged the knowledge gap between lab-scales and simplified treatment of model gases to the treatment of complex gaseous mixtures emitted from swine manure in fast-moving air. The manure emissions were treated in fast-moving air using a mobile lab equipped with UV-A and UV-C lights and photocatalytic surface coating. The percent reduction of targeted gases depended on the UV dose and wavelength. While generally mitigating targeted gases, some UV treatments resulted in CO_2 and ozone (O_3). The results proved that the UV technology was sufficiently effective in treating odorous gases, and the mobile lab was ready for farm-scale trials. The UV technology can be considered for the scaled-up treatment of emissions and air quality improvement inside livestock barns.

Citation: Lee, M.; Koziel, J.A.; Murphy, W.; Jenks, W.S.; Chen, B.; Li, P.; Banik, C. Evaluation of TiO_2 Based Photocatalytic Treatment of Odor and Gaseous Emissions from Swine Manure with UV-A and UV-C. *Animals* **2021**, *11*, 1289. https://doi.org/10.3390/ani11051289

Academic Editors: Lilong Chai and Yang Zhao

Received: 24 February 2021
Accepted: 28 April 2021
Published: 30 April 2021

Abstract: It is essential to mitigate gaseous emissions that result from poultry and livestock production to increase industry sustainability. Odorous volatile organic compounds (VOCs), ammonia (NH_3), hydrogen sulfide (H_2S), and greenhouse gases (GHGs) have detrimental effects on the quality of life in rural communities, the environment, and climate. This study's objective was to evaluate the photocatalytic UV treatment of gaseous emissions of odor, odorous VOCs, NH_3, and other gases (GHGs, O_3—sometimes considered as by-products of UV treatment) from stored swine manure on a pilot-scale. The manure emissions were treated in fast-moving air using a mobile lab equipped with UV-A and UV-C lights and TiO_2-based photocatalyst. Treated gas airflow (0.25–0.76 $m^3 \cdot s^{-1}$) simulates output from a small ventilation fan in a barn. Through controlling the light intensity and airflow, UV dose was tested for techno-economic analyses. The treatment effectiveness depended on the UV dose and wavelength. Under UV-A (367 nm) photocatalysis, the percent reduction of targeted gases was up to (i) 63% of odor, (ii) 51%, 51%, 53%, 67%, and 32% of acetic acid, propanoic acid, butanoic acid, *p*-cresol, and indole, respectively, (iii) 14% of nitrous oxide (N_2O), (iv) 100% of O_3, and 26% generation of CO_2. Under UV-C (185 + 254 nm) photocatalysis, the percent reductions of target gases were up to (i) 54% and 47% for *p*-cresol and indole, respectively, (ii) 25% of N_2O, (iii) 71% of CH_4, and 46% and 139% generation of CO_2 and O_3, respectively. The results proved that the UV technology was sufficiently effective in treating odorous gases, and the mobile lab was ready for farm-scale trials. The UV technology can be considered for the scaled-up treatment of emissions and air quality improvement inside livestock barns. Results from this study are needed to inform the experimental design for future on-farm research with UV-A and UV-C.

Keywords: air pollution control; air quality; volatile organic compounds; odor; environmental technology; advanced oxidation; UV

1. Introduction

Poor indoor air quality and gaseous emissions are undesirable side effects of livestock and poultry production. Gaseous emissions of odor, odorous volatile organic compounds (VOCs), ammonia (NH_3), hydrogen sulfide (H_2S), and greenhouse gases (GHGs) have detrimental effects on the quality of life in rural communities, the environment, and climate. Proven mitigation technologies are needed to increase the sustainability of animal agriculture. The farm-scale readiness and the effectiveness of technologies for mitigation of gaseous emissions from livestock agriculture are summarized by Maurer et al. [1]. The user-friendly description of technologies and the scientific literature database is provided by the Iowa State University Extension and Outreach website [2].

Photocatalysis with UV (ultraviolet) light has received considerable attention for special applications in indoor air quality. However, the research of UV photocatalysis in livestock agriculture applications is still limited. Costa et al. [3] and Guarino et al. [4] pioneered UV-A photocatalysis in swine weaning and farrowing units reporting mitigation of NH_3, GHGs, particulate matter (PM), and increased feed conversion efficiency. However, the previous research's technical design information, such as light dose and photocatalyst coating thickness required for application to actual farms, was not provided. Our team has been motivated by these early examples of farm-scale applications in Europe to conduct lab-to-farm-scale research to scale up and adapt UV photocatalysis to the swine farming systems prevailing in the American swine industry.

Several lessons were learned from the lab-scale to the pilot-scale progression of research. Zhu et al. [5] showed that a TiO_2 based photocatalytic coating (PureTi, Cincinnati, OH, USA) is sufficient to effectively mitigate odorous VOCs. Research showing the reduction of NH_3 and odorous VOCs with UV-C followed [6–8]. Recently, the application of UV-A photocatalysis for NH_3, odorous VOCs, ozone, and nitrous oxide (N_2O) was shown [9,10]. Testing UV-A photocatalysis (a safer bandwidth for direct human and animal exposure) showed a mitigation effect on a pilot-scale in the actual livestock farm environment [9,11].

Thus, earlier tests show practical percent reduction efficiencies for several targeted odorous air pollutants using marketed spray-on coatings for indoor building materials. Still, practical research questions must be addressed before the UV-A (or UV-C) technology can be adopted for farm-scale application to barn interiors to improve air quality inside livestock barns. Additionally, there is an interest in scaling up the UV treatment to mitigate emissions from the barn exhaust air. There is also an interest in UV-C applications to mitigate the risk of airborne pathogens from the ambient air, feed, supplies, personnel threatening farm biosecurity, and using UV to lower the pathogen load inside barns [12]. Thus, this research addresses the gap in knowledge to scale up UV-A and UV-C technology from proven performance mitigating two standard gases (NH_3, butan-1-ol) at a mobile lab-scale to the treatment of a much more complex mixture of gases released from swine manure. Comprehensive assessment of the mitigation effects for a wide range of gases is needed for scaling up technologies treating gaseous emissions of odor, odorous VOCs, NH_3, H_2S, ozone, and GHGs. This research aimed to scale up TiO_2-based photocatalysis treatment with UV-A and UV-C light to pilot-scale conditions. Specifically, the objective was to evaluate the percent reduction of gaseous emissions and investigate the required UV dose to mitigate the targeted odorous gases generated from swine manure, where the realistic mix of gases and aerosols was treated at fast-moving air and airflows consistent with those on production-scale farms. This study used a mobile UV laboratory designed and commissioned for testing with large (~1 m^3/s) airflows [13]. This study data obtained under simulated swine conditions with fast-moving airflows, like a real swine farm, is considered helpful in evaluating UV photocatalysis performance. Results from this study are needed to inform the experimental design for future on-farm research with UV-A and UV-C. Box 1 provides definitions of key acronyms used in this paper.

Box 1. Definitions of key acronyms used in this paper.

CH_4: methane
CO_2: carbon dioxide
DMDS: dimethyl disulfide
DMTS: dimethyl trisulfide
ECD: electron capture detector
FID: flame ionization detector
GC-MS: gas chromatograph-mass spectrometer
GHGs: greenhouse gases
H_2S: hydrogen sulfide
J: Joule (unit of energy)
LED: light-emitting diode
NIST: National Institute of Standards and Technology
NH_3: ammonia
N_2O: nitrous oxide
OU_E: odor unit
O_3: ozone
ppb: part per billion
ppm: part per million
SPME: solid-phase microextraction
TiO_2: titanium dioxide
UV-A: ultraviolet light (315–400 nm range)
UV-C: ultraviolet light (100–280 nm range)
VOCs: volatile organic compounds
W: watt (unit of power)

2. Materials and Methods

2.1. Materials and Methods

The mobile laboratory (7.2 × 2.4 × 2.4 m) designed and verified in the previous study [13] was used in this study. The mobile laboratory consisted of 12 chambers (7.2 × 0.9 × 2.4 m), and each chamber (0.53 × 0.9 × 2.4 m) was divided into vertical baffles. Chambers #11 and #12 were connected without a vertical baffle. Each chamber was equipped with 11 panels coated with TiO_2 (nanostructured TiO_2 anatase at 10 $\mu g/cm^2$ from PureTi, Cincinnati, OH, USA) on all sides. Two fans (I-Fan Type 40, Fancom, Panningen, the Netherlands) were installed on the mobile laboratory to control the airflow inside. The air velocity was measured with the anemometer fan (ATM, Fancom, Panningen, the Netherlands) installed in chamber #10, and the internal airflow can be controlled in real-time using the fan monitoring system (Lumina 20/21, Fancom, Panningen, the Netherlands) by controlling the two fans and the anemometer fan.

2.2. Generation of Odorous Gas Emissions from Swine Manure

A plastic drum (55 gal, ~200 L) filled with 35–40 gal of swine manure was used to generate a realistic mixture of odorous gases and aerosols and investigate UV photocatalysis performance (Figures 1 and A1). Compressed air was continuously supplied to the bottom of the manure (Figure S1), and the headspace gas was blended with ambient air. A filtration unit prevented the inflow of flies and dust into the UV mobile lab. Detailed information about the mobile laboratory and filter house has been reported in the previous study [13].

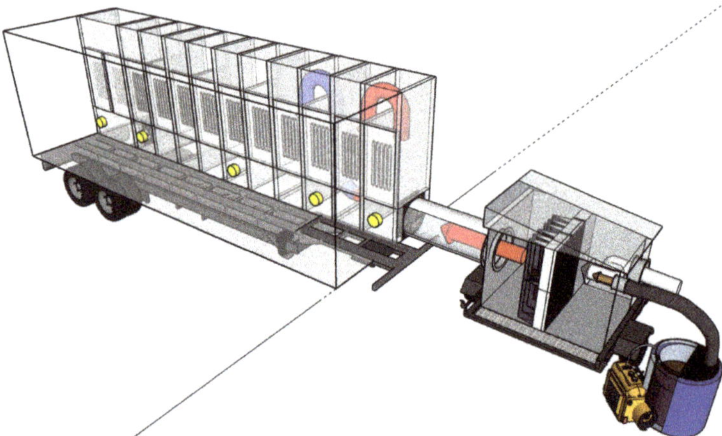

Figure 1. Schematic of mobile laboratory for UV treatment of gaseous emissions. UV dose is controlled by either adjusting treatment time (by controlled airflow rate) or adjusting irradiation (by turning lamps on/off). Brown arrow: untreated gas from the manure drum (blue barrel); white arrow: ambient air for diluting the untreated gas; red arrow: inlet air to UV treatment with reduced particle matter load (due to the filtration unit; pictured on the right); blue arrow: UV-treated air. Yellow: gas sampling ports.

2.3. Tested UV Sources

In this study, the mitigation of target gases was investigated using four different light sources (UV-A: 367 nm and UV-C: 254, 222, 185 + 254 nm, Figure A2). Two different low-pressure mercury sources were used, both of which emit strongly at 254 nm, but one additionally contains a small 185 nm component because the bulb is made from special materials that allow transmission of that line. The emission spectrum of low-pressure mercury lamps is well known, and these sources both also contained small emissions at 365 nm and other wavelengths common to all of these bulbs. Nonetheless, we refer to these as 254 nm or (185 + 254) nm light sources. An excimer source emitting at 222 nm was the third source; these three sources constitute variations on wavelengths between 222 and 365 nm. The fourth source was an LED with emission centered at 367 nm, quite near the 365 nm range that mercury lamps commonly were used for, but without many of the disadvantages of a mercury-based lamp. This is considered within the UV-A region.

Each chamber inside the mobile laboratory was equipped with 5 UV-A LED lamps (T8 LED, Eildon Technology, Shenzhen, China). An additional 100 UV-A lamps (effectively adding 20 times the light intensity, Table S1) were installed on a removable rack in each of the two chambers (#2–#3) to investigate the reduction of targeted gases according to the UV dose (Figure 2). Detailed information on UV-A lamps used in this study was reported previously [13].

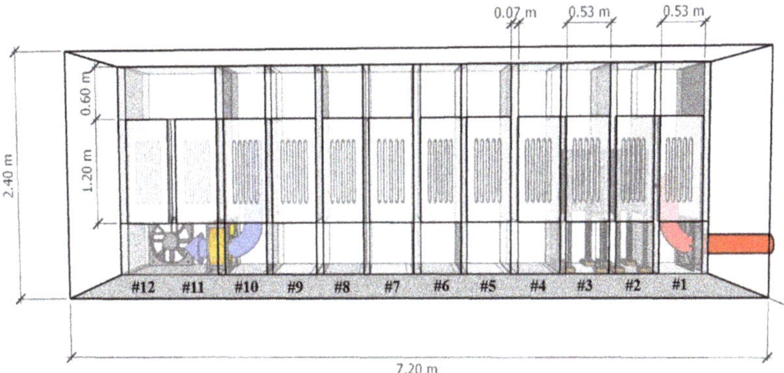

Figure 2. Detailed schematic (side view) of UV treatment inside a flow-through mobile laboratory with UV lamps. The untreated airflow is irradiated while passing through a series of chambers (#1–#12) equipped with TiO_2 photocatalytic surfaces and 5 UV-A lamps per chamber. The two chambers (#2 and #3) were equipped with additional 100 portable lamp holders to increase light intensity. Treated air moves in a serpentine pattern from the inlet (right, red) to the outlet (left, blue). UV dose is controlled by either adjusting treatment time (by controlled airflow rate) or adjusting irradiation (by turning lamps on/off).

All UV-C sources were tested inside chamber #2 while all the UV-A lights in other chambers were turned off. For UV-C (254 nm and 185 + 254 nm, American Ultraviolet Co, Lebanon, IN, USA), four lamps of each different wavelength were installed on the door in one chamber (#2). In the case of the 222 nm excimer UV-C (Ushio America Inc., Cypress, CA, USA), one lamp (Care222 Series) and power supply were installed on the door in chamber #2 (Figure A2). The effects of UV wavelength were measured locally in chamber #2 for all lamp types. The targeted gas concentrations in the untreated gas (control) were measured in the #1 chamber's sampling port. The treated gas concentration after the UV treatment was measured in the #3 chamber's sampling port (Figures 1 and 2).

2.4. The Light Intensity of Different UV Wavelength Lamps

The light intensity is needed to estimate the UV irradiation (and therefore, the dose when integrated over time). The light intensity was measured by ILT-1700 radiometer (International Light Technologies, Peabody, MA, USA) with wavelength-specific sensors and filters. The UV-C 254 nm, 222 nm, 185 nm, and UV-A 365 nm was measured by the SED240 sensor (w/NS254 filter; 254 ± 5 nm); SED240 sensor (w/NS220 filter, 220 ± 5 nm); SED185 sensor (w/NS185 filter, 185 ± 5 nm); SED033 sensor (w/NS365 filter, 365 ± 5 nm), respectively. The 222 nm sensor only imperfectly excludes light from its intended window, and non-zero artifactual measurements were seen with the two Hg sources. All UV lamps were turned on for 5 min before each measurement or experimental run to ensure stable and consistent UV irradiation. For techno-economic analysis, the electric power consumption was measured using a wattage meter (P3, Lexington, NY, USA). The summary of measured light intensity inside the mobile lab under different UV wavelengths and doses is shown in Tables S1–S7.

2.5. Measurement of Odor

Gas samples for odor analyses were collected from the inlet and outlet gas sampling ports inside the UV mobile lab into 10 L Tedlar bags using a Vac-U-Chamber and sampling pump (both from SKC Inc., Eighty-Four, PA, USA). Tedlar bags were precleaned by flushing with clean air three times before use. Gas samples were analyzed for odor using a dynamic triangular forced-choice olfactometer (St. Croix Sensory Inc., Stillwater, MN, USA). Four trained panelists at two repetitions each were used to analyze each sample, presented from low to increasingly lower dilutions to the point of consistent odor detection.

2.6. Measurement of Odorous Volatile Organic Compounds

Odorous VOCs, such as sulfur-containing VOCs, volatile fatty acids (VFAs), and phenolic compounds, are significant contributors to livestock odor [14]. VOC analysis was conducted in the same way as described in detail in the previous study [15]. Briefly, VOC samples were collected in 1 L gas sampling glass bulbs. An internal standard (hexane) was used to minimize variability in sampling and sample preparation. A 2 cm DVB/Carboxen/PDMS solid-phase microextraction (SPME) fiber (Supelco, Bellefonte, PA, USA) was used to extract VOCs from the glass bulbs for 50 min, then analyzed with a GC-MS within 12 h of sample collection. The NIST mass spectral library (with at least 80% spectral match) was used to confirm the compounds' identity. A set of 15 standards for targeted odorous VOC were used (acetic acid, propionic acid, isobutyric acid, butyric acid, isovaleric acid, valeric acid, hexanoic acid, dimethyl disulfide, diethyl disulfide, dimethyl trisulfide, guaiacol, phenol, *p*-cresol, 4-ethyl phenol, indole, and skatole) and calibrated to verify the GC retention time and MS spectral signal.

2.7. Measurement of Ozone Concentrations

Ozone is generated during UV-C irradiation of air, and thus, it was a targeted gas. On the other hand, the generated O_3 can react and mitigate odorous VOCs. In this research, the O_3 detector was connected to the monitoring system (Series 500 monitor, Aeroqual, New Zealand) and installed at the gas sampling ports when in use. The detector was factory-calibrated (Gas Sensing, Inwood, IA, USA) and certified before use. The detection range was 0–50 ppb.

2.8. Measurement of Greenhouse Gas Concentrations

UV treatment of odorous VOCs and NH_3 can result in the generation of GHGs that should be tracked. Methane (CH_4), carbon dioxide (CO_2), and nitrous oxide (N_2O) were measured. GHGs samples were collected using syringes and 5.9 mL Exetainer vials (Labco Limited, UK) and were analyzed for concentrations on a GC equipped with FID and ECD detectors (SRI Instruments, Torrance, CA, USA). Samples were analyzed on the day of collection. Standard calibrations were constructed daily using 10.3 and 20.5 ppm CH_4, 1005 and 4010 ppm CO_2, and 0.101 and 1.01 ppm N_2O. Pure helium was used to calibrate the baseline of 0 ppm (Air Liquide America, Plumsteadville, PA, USA).

2.9. Measurement of Ammonia and Hydrogen Sulfide Concentrations

Ammonia (NH_3) is a major contributor to air pollution from livestock operations. Hydrogen sulfide (H_2S) is a toxic air pollutant and a significant contributor to odor. NH_3 and H_2S concentrations were measured with a real-time analyzer (OMS-300, Smart Control & Sensing, Daejeon, Korea) calibrated with high precision standard gases (5-point dilution, $R^2 = 0.99$). The analyzer was equipped with NH_3/CR-200 and H_2S/C-50 electrochemical gas sensors (Membrapor, Wallisellen, Switzerland), NH_3/CR-200 (0 to 100 ppm), and H_2S/C-50 (0 to 50 ppm), respectively.

2.10. Evaluation of Treatment Effectiveness and Data Analysis

The overall mean percent reduction for each measured targeted gas was estimated using:

$$\% R = (C_{con} - C_{Treat})/C_{con} \times 100 \tag{1}$$

where C_{Con} and C_{Treat} are the mean measured concentrations in control and treated air, respectively. For odor and odorous VOCs, odor units ($OUE \cdot m^{-3}$) and MS detector responses (peak area counts, PAC) were used.

Emissions were calculated as a product of measured gas concentrations and the total airflow rate through the UV mobile lab, adjusted for standard conditions and dry air using collected environmental data. The overall mean emission of each measured gas was estimated as:

$$\text{Emission } (\text{g}\cdot\text{min}^{-1}) = \text{C} \times \text{V} \times (273.15\text{ K} \times \text{MW})/(273.15\text{ K} + \text{T}) \times (2.24 \times 10^4) \quad (2)$$

where C = the mean measured target gas concentration in control and treated air (mL/m^3, OU_E/m^3). V = the treated airflow rate (m^3/min). MW = the molecular weight of the targeted gas (g/mol). T = the temperature in the control and treated air. The 2.24×10^4 is an ideal gas conversion factor for L to moles at 273.15 K [13].

The electric energy consumption during UV treatment was estimated using the measured power consumption by lamps:

$$\text{EEC} = \text{P} \times t_{s}/(3600 \times 1000) \quad (3)$$

where EEC = electric energy consumption (kWh). P = measured electric power consumption for the UV lamps turned 'on' during treatment (W). t_s = treatment time for air irradiated with the UV lamps that were turned 'on' inside the mobile lab (s).

The mass of mitigated gas pollutant (M) with UV during given treatment time (t_s) was estimated by comparing gas emission rate (E) in treatment and control:

$$\text{M} = (E_{con} - E_{treat}) \times t_{s}/60 \quad (4)$$

where M = mass of mitigated gas pollutant (g). E_{con} = emission rate at the 'control' sampling location. E_{treat} = emission rate at the 'treatment' sampling location.

The electric energy of UV treatment (EE, kWh/g) was estimated as using electric energy consumption (EEC) needed to mitigate a gas pollutant mass (M):

$$\text{EE} = \text{EEC}/\text{M} \quad (5)$$

Finally, the estimated cost of electric energy (Cost) needed for UV treatment was estimated using the mean cost of rural energy in Iowa (USD 0.13/kWh):

$$\text{Cost} = \text{EE} \times \text{USD } 0.13/\text{kWh} \quad (6)$$

where Cost = estimated cost of electric energy needed for UV treatment to mitigate a unit mass of pollutants in the air (USD/g).

UV dose was estimated using measured light intensity (I) at a specific UV wavelength (mW/cm^2) and treatment time (t_s). Since the photocatalysis reaction was assumed to be the primary mechanism for the target gas mitigation, the light intensity irradiated on the TiO_2 surface was used. For lamps emitting light at multiple UV wavelengths, the UV dose was calculated using the light intensity of the primary wavelength suggested by the lamp manufacturer.

$$\text{UV dose} = \text{I} \times t_s \quad (7)$$

where UV Dose = energy of the UV light on the surface of photocatalyst (mJ/cm^2).

2.11. Statistical Analysis

The overall mean percent reduction for each measured targeted gas was estimated using the following: R studio (version 3.6.2; Boston, MA, USA) was used to analyze the target standard gases' mitigation under UV photocatalysis treatment. The UV dose and treatment time parameters between control concentration and treatment concentration were statistically analyzed using one-way ANOVA. The statistical difference was confirmed by obtaining the *p*-value through the Tukey test. A significant difference was defined for a p-value < 0.05 in this study.

3. Results

3.1. Mitigation of Targeted Gases as a Function of UV-A Dose Controlled by Light Intensity and Airflow Rate

3.1.1. Odor—Effects of UV-A Dose

The UV-A photocatalysis showed a significant percent odor reduction. The UV dose of 2.5 mJ/cm^2 was required for statistically significant odor reduction (Table 1). As the UV dose was increased, the odor reduction increased up to 63%.

Table 1. Mitigation of odor with the different UV-A doses (1.3, 2.0, 2.5, 3.9, and 5.8 mJ/cm^2) irradiating gaseous emissions from swine manure. Bold signifies statistical significance.

UV-A Dose (mJ/cm^2)	Light Intensity (mW/cm^2)	Treatment Time (s)	Control (OU$_E$/m^3)	Treatment (OU$_E$/m^3)	% Reduction (*p*-Value)
UV dose control with light intensity					
1.3	0.14	9.5	378 ± 13	229 ± 75	39.4 (0.12)
2.5	0.26	9.5	352 ± 8.0	239 ± 24	**32.2 (0.04)**
3.9	0.41	9.5	653 ± 32	277 ± 22	**57.5 (0.01)**
UV dose control with treatment time					
1.3	0.41	3.2	198 ± 59	234 ± 60	−18.6 (0.61)
2.0	0.41	4.8	212 ± 31	206 ± 75	2.9 (0.93)
UV dose control with light intensity and treatment time					
5.8 *	0.41 and 0.04	9.5 and 47.6	653 ± 9.2	243 ± 64	**62.7 (<0.01)**

Note: * Irradiation with 5 UV-A lamps per each chamber (#1–#12, a total of 60 lamps turned on) with additional 100 portable UV-A lamps turned on in chambers #2 and #3, 160 lamps turned on total. Inlet and outlet air temperature = 19 ± 2 °C and 22 ± 5 °C.

3.1.2. Volatile Organic Compounds—Effects of UV-A Dose

UV-A photocatalysis significantly mitigated selected targeted odorous VOCs while also generating a small subset of other VOCs. This is an important observation as the complex and compound-specific photocatalytic reactions can affect the overall percent reduction of odor. UV-A dose ≥ 2.5 mJ/cm^2 was required to mitigate phenolic compounds (Table 2), similarly to the findings for odor where the same UV dose resulted in significant mitigation (Table 1). As the UV dose increased up to ~3.9 mJ/cm^2, the percent reduction of VOCs and the number of mitigated (targeted) VOCs increased. The highest percent reductions were measured for acetic acid (49%), butanoic acid (53%), *p*-cresol (67%), and indole (32%). The highest dose (5.8 mJ/cm^2) did not improve the mitigation effect, suggesting that there is merit to optimizing the UV dose, especially from the techno-economic standpoint.

3.1.3. Ozone—Effects of UV-A Dose

Compared with a baseline (ambient air) amount of O_3 detected without UV irradiation, the concentration of O_3 was effectively mitigated (up to 100%) by UV-A irradiation (Table 3). This observation was consistent with our earlier UV-A research in lab-scale and pilot-scale (poultry farm) conditions [9,10]. Therefore, the treatment of the lowest UV dose (1.3 mJ·cm^{-2}) is the most economical condition if O_3 is the targeted gas. It is also important to mention that the mean O_3 concentration in the UV mobile lab outlet was 4.7 ppb. This concentration is relatively low and likely of low concern for scaling up to farm environments, where abundant VOCs are present to react with O_3 and further reduce the risk of its release to the atmosphere outside the barn.

Table 2. Mitigation of odorous VOCs with the different doses (1.3, 2.0, 2.5, 3.9, and 5.8 mJ/cm^2) irradiating gaseous emissions from swine manure. Bold signifies statistical significance.

Targeted VOCs	Percent Reduction (*p*-Value) UV-A Dose (mJ/cm^2) (UV Light Intensity, mW/cm^2 and Treatment Time, s)					
	1.3 (0.14 and 9.5)	**1.3 (0.41 and 3.2)**	**2.0 (0.41 and 4.8)**	**2.5 (0.26 and 9.5)**	**3.9 (0.41 and 9.5)**	**5.8 * (0.41 + 0.04 & 9.5 + 47.6)**
DMDS	29.3 (0.36)	54.3 (0.12)	41.9 (0.46)	59.2 (0.07)	−21.8 (0.51)	8.3 (0.92)
DMTS	−5.8 (0.96)	−5.6 (0.84)	−0.9 (0.99)	11.5 (0.35)	42.9 (0.07)	49.4 (0.22)
Acetic acid	−1.3 (0.97)	−10.8 (0.85)	23.9 (0.69)	−4.0 (0.91)	**48.6 (0.04)**	**50.5 (0.04)**
Propanoic acid	12.7 (0.36)	21.1 (0.07)	72.9 (0.30)	**50.7 (0.01)**	76.7 (0.40)	66.8 (0.45)
Isopentanoic acid	29.4 (0.39)	49.4 (0.27)	54.2 (0.23)	24.2 (0.70)	41.9 (0.10)	37.7 (0.27)
Butanoic acid	1.5 (0.95)	**36.3 (0.04)**	**35.8 (0.01)**	**44.8 (<0.01)**	**52.6 (0.04)**	**47.9 (0.04)**
Phenol	39.1 (0.15)	34.0 (0.07)	63.1 (0.26)	−7.9 (0.86)	−28.1 (0.13)	−32.1 (0.44)
p-Cresol	−3.2 (0.96)	36.1 (0.05)	−15.9 (0.84)	**41.3 (0.03)**	**66.5 (0.03)**	58.6 (0.05)
Indole	0.2 (0.99)	4.0 (0.89)	23.3 (0.61)	21.5 (0.58)	**32.3 (0.02)**	**20.0 (0.03)**
Skatole	−9.2 (0.89)	17.6 (0.06)	6.4 (0.93)	6.4 (0.93)	70.0 (0.37)	64.6 (0.45)

Note: DMDS = dimethyl disulfide, DMTS = dimethyl trisulfide; values in table report percent reduction (*p*-values). * Irradiation with 5 UV-A lamps per each chamber (#1–#12, a total of 60 lamps turned on) with additional 100 portable UV-A lamps turned on in chambers #2 and #3, 160 lamps turned on total. Inlet and outlet air temperature = 19 ± 2 °C and 22 ± 5 °C.

Table 3. Mitigation of O_3 with the different UV doses (1.3, 2.0, 2.5, 3.9, and 5.8 mJ/cm^2) irradiating gaseous emissions from swine manure. Bold signifies statistical significance.

UV-A Dose (mJ/cm^2)	Light Intensity (mW/cm^2)	Treatment Time (s)	Control (ppb)	Treatment (ppb)	% Reduction (*p*-Value)
		UV dose control with UV light intensity			
1.3	0.14	9.5	2.9 ± 0.3	0.0 ± 0.0	**100 (<0.01)**
2.5	0.26	9.5	3.8 ± 0.1	0.0 ± 0.0	**100 (<0.01)**
3.9	0.41	9.5	9.5 ± 1.2	0.0 ± 0.0	**100 (<0.01)**
		UV dose control with treatment time			
1.3	0.41	3.2	5.8 ± 2.0	0.0 ± 0.0	**100 (<0.01)**
2.0	0.41	4.8	2.3 ± 0.2	0.0 ± 0.0	**100 (<0.01)**
		UV dose control with UV light intensity and treatment time			
5.8 *	0.41 and 0.04	9.5 and 47.6	3.0 ± 0.2	0.0 ± 0.0	**100 (<0.01)**

Note: * Irradiation with 5 UV-A lamps per each chamber (#1–#12, a total of 60 lamps turned on) with additional 100 portable UV-A lamps in chambers #2 and #3 turned on, 160 lamps total (installed and portable). Inlet and outlet air temperature = 19 ± 2 °C and 22 ± 5 °C.

3.1.4. Greenhouse Gases—Effects of UV-A Dose

The N_2O concentrations were significantly reduced (by 4–14%) with UV-A photocatalysis for 2.5 mJ/cm^2 or higher doses (Table 4). However, there was no statistically significant change in CH_4 concentrations (Table S8), and there was a significant generation of CO_2 (up to −26%) (Table S9) that increased with the UV dose.

Table 4. Mitigation of N_2O with the different UV doses (1.3, 2.0, 2.5, 3.9, and 5.8 mJ/cm^2) irradiating gaseous emissions from swine manure. Bold signifies statistical significance.

UV-A Dose (mJ/cm^2)	Light Intensity (mW/cm^2)	Treatment Time (s)	Control (ppm)	Treatment (ppm)	% Reduction (p-Value)
		UV dose control with UV light intensity			
1.3	0.14	9.5	0.4 ± 0.0	0.4 ± 0.0	3.5 (0.22)
2.5	0.26	9.5	0.4 ± 0.0	0.3 ± 0.0	**9.0 (<0.01)**
3.9	0.41	9.5	0.3 ± 0.0	0.3 ± 0.0	**4.3 (0.02)**
		UV dose control with treatment time			
1.3	0.41	3.2	0.3 ± 0.0	0.3 ± 0.0	0.4 (0.85)
2.0	0.41	4.8	0.3 ± 0.0	0.2 ± 0.0	17.1 (0.09)
		UV dose control with UV light intensity and treatment time			
5.8 *	0.41 and 0.04	9.5 and 47.6	0.3 ± 0.0	0.2 ± 0.0	**14.2 (0.03)**

Note: * Irradiation with 5 UV-A lamps per each chamber (#1–#12, a total of 60 lamps turned on) with additional 100 portable UV-A lamps in chambers #2 and #3 turned on, 160 lamps total (installed and portable). Inlet and outlet air temperature = 19 ± 2 °C and 22 ± 5 °C.

3.1.5. Ammonia and Hydrogen Sulfide—Effects of UV-A Dose

Significant percent reduction of NH_3 concentrations was measured only for the maximum UV-A dose (5.8 mJ/cm^2; Table 5). The treatment efficiency was low (6%), similar to the previous reports [4,9,10,13]. The mean NH_3 concentration in control was 5.4 ppm. No steady concentration of H_2S was measured in control (likely due to the limited supply of it in manure). The H_2S was typically detectable at the start of the experiment, but its concentration in control was rapidly diminishing, preventing reproducible measurements after UV-A treatment. This limitation will be addressed in farm-scale trials, where H_2S in barn air or barn exhaust is continuously present, the mitigation of H_2S can be objectively tested.

Table 5. Mitigation of NH_3 with the different UV doses (1.3, 2.0, 2.5, 3.9, and 5.8 mJ/cm^2) irradiating gaseous emissions from swine manure. Bold signifies statistical significance.

UV-A Dose (mJ/cm^2)	Light Intensity (mW/cm^2)	Treatment Time (s)	Control (ppm)	Treatment (ppm)	% Reduction (p-Value)
		UV dose control with UV light intensity			
1.3	0.14	9.5	4.6 ± 0.1	4.5 ±0.1	1.0 (0.33)
2.5	0.26	9.5	5.8 ± 0.1	5.7 ±0.1	1.3 (0.21)
3.9	0.41	9.5	5.5 ± 0.2	5.3 ±0.2	2.1 (0.38)
		UV dose control with treatment time			
1.3	0.41	3.2	4.3 ± 0.4	4.2 ± 0.4	4.3 (0.41)
2.0	0.41	4.8	6.3 ± 0.1	6.2 ± 0.1	2.9 (0.93)
		UV dose control with UV light intensity and treatment time			
5.8 *	0.41 and 0.04	9.5 and 47.6	6.0 ± 0.3	5.6 ± 0.3	**6.1 (0.04)**

Note: * Irradiation with 5 UV-A lamps per each chamber (#1–#12, a total of 60 lamps turned on) with additional 100 portable UV-A lamps in chambers #2 and #3 turned on, 160 lamps total (installed and portable). Inlet and outlet air temperature = 19 ± 2 °C and 22 ± 5 °C.

3.2. *Comparison of the Mitigation of Targeted Gases as a Function of UV Wavelength*

The comparison of UV-A and UV-C photocatalysis treatment was conducted in only one chamber (#2) due to the limited number of available UV-C lamps that are more costly than UV-A. The results are summarized below. Testing conditions were the same for all lamps to enable a fair side-by-side comparison.

3.2.1. Odor—Effects of UV Wavelength

The short UV-C wavelength (185 + 254 nm) resulted in a 44% reduction of overall detected odor. This was a remarkable mitigation effect, considering that the UV dose was the lowest among all tested (Table 6). However, odor reduction was not significant for all treatments ($0.09 < p$-value < 0.94). This was likely due to the variability of control used for just one treatment chamber tested (Table 6). Odor measurements via dilution olfactometry and human panelists are inherently more variable than chemical analyses. This limitation could be addressed by refurbishing the entire UV mobile laboratory with one type of lamp, effectively allowing a more extensive range of doses to be tested (e.g., lower variability reported for UV treatment using an entire mobile lab with 12 chambers facilitating treatment, Table 1).

Table 6. Mitigation of odor with different UV wavelengths irradiating gaseous emissions inside the #2 chamber.

UV Wavelengths (nm)	UV Dose (μJ/cm^2)	Light Intensity (μW/cm^2)	Control (OU$_E$/m^3)	Treatment (OU$_E$/m^3)	% Reduction (*p*-Value)
185 + 254	0.03	0.01		182 ± 73	43.9 (0.09)
222	2.83	0.59		262 ± 22	19.5 (0.30)
254	1.78	0.37	325 ± 78	290 ± 55	10.6 (0.19)
367	192	40		332 ± 58	−2.2 (0.94)
	1968	410		270 ± 38	17.0 (0.11)

Note: Treatment time = 4.8 s (airflow = 0.25 m^3/s), inlet and outlet air temp. = 16 ± 1 °C and 19 ± 2 °C.

3.2.2. Volatile Organic Compounds—Effects of UV Wavelength

The phenolic compounds of *p*-cresol and indole were effectively treated with UV-C (185 + 254 nm) with a statistically significant percent reduction at 47 and 54%, respectively (Table 7). *p*-Cresol and indole are often referred to as the 'signature' barnyard odors and potent odorants; thus, their mitigation is consistent with the results for overall odor reduction (Table 6). The UV-C (185 + 254 nm) dose was the lowest tested, yet the percent reductions for other targeted VOCs were notable and ranged from 10 to 59%. The 185 + 254 nm light source is essentially identical to the 254 nm light source, save that the 'glass' of the lamp itself additionally transmits a small amount of very high energy 185 nm photons. The additional 185 nm irradiation (when part of 185 + 254 nm treatment) results in effective reduction of targeted VOCs. The UV-C (254 and 222 nm) sources also effectively mitigated much targeted VOCs (from 15 to 70%), although there is no significant statistical mitigation. The use of longer-wavelength UV-A (367 nm) and the highest dose resulted in a statistically significant reduction for acetic acid (57%) and butanoic acid (33%).

3.2.3. Ozone—Effects of UV Wavelength

O_3 was reduced at all wavelengths except for 185 + 254 nm. Specifically, complete mitigation (below detection limits) was measured for 222 and 254 nm treatments. The percent reduction increased from 30 to 97% as the UV dose increased for the 367 nm wavelength. O_3 increased by ~140% (Table 8) for the 185 + 254 nm treatment. This is due to the direct photolysis of O_2 in the air, which leads to O_3 formation.

Table 7. Mitigation of odorous VOCs with different UV wavelength irradiating gaseous emissions inside the #2 chamber. Bold signifies statistical significance.

Targeted VOCs	Percent Reduction (p-Value)				
	UV-C Dose, μJ/cm² (Light Intensity, μW/cm²)			UV-A Dose (Light Intensity, μW/cm²)	
	185 + 254 nm 0.03 (0.01)	222 nm 2.83 (0.59)	254 nm 1.78 (0.37)	367 nm 192 (40)	367 nm 1968 (410)
DMDS	59.3 (0.14)	59.8 (0.27)	14.8 (0.70)	−14.7 (0.87)	−4.1 (0.93)
DMTS	56.4 (0.15)	67.1 (0.12)	21.7 (0.50)	−6.2 (0.63)	5.8 (0.86)
Acetic acid	10.0 (0.68)	−12.2 (0.72)	−12.1 (0.70)	0.4 (0.99)	**57.2 (0.04)**
Propanoic acid	13.3 (0.83)	37.4 (0.47)	23.1 (0.64)	−32.6 (0.54)	36.2 (0.49)
Isopentanoic acid	24.9 (0.72)	60.5 (0.39)	70.0 (0.33)	−13.6 (0.85)	18.1 (0.80)
Butanoic acid	10.3 (0.66)	27.8 (0.08)	−15.5 (0.79)	21.8 (0.39)	**33.4 (0.03)**
Phenol	43.4 (0.08)	32.4 (0.26)	7.8 (0.79)	23.2 (0.53)	24.7 (0.52)
p-Cresol	**47.1 (0.04)**	46.9 (0.09)	29.2 (0.32)	8.9 (0.75)	46.8 (0.05)
Indole	**54.2 (0.01)**	19.2 (0.48)	16.9 (0.51)	−14.5 (0.64)	46.6 (0.17)
Skatole	35.1 (0.35)	55.8 (0.14)	64.6 (0.11)	3.7 (0.83)	56.5 (0.09)

Note: DMDS = dimethyl disulfide, DMTS = dimethyl trisulfide. Treatment time = 4.8 s (airflow = 0.25 m^3/s), inlet and outlet air temperature = 16 ± 1 °C and 19 ± 2 °C.

Table 8. Mitigation of O_3 concentration with different UV wavelength irradiating gaseous emissions inside the #2 chamber. Bold signifies statistical significance.

UV Wavelength (nm)	UV Dose (μJ/cm²)	Light Intensity (μW/cm²)	Control (ppb)	UV Treatment (ppb)	% Reduction (p-Value)
185 + 254	0.03	0.01	14.6 ± 4.2	34.8 ± 5.7	**−139 (<0.01)**
222	2.83	0.59	18.6 ± 2.2	0.0 ± 0.0	**100 (<0.01)**
254	1.78	0.37	16.8 ± 2.9	0.0 ± 0.0	**100 (<0.01)**
367	192	40	10.9 ± 2.1	7.7 ± 2.4	**30 (0.02)**
	1968	410	6.4 ± 1.6	0.2 ± 0.7	**97 (<0.01)**

Note: Treatment time = 4.8 s (airflow = 0.25 m^3/s), inlet and outlet air temp. = 16 ± 1 °C and 19 ± 2 °C.

3.2.4. Greenhouse Gases—Effects of UV Wavelength

Significant mitigation was measured for CH_4, with the (185 + 254) nm lamps. However, other wavelength lamps did not show statistically significant reduction (Table 9). CO_2 concentrations increased for all UV wavelengths tested (Table S10) and were statistically significant for 185 + 254 nm and 367 nm (high UV dose). N_2O was mitigated at statistically significant levels (from 8 to 25%) for all treatments except for the low 367 nm dose (Table 10). The highest percent reduction for CH_4 and N_2O resulted from the 185 + 254 nm treatment.

Table 9. Mitigation of CH_4 concentration with different UV wavelength irradiating gaseous emissions inside the #2 chamber. Bold signifies statistical significance.

UV Wavelength (nm)	UV Dose (μJ/cm²)	Light Intensity (μW/cm²)	Control (ppm)	Treatment (ppm)	% Reduction (p-Value)
185 + 254	0.03	0.01		1.1 ± 0.0	**70.9 (0.04)**
222	2.83	0.59		1.7 ± 0.0	55.7 (0.06)
254	1.78	0.37	3.7 ± 0.9	1.4 ± 0.2	63.6 (0.06)
367	192	40		3.9 ± 0.5	−3.7 (0.77)
	1968	410		3.2 ± 1.3	15.9 (0.59)

Note: Treatment time = 4.8 s (airflow = 0.25 m^3/s), inlet and outlet air temp. = 16 ± 1 °C and 19 ± 2 °C.

Table 10. Mitigation of N_2O concentration with different UV wavelength irradiating gaseous emissions inside the #2 chamber. Bold signifies statistical significance.

UV Wavelength (nm)	UV Dose (μJ/cm^2)	Light Intensity (μW/cm^2)	Control (ppm)	Treatment (ppm)	% Reduction (p-Value)
185 + 254	0.03	0.01		0.18 ± 0.00	**25.4 (<0.01)**
222	2.83	0.59		0.22 ± 0.00	**8.1 (0.01)**
254	1.78	0.37	0.24 ± 0.01	0.21 ± 0.01	**13.6 (0.01)**
367	192	40		0.23 ± 0.01	5.9 (0.17)
	1968	410		0.21 ± 0.00	**13.5 (0.02)**

Note: Treatment time = 4.8 s (airflow = 0.25 m^3/s), inlet and outlet air temp. = 16 ± 1 °C and 19 ± 2 °C.

3.2.5. Ammonia and Hydrogen Sulfide—Effects of UV Wavelength

There was no statistically significant reduction for all UV wavelengths tested, and the percent reduction ranged from 0.3 to 2.1% (Table 11). The average concentration of the control group was 3.1 ppm (Table 11). H_2S concentrations in control were not stable enough to warrant reporting the effect.

Table 11. Mitigation of NH_3 concentration with different UV wavelength irradiating gaseous emissions inside the #2 chamber.

UV Wavelength (nm)	UV Dose (μJ/cm^2)	Light Intensity (μW/cm^2)	Control (ppm)	Treatment (ppm)	% Reduction (p-Value)
185 + 254	0.03	0.01	3.3 ± 0.1	3.3 ± 0.0	0.5 (0.55)
222	2.83	0.59	3.2 ± 0.1	3.2 ± 0.1	1.3 (0.35)
254	1.78	0.37	3.1 ± 0.1	3.0 ± 0.1	1.4 (0.22)
367	192	40	2.5 ± 0.1	2.5 ± 0.1	0.3 (0.71)
	1968	410	2.6 ± 0.1	2.6 ± 0.1	2.1 (0.16)

Note: Treatment time = 4.8 s (airflow = 0.25 m^3/s), inlet and outlet air temp. = 16 ± 1 °C and 19 ± 2 °C.

4. Discussion

4.1. Summary of the UV-A Photocatalysis—Comparison with Previous Research

UV photocatalysis can be considered a potential technology to reduce odorous gases and improve air quality. This research provides mitigation data for a more extensive set of odorants and air pollutants compared with the state-of-the-art. UV-A photocatalysis reduced several of the targeted odorous gases (Table 12) with statistical significance. The reproducibility of mitigation with UV-A photocatalysis warrants further scaling up into larger volumetric flowrates common for farm applications.

This research provided data that can be considered for early assessment and extrapolating the techno-economic analysis of the UV-A treatment to practical scales (Table 13).

Caution needs to be exercised when extrapolating pilot-scale data. However, several major recommendations can be made. The UV-A treatment does not appear to be effective for farm-scale mitigation of NH_3, considering that the mitigation effect was rather small (Tables 5 and 11). Thus, effective reduction of kg/day quantities of NH_3 from typical swine farms with UV-A appears to be too costly.

Table 12. Summary and comparison of the % reduction of targeted gases with UV-A photocatalysis. Bold signifies statistical significance.

Reference	Targeted Gas	UV Dose (mJ/cm^2)	Target Gas Concentration (ppm, O_3 = ppb, Odor = OU_E/m^3)		% Reduction
			Control	UV Treatment	
[13] (pilot-scale treating standard gases)	NH_3	3.9	67.4 ± 0.36	61.1 ± 0.30	**9**
		5.8	68.9 ± 0.68	61.1 ± 0.70	**11**
	Butan-1-ol	2.5	0.66 ± 0.02	0.53 ± 0.06	**19**
		3.9	0.65 ± 0.03	0.43 ± 0.04	**34**
		5.8	0.69 ± 0.02	0.41 ± 0.07	**41**
This study (pilot-scale with swine manure)	NH_3	5.8	5.98 ± 0.28	5.62 ± 0.34	6
	N_2O	3.9	0.29 ± 0.01	0.28 ± 0.00	4
	N_2O	5.8	0.29 ± 0.00	0.25 ± 0.01	**14**
	O_3	1.3	0.34 ± 0.03	0	**100**
	O_3	5.8	0.31 ± 0.02	0	**100**
	VOCs	2.5	N/A	N/A	**PA(51), BA(45), *p*-Cresol(41)**
		3.9	N/A	N/A	**AA(49), BA(53), *p*-Cresol (67), Indole(32)**
		5.8	N/A	N/A	**AA(51), BA(48), Indole(20)**
	Odor	2.5	352 ± 7.98	239 ± 24.4	**32**
		3.9	653 ± 32.1	277 ± 21.7	**58**
		5.8	653 ± 9.25	243 ± 64.4	**63**

Note: DMDS = dimethyl disulfide, acetic acid = AA, propanoic acid = PA, isobutyric acid = IA, and butanoic acid = BA, N/A = not available.

Table 13. Techno-economic analysis of mitigating target gases with UV-A photocatalysis.

Reference	Targeted Gas	UV Dose (mJ/cm^2)	Target Gas Emission (*E*, mg/min, Odor, OU_E/min)		Cost [1] (USD/kg for NH_3, USD/g for butan-1-ol, N_2O, O_3 USD/ton of OU_E for Odor)
			Control	UV Treatment	
[13] (pilot-scale treating standard gases)	NH_3	3.9	746	676	53.4
	NH_3	5.8	763	676	62.5
	Butan-1-ol	2.5	31.5	25.3	442
		3.9	30.9	20.3	352
		5.8	32.9	19.4	403
This study (pilot-scale treating emissions from swine manure)	NH_3	5.8	64.2	60.2	1260
	N_2O	3.9	8.14	7.79	10.6
	N_2O	5.8	8.06	6.92	4.72
	O_3	1.3	0.01	0.00	18.9
	O_3	5.8	0.01	0.00	60.0
	Odor	3.9	9200	3910	0.71
		5.8	9200	3430	0.94

Note: [1] electric energy needed for UV treatment to mitigate a unit mass of pollutants in the air (USD/g).

On the other hand, mitigation of several targeted air pollutants is worth considering. For example, mitigation of N_2O (the most potent GHG, Tables 4 and 10) might be further exploited for farm income generation that uses subsidies and programs focused on reducing GHGs emissions and mitigating climate change. Direct emissions of O_3 from farms have not been a concern, as opposed to the secondary pollutant generation of O_3 as a by-product of emitted VOCs and their atmospheric chemistry. Thus, the incentivization and credit taking for the at-source mitigation of O_3 might be considered (Tables 3 and 8). Finally, the significant reduction of odor and odorous VOCs is encouraging at this scale (Tables 1, 2, 6 and 7). Of course, planned farm-scale trials can provide a more realistic techno-economic assessment of UV-A cost. Farm-scale trials with the UV-A photocatalysis installed inside barns to mitigate indoor air quality and the pathogen load are warranted.

4.2. Summary of the UV-C Photocatalysis

The effects of UV wavelength were only tested in one chamber inside the UV mobile lab due to increasing UV-C dose limitation. Thus, the comparison is somewhat limited (Table 14). Therefore, the results show the UV-C's future potential that still needs to be tested on a larger scale. The UV-C can efficiently reduce odorous VOCs with a lower dose (compared to UV-A). One caveat to UV-C use is risks associated with direct skin and eye tissue exposure and O_3 generation.

Table 14. Summary of the % reduction of targeted gases with UV-C photocatalysis. Bold signifies statistical significance.

UV Wavelength, nm (UV dose, µJ/cm²)	Targeted Gas	Target Gas Concentration (ppm; O_3 = ppb)		% Reduction
		b	UV Treatment	
185 + 254 (0.03)	VOCs	N/A	N/A	***p*-Cresol (47); Indole (54)**
	CH_4	3.7 ± 0.9	1.1 ± 0.0	**71**
	N_2O	0.2 ± 0.0	0.2 ± 0.0	**25**
222 (2.80)	N_2O	0.2 ± 0.0	0.2 ± 0.0	**8**
	O_3	18.6 ± 2.2	0.0 ± 0.0	**100**
254 (1.76)	N_2O	0.2 ± 0.0	0.2 ± 0.0	**14**
	O_3	16.8 ± 2.9	0.0 ± 0.0	**100**

Note: N/A = not available.

4.3. Evaluation of UV Photocatalysis Based on TiO_2 in the Livestock Environment

We summarized the percent mitigation of targeted gases in the previous studies and this mobile lab research series to show the UV photocatalysis performance (Lee et al., 2021; Table 14) [16]. The mitigation of selected target gases via photocatalysis with UV-A and UV-C in livestock-relevant environmental conditions can be considered as an effective method to mitigate the odorous gases.

The TiO_2 based photocatalysis with UV-A yields significant reductions of NH_3 (~31%), H_2S (~40%), CH_4 (~27%), N_2O (~14%), O_3 (~100%), Odorous VOCs (~100%), and odor (~63%) [16]. In the case of CO_2, generation has been reported after UV-A photocatalysis in previous studies. CO_2 is the oxidative endpoint for photocatalytic oxidation of virtually all carbon-containing compounds, and thus its mitigation would not derive from its chemical removal. The percent reduction for the targeted gas showed a difference depending on the coating thickness and UV dose.

The UV-C photocatalyst showed a higher mitigation effect at a lower dose than UV-A photocatalysis. In particular, it showed a significant reduction in H_2S (~100%), CH_4 (~40%) and VOCs (~100%) even after irradiation for a relatively short time (1 s) [6,17,18]. Additionally, it is encouraging that it can effectively reduce H_2S, which is harmful to farms, among the compounds generated in swine barns. In previous research results, it was reported that UV photocatalysis showed high efficiency compared to other mitigation technologies in economic analysis (estimated average electricity cost of UV treatment per pig was USD 0.15–0.23) [6,18].

5. Conclusions

This study evaluated the photocatalytic UV-A and UV-C treatment of gaseous emissions of odor, odorous VOCs, NH_3, and other gases (GHGs, O_3) from stored swine manure on a pilot-scale. To our knowledge, this is the first study of this scope that was needed for scaling up technologies treating gaseous emissions of odor, odorous VOCs, NH_3, H_2S, ozone, and GHGs. The study bridged the knowledge gap between lab-scales and simplified treatment of model gases to the treatment of complex gaseous mixtures emitted from swine manure in fast-moving air. The results showed that the proposed UV technology is ready for the next stage of testing and mitigation of emissions from swine farms. The treatment effectiveness depended on the UV dose and wavelength. Specific findings are summarized below.

Under UV-A (367 nm) photocatalysis, the percent reduction of targeted gases was up to:

- 63% of odor,
- 51%, 51%, 53%, 67%, and 32% of acetic acid, propanoic acid, butanoic acid, *p*-cresol, and indole, respectively,
- 14% of nitrous oxide (N_2O),
- 100% of O_3, 6% of NH_3, and
- 26% generation of CO_2.

Under UV-C (185 + 254 nm) photocatalysis, the percent reduction of target gases was up to:

- 54% and 47% for *p*-cresol and indole, respectively,
- 25% of N_2O,
- 71% of CH_4, and
- 46% and 139% generation of CO_2 and O_3, respectively.

UV-C (222 nm) photocatalysis showed a reduction of 8% for N_2O, 100% for O_3. Lastly, UV-C (254 nm) photocatalysis showed a reduction of 14% for N_2O, 100% for O_3. The UV-A photocatalysis (367 nm) was not economical to reduce NH_3; while it appeared to be economical and effective in mitigating odor and VOC. The 2.5 mJ/cm^2 dose is required to significantly reduce odor. UV-C photocatalysis (185 + 254 nm) was shown to be more efficient than UV-A photocatalysis by significantly reducing several target gases with a low dose, but additional research is needed because there was a limit to the dose control of UV-C in this study. The results proved that the UV technology was sufficiently effective in treating odorous gases in a simulated swine emissions environment, and the mobile lab was ready for farm-scale trials. The UV technology can be considered for the scaled-up treatment of emissions and air quality improvement inside livestock barns.

Supplementary Materials: The following are available online at https://www.mdpi.com/article/10.3390/ani11051289/s1, Figure S1: Generation of gaseous emissions from swine manure. Compressed air is fed into the bottom of the swine manure storage vessel. Gaseous emissions from the vessel's headspace are then blended with clean air, Table S1: Measurement of UV-A light intensity according to the number of lamps in the chamber of the mobile laboratory, Table S2: The measured light intensity (μW/cm^2) with 254 nm lamp irradiation in chamber #2, Table S3: The measured light intensity (μW/cm^2) of photolysis with UV-C excimer (222 nm) irradiation in chamber #2, Table S4: The measured light intensity (μW/cm^2) with UV-C fluorescent (185+254 nm) irradiation in chamber #2, Table S5: UV-C fluorescent (254 nm) light intensity (μW/cm^2) at 11 panels in #2 chamber (Top, Bottom, Front Top, Front Bottom, Left Top, Left Bottom, Right Top, Right Bottom, Back Top, Back Middle, and Back Bottom), Table S6: UV-C excimer (222 nm) light intensity (μW/cm^2) at 11 panels in #2 chamber (Top, Bottom, Front Top, Front Bottom, Left Top, Left Bottom, Right Top, Right Bottom, Back Top, Back Middle, and Back Bottom), Table S7: UV-C fluorescent (185 + 254 nm) light intensity (μW/cm^2) at 11 panels in #2 chamber (Top, Bottom, Front Top, Front Bottom, Left Top, Left Bottom, Right Top, Right Bottom, Back Top, Back Middle, and Back Bottom), Table S8: Mitigation of CH4 with the different UV doses (1.3, 2.0, 2.5, 3.9, and 5.8 mJ/cm^2) irradiating gaseous emissions from swine manure, Table S9: Mitigation of CO2 with the different UV doses (1.3, 2.0, 2.5, 3.9, and 5.8 mJ/cm^2) irradiating gaseous emissions from swine manure. Bold signifies statistical significance, Table S10: Mitigation of CO2 concentration with different UV wavelength irradiating gaseous emissions inside the #2 chamber. Bold signifies statistical significance.

Author Contributions: Conceptualization, J.A.K., W.S.J.; methodology, J.A.K.; validation, M.L., W.M., and J.A.K.; formal analysis, M.L.; investigation, M.L., W.M., B.C., P.L. and C.B.; resources, M.L., W.M., B.C., P.L., C.B. and J.A.K.; data curation, M.L., W.M. and J.A.K.; writing—original draft preparation, M.L.; writing—review and editing, M.L., J.A.K. and W.S.J.; visualization, M.L.; supervision, J.A.K.; project administration, J.A.K., W.S.J.; funding acquisition, J.A.K., W.S.J. All authors have read and agreed to the published version of the manuscript.

Funding: This research was supported by Iowa Pork Producers Association Project #18-089 "Employing environmental mitigation technology and/or practices: Treating swine odor and improving air quality with black light." In addition, this research was partially supported by the Iowa Agriculture and Home Economics Experiment Station, Ames, Iowa. Project no. IOW05556 (Future Challenges in Animal Production Systems: Seeking Solutions through Focused Facilitation) sponsored by Hatch Act and State of Iowa funds.

Institutional Review Board Statement: Not applicable.

Informed Consent Statement: Not applicable.

Data Availability Statement: The original contributions presented in the study are included in the article/Supplementary Materials; further inquiries can be directed to the corresponding author.

Acknowledgments: The authors gratefully acknowledge Woosang Lee (Smart Control & Sensing Inc.) for his help with the NH_3 and H_2S monitoring system.

Conflicts of Interest: The author does not declare a conflict of interest. The funders did not play any role in the study design, data collection, analysis, interpretation, and decision to write a manuscript or present results.

Appendix A

Figure A1. Picture of UV mobile laboratory (**back**) and filter house mounted on a trailer (**front**) with manure drum and gaseous emission generation system.

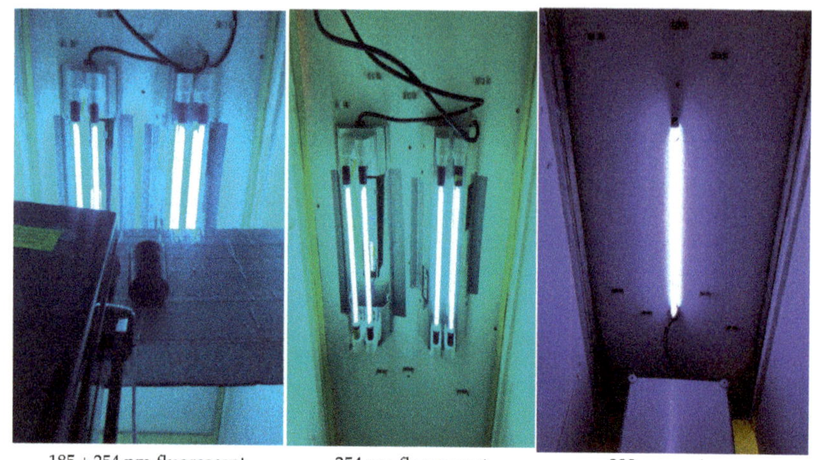

Figure A2. Picture of UV-C lamps installed inside Chamber #2. Caution—colors should be interpreted as actual UV output.

References

1. Maurer, D.L.; Koziel, J.A.; Harmon, J.D.; Hoff, S.J.; Rieck-Hinz, A.M.; Andersen, D.S. Summary of performance data for technologies to control gaseous, odor, and particulate emissions from livestock operations: Air management practices assessment tool (AMPAT). *Data Brief* **2016**, *7*, 1413–1429. [CrossRef] [PubMed]
2. Air Management Practices Assessment Tool. Available online: https://www.extension.iastate.edu/ampat/animal-housing (accessed on 24 February 2021).
3. Costa, A.; Chiarello, G.L.; Selli, E.; Guarino, M. Effects of TiO_2 based photocatalytic paint on concentrations and emissions of pollutants and on animal performance in a swine weaning unit. *J. Environ. Manag.* **2012**, *96*, 86–90. [CrossRef] [PubMed]
4. Guarino, M.; Costa, A.; Porro, M. Photocatalytic TiO_2 coating—To reduce ammonia and greenhouse gases concentration and emission from animal husbandries. *Bioresour. Technol.* **2008**, *99*, 2650–2658. [CrossRef] [PubMed]
5. Zhu, W.; Koziel, J.A.; Maurer, D.L. Mitigation of livestock odors using black light and a new titanium dioxide-based catalyst: Proof-of-concept. *Atmosphere* **2017**, *8*, 103. [CrossRef]
6. Koziel, J.A.; Yang, X.; Cutler, T.; Zhang, S.; Zimmerman, J.J.; Hoff, S.J.; Jenks, W.S.; Laor, Y.; Ravid, U.; Armon, R. Mitigation of odor and pathogens from CAFOs with UV/TiO_2: Exploring the cost effectiveness. In Proceedings of the Mitigating Air Emissions from Animal Feeding Operations Conference, Des Moines, IA, USA, 19–21 May 2008; pp. 169–173.
7. Rockafellow, E.M.; Koziel, J.A.; Jenks, W.S. Laboratory-Scale Investigation of UV Treatment of Ammonia for Livestock and Poultry Barn Exhaust Applications. *J. Environ. Qual.* **2012**, *41*, 281–288. [CrossRef] [PubMed]
8. Yang, X.; Koziel, J.A.; Laor, Y.; Zhu, W.; van Leeuwen, J.H.; Jenks, W.S.; Hoff, S.J.; Zimmerman, J.; Zhang, S.; Ravid, U. VOC Removal from Manure Gaseous Emissions with UV Photolysis and UV-TiO_2 Photocatalysis. *Catalysts* **2020**, *10*, 607. [CrossRef]
9. Lee, M.; Li, P.; Koziel, J.A.; Ahn, H.; Wi, J.; Chen, B.; Meiirkhanuly, Z.; Banik, C.; Jenks, W.S. Pilot-scale testing of UV-A light treatment for mitigation of NH_3, H_2S, GHGs, VOCs, odor, and O_3 inside the poultry barn. *Front. Chem.* **2020**, *8*, 613. [CrossRef] [PubMed]
10. Lee, M.; Wi, J.; Koziel, J.A.; Ahn, H.; Li, P.; Chen, B.; Meiirkhanuly, Z.; Banik, C.; Jenks, W. Effects of UV-A Light Treatment on Ammonia, Hydrogen Sulfide, Greenhouse Gases, and Ozone in Simulated Poultry Barn Conditions. *Atmosphere* **2020**, *11*, 283. [CrossRef]
11. Maurer, D.L.; Koziel, J.A. On-farm pilot-scale testing of black ultraviolet light and photocatalytic coating for mitigation of odor, odorous VOCs, and greenhouse gases. *Chemosphere* **2019**, *221*, 778–784. [CrossRef] [PubMed]
12. Li, P.; Koziel, J.A.; Zimmerman, J.J.; Zhang, J.; Cheng, T.Y.; Yim-Im, W.; Hoff, S.J. Mitigation of Airborne PRRSV Transmission with UV Light Treatment: Proof-of-Concept. *Agriculture* **2021**, *11*, 259. [CrossRef]
13. Lee, M.; Koziel, J.A.; Murphy, W.; Jenks, W.S.; Fonken, B.; Storjohann, R.; Chen, B.; Li, P.; Banik, C.; Wahe, L. Design and testing of mobile laboratory for mitigation of gaseous emissions from livestock agriculture with photocatalysis. *Int. J. Environ. Res. Public Health* **2021**, *18*, 1523. [CrossRef] [PubMed]
14. Lo, Y.C.M.; Koziel, J.A.; Cai, L.; Hoff, S.J.; Jenks, W.S.; Xin, H. Simultaneous Chemical and Sensory Characterization of Volatile Organic Compounds and Semi-Volatile Organic Compounds Emitted from Swine Manure using Solid Phase Microextraction and Multidimensional Gas Chromatography–Mass Spectrometry–Olfactometry. *J. Environ. Qual.* **2008**, *37*, 521–534. [CrossRef] [PubMed]

15. Chen, B.; Koziel, J.A.; Banik, C.; Ma, H.; Lee, M.; Wi, J.; Meiirkhanuly, Z.; Andersen, D.S.; Białowiec, A.; Parker, D.B. Emissions from swine manure treated with current products for mitigation of odors and reduction of NH_3, H_2S, VOC, and GHG emissions. *Data* **2020**, *5*, 54. [CrossRef]
16. Lee, M.; Koziel, J.A.; Murphy, W.; Jenks, W.S.; Chen, B.; Li, P.; Banik, C. Mitigation of Odor and Gaseous Emissions from Swine Barn with UV-A and UV-C Photocatalysis. *Preprints* **2021**, 2021030629. [CrossRef]
17. Yao, H.; Feilberg, A. Characterisation of photocatalytic degradation of odorous compounds associated with livestock facilities by means of PTR-MS. *Chem. Eng. J.* **2015**, *277*, 341–351. [CrossRef]
18. Liu, Z.; Murphy, P.; Maghirang, R.; DeRouchey, J. Mitigation of air emissions from swine buildings through the photocatalytic technology using UV/TiO_2. In Proceedings of the 2015 ASABE Annual International Meeting, New Orleans, LA, USA, 26–29 July 2015. [CrossRef]

Article

Effects of a Partially Perforated Flooring System on Ammonia Emissions in Broiler Housing—Conflict of Objectives between Animal Welfare and Environment?

Carolin Adler [1,*], Alexander J. Schmithausen [2], Manfred Trimborn [1], Sophia Heitmann [3], Birgit Spindler [3], Inga Tiemann [1,4], Nicole Kemper [3] and Wolfgang Büscher [1]

1 Institute of Agricultural Engineering, University of Bonn, 53115 Bonn, Germany; m.trimborn@uni-bonn.de (M.T.); inga.tiemann@uni-bonn.de (I.T.); buescher@uni-bonn.de (W.B.)
2 Corteva Agriscience, Riedenburger Straße 7, 81677 München, Germany; alexander.schmithausen@corteva.com
3 Institute of Animal Hygiene, Animal Welfare and Farm Animal Behaviour, University of Veterinary Medicine Hannover, Foundation, 30559 Hannover, Germany; sophia.heitmann@tiho-hannover.de (S.H.); birgit.spindler@tiho-hannover.de (B.S.); nicole.kemper@tiho-hannover.de (N.K.)
4 Institute of Animal Science, University of Bonn, 53115 Bonn, Germany
* Correspondence: c.adler@uni-bonn.de

Simple Summary: Previous studies have shown positive effects of a partially perforated flooring system on animal welfare in broiler housing. Towards the end of the fattening periods, the present study showed a higher ammonia emission rate (NH_3 ER) for a partially perforated flooring system compared with a littered control barn. Nevertheless, the measured NH_3 concentrations were below 20 ppm, except during a mechanical litter treatment in the winter fattening period. Furthermore, the system offers the possibility of applying practical solutions that were not feasible before. By using underfloor air extraction, manure belts, or acidification systems underneath the elevated perforated area, NH_3 concentrations and the resulting NH_3 ER could be reduced. Thus, with some optimization, the partially perforated flooring system could be used to contribute to an increase in animal welfare and environmental protection at the same time.

Abstract: A partially (50%) perforated flooring system showed positive effects on health- and behavior-based welfare indicators without affecting production performance. Ammonia (NH_3) is the most common air pollutant in poultry production, with effects on animal welfare and the environment. The objectives of animal welfare and environmental protection are often incompatible. Therefore, this study addresses the question of how a partially perforated flooring system affects NH_3 emissions. According to German regulations, three fattening periods were carried out with 500 Ross 308 broilers per barn (final stocking density: 39 kg m^{-2}). The experimental barn was equipped with an elevated perforated area in the supply section, accessible by perforated ramps. The remaining area in the experimental barn and the control barn were equipped with wood shavings (600 g m^{-2}). Besides the different floor types, management was identical. Air temperature (Temp), relative air humidity (RH), NH_3 concentration, and ventilation rate (VR) were measured continuously. Furthermore, dry matter (DM) content, pH, and litter quality were assessed. Towards the end of the fattening periods, the NH_3 emission rate (ER) of the partially perforated flooring system was higher compared with that of the littered control barn (all $p < 0.001$). This effect is mainly caused by the higher NH_3 concentrations, which are promoted by the lack of compaction underneath the elevated perforated area and the increase in pH value under aerobic conditions. Nevertheless, the partially perforated flooring system offers different approaches for NH_3 reduction that were previously not feasible, potentially contributing equally to animal welfare and environmental protection.

Keywords: broiler production; alternative flooring; ammonia emissions; animal welfare; environmental impact

Citation: Adler, C.; Schmithausen, A.J.; Trimborn, M.; Heitmann, S.; Spindler, B.; Tiemann, I.; Kemper, N.; Büscher, W. Effects of a Partially Perforated Flooring System on Ammonia Emissions in Broiler Housing—Conflict of Objectives between Animal Welfare and Environment?. *Animals* **2021**, *11*, 707. https://doi.org/10.3390/ani11030707

Academic Editors: Yang Zhao and Lilong Chai

Received: 11 February 2021
Accepted: 1 March 2021
Published: 5 March 2021

1. Introduction

In Germany, broilers are conventionally kept on concrete floors equipped with organic bedding materials [1]. From day 14, about 80% of the litter's dry matter (DM) consists of excrements and feed residues [2]. If the drying conditions are unfavorable, footpad dermatitis, hock burn, and plumage contamination occur [3–5]. As the abovementioned aspects lead to pain and secondary diseases, the incidence and severity are used as animal-based indicators to record animal welfare [6–8]. In addition, the litter provides a potential reservoir for antibiotic-resistant bacteria [9] that can be transmitted to humans via the food chain [10,11]. Two studies were carried out in the past comparing an innovative partially (50%) perforated flooring system with a littered flooring system [12,13]. One feature of the partially perforated flooring system is an elevated perforated area in the section of feed and water supply, accessible by perforated ramps. Next to the elevated perforated area, littered areas are available. The system provides access to two different floor types at different height levels and promotes the animals' natural behavior such as perching and resting on elevated levels [14,15] or pecking, scratching, and dustbathing in contact with litter [16,17]. At the same time, animals are separated from at least 50% of the litter, which contains excrements, moisture, and bacteria. Adler et al. [12] examined the effect of the partially perforated flooring system on animal-based welfare indicators and production performance. The creation of different functional areas at different heights enriched the husbandry environment, increased the environmental complexity, and reduced animals' general fear response, as confirmed by [18,19]. Furthermore, the separation of the animals from at least 50% of the litter had a positive influence on foot pad dermatitis and hock burn. Production performance was not affected by the floor type [12]. Heitmann et al. [13] studied the effect of the partially perforated flooring system on the occurrence of bacteria. A tendency was shown for a higher content of *Escherichia coli* (*E. coli*) in the supply area of the partially perforated system compared with the supply area of the littered flooring system. Owing to the elevated perforated area, the animals did not come into contact with the material containing *E. coli* underneath the perforated floor. Regarding the total bacteria count, a tendency for lower contents in the littered side areas of the partially perforated flooring system was found compared with the littered control barn. In summary, the partially perforated flooring system has a positive effect on health- and behavior-based welfare indicators without a reduction in production performance [12].

Due to the conflict of objectives between animal welfare and the environment, it is also important to consider the influence of the partially perforated flooring system on environmental aspects [20]. Ammonia (NH_3) is the most common air pollutant emitted by poultry production [21]. Several studies revealed that NH_3 has negative effects on human and animal health as well as the environment. NH_3 inside the barn is known to irritate the mucous membranes and damage the respiratory tract of humans and animals [22–24]. Negative effects on broilers' production performance as a result of NH_3 concentrations above 30 ppm were also observed [25,26]. Furthermore, NH_3 released in the air from poultry houses is able to contribute to the production of acid rain [27] and therefore nitrogen (N) deposition in the ecosystem [28]. Potential consequences of N deposition are eutrophication, acidification, less biodiversity, and nitrification of groundwater [29].

Several studies have been carried out on different flooring systems with regard to NH_3. For example, Boggia et al. [30] showed a reduction in the NH_3 concentration using a no-litter flooring system for broilers compared with conventional litter flooring. Almeida et al. [31] used a totally (100%) perforated flooring system and showed a reduction in the NH_3 concentration, if manure was continuously removed during the fattening period. In the case of manure storage underneath a totally (100%) perforated flooring system over several fattening periods, the NH_3 concentrations and the resulting NH_3 emissions were higher compared with litter flooring [32].

It is known that the partially perforated flooring system has positive effects on animal-based welfare indicators [12]. So far it is not known how the partially perforated flooring system affects NH_3 emissions and whether harmful NH_3 concentrations are to be expected.

Therefore, the aim of this study was to investigate the effect of the partially perforated flooring system on NH_3 emissions compared with a littered system.

2. Materials and Methods

2.1. Animals and Housing

This case-control study was carried out at the Educational and Research Center Frankenforst of the Faculty of Agriculture, University of Bonn (Königswinter, Germany; 55°42′55 N and 7°12′26 E). The experiments were performed in accordance with German regulations and approved by the relevant authority (Landesamt für Natur-, Umwelt- und Verbraucherschutz Nordrhein-Westfalen, Recklinghausen; 81.02.04.2018.A057). A total of three fattening periods were carried out in three different seasons from August 2018 to June 2019. Due to specifications of the slaughterhouse, each fattening period lasted 31 to 32 days. The study was carried out according to German regulations by housing 500 Ross 308 broilers per barn and fattening period to achieve a final stocking density of 39 kg m^{-2}. Two identical barns were used to fulfill the conditions of a case-control study (experimental vs. control barn). Both barns were automatically ventilated by negative pressure ventilation, regulated identically via climate computers (PL-9400, Stienen Bedrijfselektronica B.V., RT Nederweert, The Netherlands). More information on management, feeding, lighting program, and vaccination can be found in the previous study by Adler et al. [12].

2.2. Floor Design and Litter Management

Figure 1 illustrates the two different flooring systems. The experimental barn was equipped with an elevated perforated floor in the area of feed and water supply, accessible by perforated ramps. Conventional wood shavings (600 g m^{-2}) were used in the concrete floor areas next to the feed and water supply. The control barn was completely equipped with wood shavings (600 g m^{-2}).

Figure 1. Illustration of the partially perforated flooring system, with an elevated perforated area in the section of feed and water supply, and the littered flooring system.

After each fattening period, both barns were cleaned, disinfected, and provided with new wood shavings. During the winter fattening period, the litter was additionally mechanically treated with a rake on day 11 to ensure an even distribution in all areas of the barn, especially in the area near the ramps.

2.3. Indoor Environmental Factors

Inside the barn, air temperature (Temp) and relative air humidity (RH) were measured every three minutes using data loggers (Tinytag Plus 2—TGP-4500 loggers, Gemini Data Loggers Ltd., Chichester, West Sussex, UK). A total of three data loggers per barn were placed in the same positions and at a height of 55 cm (Figure 2). Data regarding Temp and RH in the environment outside the barns were provided every 10 min by the nearby weather station at a height of 2.0 m (Königswinter, Germany; 50°42′54 N and 7°12′31 E).

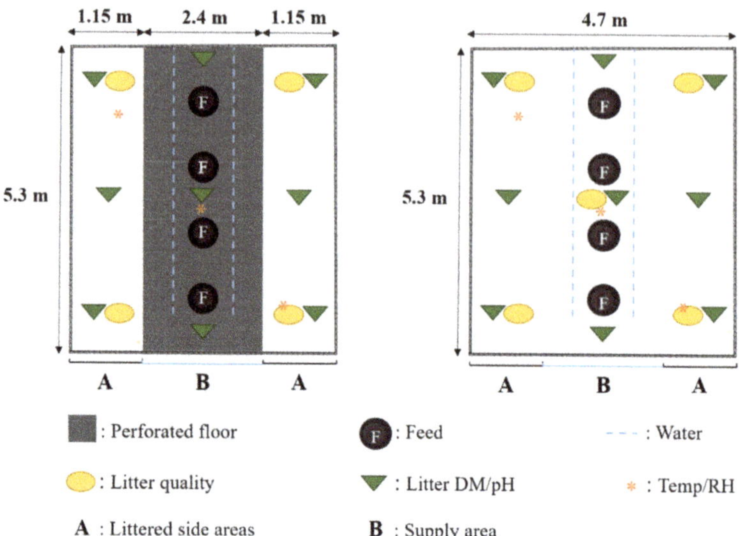

Figure 2. Floor plan view of the partially perforated flooring system, with an elevated perforated area in the section of feed and water supply, and the littered flooring system, including measurement positions for litter quality, litter dry matter (DM), litter pH (pH), air temperature (Temp), and relative air humidity (RH).

2.4. Ammonia Concentration

The NH_3 concentrations were measured continuously via photoacoustic infrared spectroscopy using a Photoacoustic Gas Monitor INNOVA 1412 in combination with a Multipoint Sampler INNOVA 1309 (LumaSense Technologies SA, Ballerup, Dennmark). The measurement setup was based on studies carried out by Schmithausen et al. [33,34]. Additionally, the gas monitor was calibrated by the manufacturer and regularly checked in the measuring laboratory between the fattening periods. In total, three sampling points were installed with filter orifices to protect the technique from dust. Two sample points were installed in the exhaust chimneys of the barns. A third sampling point was installed outside the barns to measure the background concentration. The air of each sampling point was collected continuously by vacuum pumps (ME 2C, Vacuubrand GmbH + Co. KG, Wertheim, Germany) through polytetrafluoroethylene tubes in three separate sample bottles (600 mL). The tubes were equipped with heating cables (A. Rak Wärmetechnik GmbH, Frankfurt am Main, Germany) to avoid water condensation in the tubes with temperature decrease. Due to the continuously flushed sample bottles and tube system, there was always actual sample air available. The tubes between the sample bottles and the Multipoint Sampler were as short as possible (<0.5 m). This measurement setup is useful to ensure a small distance from the sampling point until the analysis. One measuring cycle lasted three minutes with a measuring time of 60 s per sampling point.

2.5. Ventilation Rate

The ventilation rate (VR) in both exhaust chimneys was estimated using ProVent measurement fans (Reventa GmbH, Horstmar, Germany) of the same diameter as the exhaust chimney. The measuring fans were calibrated in a wind tunnel by the manufacturer. There was a calming distance of 2.0 m between the air outlet and the measuring fan to fulfill laminar air flow conditions. These conditions are useful to increase the measurement accuracy of the measurement fans. The data were recorded every minute by Almemo 2590 data loggers (Ahlborn Mess- und Regelungstechnik GmbH, Holzkirchen, Germany).

2.6. Litter Analysis

At the end of each fattening period on day 31 to 32, representative litter samples were taken to determine the litter dry matter (DM). In the spring fattening period, further litter samples were taken on days 7, 14, 21, and 28. Litter samples were taken in nine different positions: a total of six positions in the littered side areas and three positions in the supply area of each barn (Figure 2). In each position, a sample of the entire depth of the litter was taken. The litter samples were weighed before and after being oven-dried at 105 °C for 24 h.

In addition, the litter samples of the spring fattening period were analyzed regarding the pH value. A total amount of 20 g of each litter sample was mixed with 300 g deionized water. The samples were then shaken with an overhead shaker (Reax 20 overhead shaker, Heidolph Instruments GmbH & Co. KG, Schwabach, Germany) for a total time of one hour (30 turns m^{-1}). After shaking, the pH was measured using a pH electrode (InLab Max Pro-ISM electrode, Mettler Toledo, OH, USA). The pH electrode was calibrated using a buffer solution for pH 4 and 7.

2.7. Litter Quality

In all three fattening periods, litter quality was assessed on days 7, 14, 21, and 28 using the scoring system developed by Welfare Quality® [35]. Figure 2 shows the positions where litter quality was evaluated. A total of five litter samples were taken in the littered control barn, with four samples in the littered side areas and one sample in the supply area. In the experimental barn, four litter samples were taken in the littered side areas. No litter quality assessment was performed underneath the perforated area in the supply area of the experimental barn. Underneath the perforated area, mainly excrements are stored, which cannot be defined and evaluated as litter. A scoring system from 0 to 4 was used to evaluate the litter quality [35]. Figure 3 illustrates the images of the different scores. Score 0 was equal to "completely dry and flaky, that is, moves easily with the foot"; score 1 was equal to "dry but not easy to move with foot"; score 2 was equal to "leaves imprint of foot and will form a ball if compacted, but ball does not stay together well"; score 3 was equal to "sticks to boots and sticks readily in a ball if compacted"; and score 4 was equal to "sticks to boots once the cap or compacted crust is broken."

Figure 3. Illustration of the scores used during the litter quality evaluation.

2.8. Data Processing and Statistical Analysis

For statistical analysis, SPSS®® Statistics 25 (IBM Corporation, Armonk, NY, USA) was used. Graphical presentation was done with SigmaPlot 14.0 (Systat Software Inc., Chicago, IL, USA). The daily emission rate (ER) was calculated using the average hourly NH_3 concentrations (n = 20 values per hour and sample point) and the average hourly ventilation rates (n = 60 values per hour and barn) using the following equation:

$$ER = (((C_{inside} - C_{outside}) \times VR)/N) \times 24 \quad (1)$$

where:

ER = emission rate (g d^{-1} $bird^{-1}$)

C_{inside} = inside ammonia concentration (g m^{-3})

$C_{outside}$ = outside ammonia concentration (g m^{-3})

VR = ventilation rate (m^3 h^{-1})

N = actual number of birds per day

For each fattening period, the hourly values for NH_3 ER, NH_3 concentration, VR, air Temp, and RH were divided into three sections according to the feeding program: start (day 0 to 6), middle (day 7 to 27), and end of the fattening period (day 28 to 31 or 32). Data were analyzed using general linear models (GLMs). For litter quality, the link function was Poisson distributed, otherwise it was linear distributed. In the first step, univariate GLMs were used to select the significant main effects with NH_3 ER, NH_3 concentration, VR, air Temp, RH, litter quality, DM, and pH as response variables. Significant main effects were then analyzed by a multifactorial GLM. After backward selection, the final GLMs were interpreted with interaction terms (Fattening period × Section × Floor type, Area × Fattening period, or Area × Day). The *p*-values were corrected by Bonferroni. Differences of $p \leq 0.05$ were considered statistically significant, and differences of $0.05 \leq p \leq 0.10$ were considered a tendency.

3. Results

3.1. Indoor Environmental Factors

Figure 4 shows typical Temp curves with a decreasing Temp towards the end of the fattening periods (all $p < 0.001$). During the summer fattening period, no additional heating sources were used. Therefore, the barn Temp of the summer fattening period was more oriented to the outside Temp. The initial Temp differences between both floor types in the summer fattening period are due to the settings of the climate computer ($p < 0.01$).

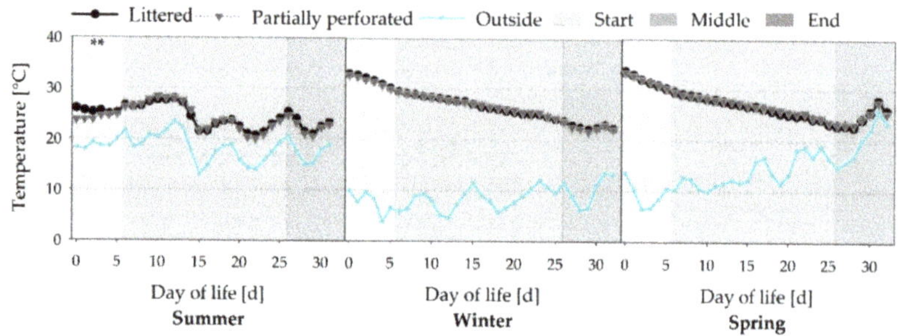

Figure 4. Average daily air temperature inside the broiler houses with two different floor types, and the outside environment air temperature measured over three different fattening periods ($n = 144$ values per day and barn). Significant differences between both floor types within the three different sections (start, middle, end of the fattening period) are marked by asterisks: ** $p < 0.01$.

An increase in RH from the start to the end of the three fattening periods is presented in Figure 5 (all $p < 0.001$). Analogous to the Temp, the summer fattening period is oriented to the outside RH. Differences in RH between the flooring systems in the start and middle sections of the summer fattening period are due to the settings of the climate computer (all $p < 0.05$). On day 11 of the winter fattening period, the litter was mechanically treated, resulting in an RH peak. At this time, the RH was higher for the littered control barn compared with the partially perforated flooring system ($p < 0.001$).

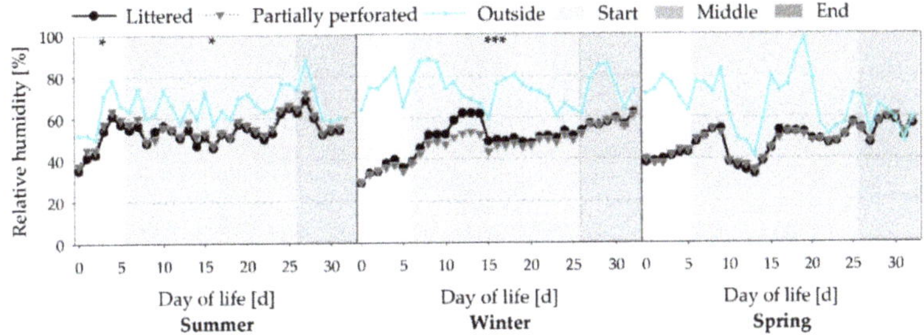

Figure 5. Average daily relative air humidity inside the broiler houses with two different floor types, and the outside environment RH measured over three different fattening periods (n = 144 values per day and barn). Significant differences between both floor types within the three different sections (start, middle, end of the fattening period) are marked by asterisks: * $p \leq 0.05$, *** $p < 0.001$.

3.2. Ammonia Concentration

Figure 6 shows a typical increase in NH_3 concentration towards the end of the fattening periods (all $p < 0.01$). Higher ventilation rates in the summer fattening period reflect lower NH_3 concentrations (all $p < 0.001$). Owing to the mechanical litter treatment on day 11 of the winter fattening period, an NH_3 concentration peak occurred. This peak was higher in the littered control barn than in the barn with the partially perforated flooring system ($p < 0.001$). Towards the end of the winter and spring fattening periods, NH_3 concentrations of the partially perforated flooring system were higher compared with the littered control barn (all $p < 0.01$).

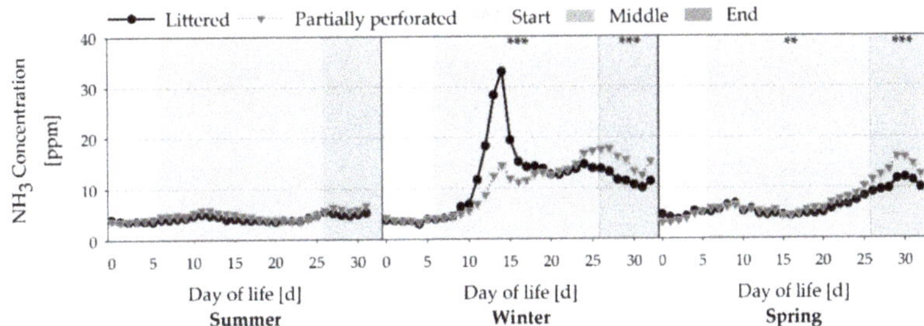

Figure 6. Average daily ammonia (NH_3) concentrations from broiler houses with two different floor types measured over three different fattening periods (n = 480 values per day and barn). Significant differences between both floor types within the three different sections (start, middle, end of the fattening period) are marked by asterisks: ** $p < 0.01$, *** $p < 0.001$.

3.3. Ventilation Rate

A typical course with an increasing VR towards the end of the fattening periods is shown in Figure 7 (all $p < 0.001$). Depending on the seasons, the highest VR is shown in the summer compared with the winter and spring fattening periods (all $p < 0.001$). Differences between both floor types in the middle section of the summer fattening period are due to the settings of the climate computer ($p < 0.001$). The VR is a preset parameter of the climate computer, resulting in parallel VR curves.

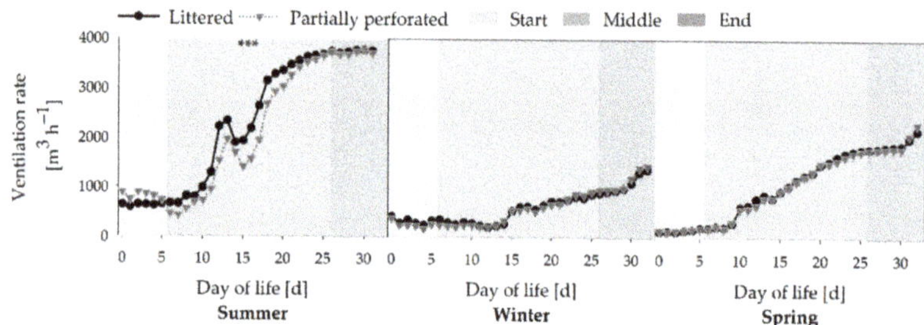

Figure 7. Average daily ventilation rate from broiler houses with two different floor types measured over three different fattening periods (n = 1.440 values per day and barn). Significant differences between both floor types within the three different sections (start, middle, end of the fattening period) are marked by asterisks: *** $p < 0.001$.

3.4. Ammonia Emission Rate

Figure 8 shows a common NH_3 ER increase over the time for all fattening periods (all $p < 0.001$). Overall, the average NH_3 ER of the three fattening periods was lower for the littered control barn (0.15 ± 0.12 g day^{-1} bird^{-1}) compared with the partially perforated flooring system (0.19 ± 0.19 g day^{-1} bird^{-1}). These differences were especially observed towards the end of the fattening periods, with a lower NH_3 ER for the littered control barn than for the partially perforated flooring system (all $p < 0.001$). Drifting apart of the NH_3 ER between the two flooring systems began on days 26, 22, and 18 for the summer, winter, and spring fattening periods, respectively. Furthermore, a stagnation of the NH_3 ER could be observed in the end section of the littered control barn in the summer and winter fattening periods.

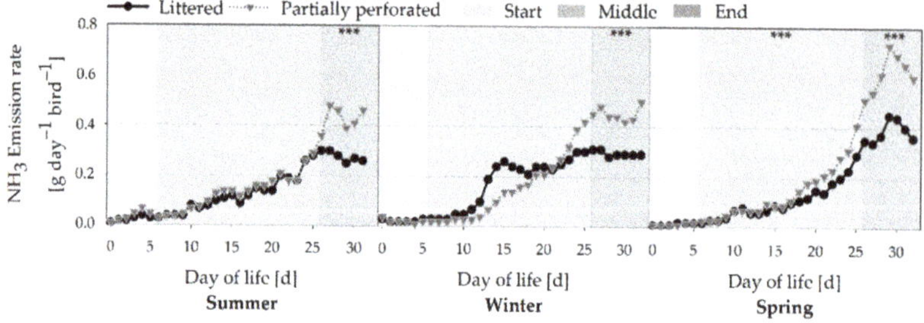

Figure 8. Average daily ammonia (NH_3) emission rate from broiler houses with two different floor types measured over three different fattening periods (n = 24 values per day and barn). Significant differences between both floor types within the three different sections (start, middle, end of the fattening period) are marked by asterisks: *** $p < 0.001$.

3.5. Litter Analysis

The average litter DM at the end of the fattening periods is shown in Figure 9. At the end of the summer fattening period, litter DM was higher in the barn with the partially perforated flooring system compared with litter flooring ($p = 0.004$). In the supply area, litter DM was always lower for the partially perforated flooring system (all $p < 0.032$).

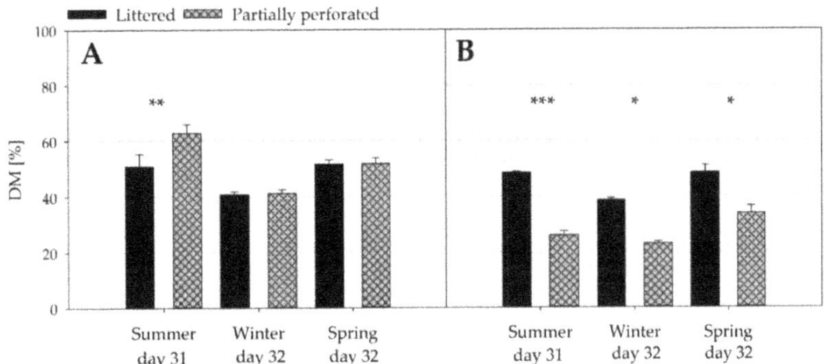

Figure 9. Average dry matter (DM) contents of litter samples from (**A**) the littered side areas and (**B**) the supply area from broiler houses with two different floor types at the end of each of the three different fattening periods ($n = 6$ values per fattening period and barn in area (**A**); $n = 3$ values per fattening period and barn in area (**B**)). Significant differences are marked by asterisks: * $p \leq 0.05$, ** $p < 0.01$, *** $p < 0.001$.

The effect of a lower litter DM in the supply area of the experimental barn compared with the control barn is also reflected in Figure 10 over the whole spring fattening period (all $p \leq 0.01$).

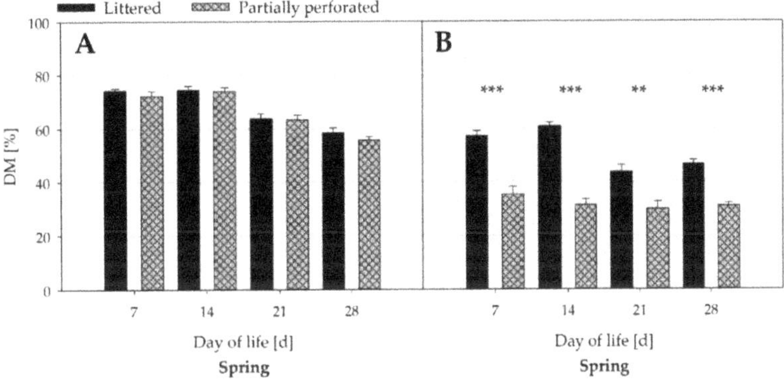

Figure 10. Average dry matter (DM) contents of litter samples from (**A**) the littered side areas and (**B**) the supply area ($n = 6$ values per day and barn in area (**A**); $n = 3$ values per day and barn in area (**B**)) from broiler houses with two different floor types measured over the spring fattening period. Significant differences are marked by asterisks: ** $p < 0.01$, *** $p < 0.001$.

Figure 11 presents the average daily litter pH values of the spring fattening period. In the littered side areas, litter pH increased from day 1 to 32 for both floor types (all $p < 0.001$). On day 21, litter pH in the littered side areas was higher in the control barn compared with the partially perforated flooring system ($p = 0.032$). In the supply area of the control barn, litter pH increased from day 14 to 28 and dropped back to the starting level by day 32 (all $p < 0.015$). By contrast, there was a continuous increase in litter pH from day 14 in the supply area of the barn with the partially perforated flooring system (all $p < 0.001$). From day 28, litter pH in the supply area was higher in the barn with the partially perforated flooring system compared with litter flooring (all $p < 0.015$).

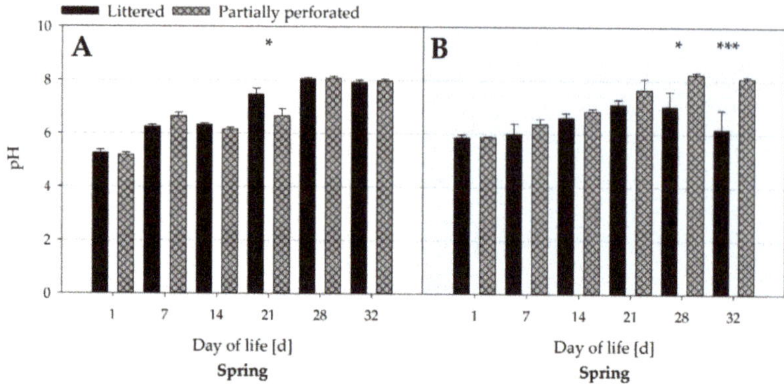

Figure 11. Average pH values of litter samples from (**A**) the littered side areas and (**B**) the supply area from broiler houses with two different floor types measured over the spring fattening period (n = 6 values per day and barn in area (**A**); n = 3 values per day and barn in area (**B**)). Significant differences are marked by asterisks: * $p \leq 0.05$, *** $p < 0.001$.

3.6. Litter Quality

Figure 12 presents a decrease in the litter quality of the littered side areas from the start to the end of the fattening periods ($p < 0.001$). For the littered control barn, litter quality was higher in the littered side areas compared with the supply area ($p = 0.002$). In the supply area of the littered control barn, the highest litter quality score was achieved from day 14 with a fully compacted litter surface. This score was never found in the littered side areas for both floor types. No litter quality assessment was carried out in the supply area of the experimental barn, as the accumulated excrements underneath the perforated area cannot be scored as litter.

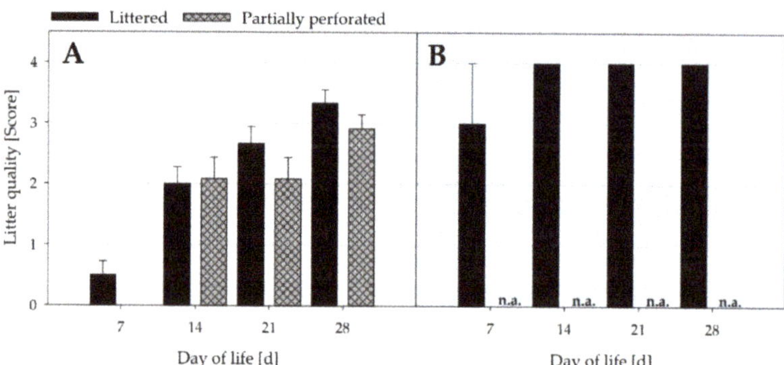

Figure 12. Mean scores of litter quality from (**A**) the littered side areas (n = 12 values per day and barn) and (**B**) the supply area (n = 3 values per day and barn) from broiler houses with two different floor types. n.a. = not assessed (no litter quality assessment underneath the perforated area in the supply area of the experimental barn).

4. Discussion

A multifactorial evaluation of an innovative flooring system in broiler production is important to ensure the compatibility of animal welfare and the environment. Previous studies have shown positive effects of a partially perforated flooring system on animal-based welfare indicators, without affecting production performance or the occurrence of bacteria [12,13]. Positive effects of elevated perforated areas on health- and behavior-based welfare indicators are confirmed by numerous studies [18,19,36–39]. Conflicting information is available regarding the environmental impact of a totally perforated flooring system, with a focus on NH_3 concentration and ER [31,32]. No results are published on the effect of a partially perforated flooring system on NH_3 concentration and ER.

By measuring the NH_3 concentrations and VR, the NH_3 ER was calculated. Results showed that the NH_3 ER of the partially perforated flooring system was higher compared with the littered control barn towards the end of the fattening periods. Since no differences occurred in the VR between both flooring systems at that time, the higher NH_3 ER is explained by the higher NH_3 concentrations. As reported by Groot Koerkamp [40], NH_3 is mainly generated from the decomposition of uric acid and undigested protein excreted by the broilers. NH_3 volatilization depends on several factors such as VR, indoor Temp, and RH, as well as litter characteristics such as DM and pH [41]. In the current study, no differences in Temp and RH were measured towards the end of the fattening periods. Litter DM in the supply area was higher in the control compared with the experimental barn. From day 28, the pH in the supply area of the control barn was in an acidic range, and in the supply area of the experimental barn the pH was in an alkaline range. In addition, litter quality assessment revealed cake formation in the supply area of the control barn from day 14.

A decrease in litter DM increases the NH_3 volatilization [42]. Groot Koerkamp [40] reported that 40–60% litter DM provides the best conditions for microbial growth and thus for the release of NH_3. Therefore, Miles et al. [43] stated that 54–55% of the total NH_3 generated by the litter can be expected near the water lines promoted by the low litter DM level. The dissociation equilibrium between ammonium (NH_4^+) and NH_3 is strongly pH-dependent. With an increase in the litter pH above pH 7, N is increasingly released as NH_3 due to the microbial implementation processes [44–47]. This is supported by a function set up by Kirchmann and Lundvall [48], explaining 79% of the variance in the extent of the NH_3 volatilization from animal wastes due to an increase in pH value. Various studies on the storage of animal excrements showed that the NH_3 release is promoted by aerobic conditions and inhibited by anaerobic conditions [49–51]. Anaerobic conditions during farmyard manure storage cause the degradation of organic material into volatile organic acids, resulting in a decrease in pH [52,53]. Under aerobic litter conditions, the volatile organic acids are degraded to CO_2, which increases the pH with time [54]. From day 14, cake formation occurred in the supply area of the littered control barn. Cake formation is described as compacted litter at a height of 5 to 10 cm on the surface of the litter [55]. Chadwick [56] demonstrated an NH_3 reducing effect of 50–90% by using compaction and covering cattle manure heaps. The opposite effect is shown during the turning of cattle manure, where the air exchange and the increase in the emitting surface area caused an increase in the NH_3 concentration [57]. In the present study, loosening of litter also increased the NH_3 concentration, as presented by the post-litter treatment NH_3 concentration peak on day 11 of the winter fattening period.

In summary, the animals compacted the litter in the supply area of the littered control barn, resulting in cake formation from day 14. Cake formation led to anaerobic conditions, a decrease in litter pH, and thus an inhibition of the NH_3 release. In contrast, animals were separated from their excrements by the elevated perforated floor in the supply area of the experimental barn. As a result, no compaction by the animals took place and aerobic conditions occurred. As mentioned above, aerobic conditions increased the pH to an alkaline range, which promoted the NH_3 release and thus the NH_3 concentration and NH_3 ER.

Despite the higher NH_3 concentrations and NH_3 ER, the partially perforated flooring system promotes animal welfare as indicated by animal-based welfare indicators without affecting production performance [12]. Such conflicting goals between animal welfare vs. environmental aspects are already known in livestock husbandry [58]. For example, free-range and organic systems in broiler production are often associated with longer fattening periods in combination with higher feed consumptions and nutrient excretions. These systems promote animal welfare but have a negative impact on the environment [59,60].

Compared with previous systems, the partially perforated flooring system offers the possibility of implementing reduction measures that were not implementable before. Solution approaches can be used in the supply area, where more than 50% of the total NH_3 generation is expected [43]. For example, underfloor air extraction has become established in pig production to eliminate released NH_3 directly at the source [61,62]. Such a system could also be tested in combination with the partially perforated flooring system. Almeida et al. [31] mentioned a decrease in NH_3 concentration of a totally (100%) perforated flooring system in combination with regular manure removal during the fattening period. It is known from aviary systems in laying hens that regular manure removal with manure belts reduces NH_3 concentrations [40,63,64]. Manure belts underneath the elevated partially perforated area could be tested for regular manure removal in the supply area. The system could also be used for catching animals for slaughter and final manure removal of the whole barn. In addition, the use of acids in pig and cattle slurry has proven to be effective in reducing the pH and thus the NH_3 release [65,66]. Similar approaches exist with pH-lowering litter additives in poultry production [67,68]. The partially perforated flooring system offers the possibility of testing a system to acidify the material below the elevated perforated area and thus reduce NH_3 volatilization. The advantage is that the animals are separated from the acidified material due to the elevated perforated area. With these measures, the increased NH_3 volatilization underneath the perforated area could be compensated for and largely avoided. Further research approaches are necessary to reduce the NH_3 concentrations inside the barn as well as the resulting NH_3 ER, contributing equally to animal welfare and environmental protection.

5. Conclusions

This study examined how a partially perforated (50%) flooring system affects NH_3 emissions compared with a littered system in broiler housing. Compared with the littered flooring system, the material underneath the elevated perforated area was not compacted during the fattening periods. Therefore, aerobic conditions were present, which increased the litter pH and thus the NH_3 concentration and NH_3 ER in the final phase of the fattening period. Nevertheless, with the exception of the mechanical litter treatment in the winter fattening period, NH_3 concentrations in the exhaust air were below 20 ppm for both floor types. Underfloor air extraction, manure belts, or acidification systems below the elevated perforated area are potential solution approaches to inhibit NH_3 release that were not feasible before. Previous studies showed a positive effect of the partially (50%) perforated flooring system on health- and behavior-based welfare indicators without affecting production performance. In summary, with optimization measures, the partially perforated flooring system could contribute to an improvement in animal welfare and environmental protection at the same time. Further studies should be carried out with a focus on the partially perforated flooring system combined with NH_3 reduction strategies from agricultural practice.

Author Contributions: Conceptualization, C.A., A.J.S., M.T., S.H., B.S., I.T., N.K., and W.B.; methodology, C.A., A.J.S., I.T., and W.B.; software, C.A., A.J.S., and I.T.; validation, C.A., A.J.S., and I.T.; formal analysis, C.A., A.J.S., and M.T.; investigation, C.A. and A.J.S.; resources, I.T. and W.B.; data curation, C.A. and S.H.; writing—original draft preparation, C.A.; writing—review and editing, C.A., A.J.S., M.T., I.T., and W.B.; visualization, C.A.; supervision, I.T. and W.B.; project administration, N.K. and W.B.; funding acquisition, B.S., N.K., and W.B. All authors have read and agreed to the published version of the manuscript.

Funding: This research was funded by the Federal Ministry of Food and Agriculture and supported by the Federal Office for Agriculture and Food (2817700214). The authors thank the Verein zur Förderung der Landtechnik Bonn und Haushaltstechnik Bonn e. V. for their contribution to covering publication costs.

Institutional Review Board Statement: The study was carried out at the Educational and Research Center Frankenforst of the Faculty of Agriculture, University of Bonn (Königswinter, Germany;

55°42′55 N and 7°12′26 E). The experiments were performed in accordance with German regulations and approved by the relevant authority (Landesamt für Natur-, Umwelt- und Verbraucherschutz Nordrhein-Westfalen, Recklinghausen; 81.02.04.2018.A057).

Data Availability Statement: The data presented in this study are available on request from the corresponding author.

Acknowledgments: The authors wish to acknowledge the Educational and Research Center Frankenforst of the Faculty of Agriculture, University of Bonn (Königswinter, Germany) for the care of the animals and their comprehensive support. Also acknowledged are the team of the Institute of Agricultural Engineering, University of Bonn, and many students for their great help in data collection.

Conflicts of Interest: The authors declare no conflict of interest.

References

1. Bergmann, S.; Schwarzer, A.; Wilutzky, K.; Louton, H.; Bachmeier, J.; Schmidt, P.; Erhard, M.; Rauch, E. Behavior as welfare indicator for the rearing of broilers in an enriched husbandry environment—A field study. *J. Vet. Behav.* **2017**, *19*, 90–101. [CrossRef]
2. Kamphues, J.; Youssef, A.; Abd El-Wahab, B.; Üffing, B.; Witte, M.; Tost, M. Influence of feeding and housing on foot pad health in hens and turkeys. *Übers. Tierernährg.* **2011**, *39*, 147–195.
3. Youssef, I.M.I.; Beineke, A.; Rohn, K.; Kamphues, J. Experimental study on effects of litter material and its quality on foot pad dermatitis in growing turkeys. *Int. J. Poult. Sci.* **2010**, *9*, 1125–1135. [CrossRef]
4. De Jong, I.C.; van Harn, J.; Gunnink, H.; Hindle, V.A.; Lourens, A. Footpad dermatitis in Dutch broiler flocks: Prevalence and factors of influence. *Poult. Sci.* **2012**, *91*, 1569–1574. [CrossRef] [PubMed]
5. Haslam, S.M.; Knowles, T.G.; Brown, S.N.; Wilkins, L.J.; Kestin, S.C.; Warriss, P.D.; Nicol, C.J. Factors affecting the prevalence of foot pad dermatitis, hock burn and breast burn in broiler chicken. *Br. Poult. Sci.* **2007**, *48*, 264–275. [CrossRef] [PubMed]
6. Martland, M.F. Ulcerative dermatitis dm broiler chickens: The effects of wet litter. *Avian Pathol.* **1985**, *14*, 353–364. [CrossRef]
7. Michel, V.; Prampart, E.; Mirabito, L.; Allain, V.; Arnould, C.; Huonnic, D.; Le Bouquin, S.; Albaric, O. Histologically-validated footpad dermatitis scoring system for use in chicken processing plants. *Br. Poult. Sci.* **2012**, *53*, 275–281. [CrossRef] [PubMed]
8. Bessei, W. Welfare of broilers: A review. *Worlds Poult. Sci. J.* **2006**, *62*, 455–466. [CrossRef]
9. Furtula, V.; Farrell, E.G.; Diarrassouba, F.; Rempel, H.; Pritchard, J.; Diarra, M.S. Veterinary pharmaceuticals and antibiotic resistance of Escherichia coli isolates in poultry litter from commercial farms and controlled feeding trials. *Poult. Sci.* **2010**, *89*, 180–188. [CrossRef] [PubMed]
10. Kemper, N. Veterinary antibiotics and their possible impact on resistant bacteria in the environment. In *Antibiotic Resistance: Causes and Risk Factors, Mechanisms and Alternatives*, 2nd ed.; Bonilla, A.R., Muniz, K.P., Eds.; Nova Science Publishers, Inc.: New York, NY, USA, 2009; pp. 467–495. ISBN 9781607419730.
11. Amador, P.; Fernandes, R.; Prudêncio, C.; Duarte, I. Prevalence of antibiotic resistance genes in multidrug-resistant enterobacteriaceae on portuguese livestock manure. *Antibiotics* **2019**, *8*, 23. [CrossRef] [PubMed]
12. Adler, C.; Tiemann, I.; Hillemacher, S.; Schmithausen, A.J.; Müller, U.; Heitmann, S.; Spindler, B.; Kemper, N.; Büscher, W. Effects of a partially perforated flooring system on animal-based welfare indicators in broiler housing. *Poult. Sci.* **2020**, *99*, 3343–3354. [CrossRef]
13. Heitmann, S.; Stracke, J.; Adler, C.; Ahmed, M.F.E.; Schulz, J.; Büscher, W.; Kemper, N.; Spindler, B. Effects of a slatted floor on bacteria and physical parameters in litter in broiler houses. *Vet. Anim. Sci.* **2020**, *9*, 100115. [CrossRef] [PubMed]
14. Wood, D.J.; van Heyst, B.J. A review of ammonia and particulate matter control strategies for poultry housing. *Trans. ASAE* **2016**, *59*, 329–344. [CrossRef]
15. Schrader, L.; Müller, B. Night-time roosting in the domestic fowl: The height matters. *Appl. Anim. Behav. Sci.* **2009**, *121*, 179–183. [CrossRef]
16. Larsen, B.; Vestergaard, K.S.; Hogan, J.A. Development of dustbathing behavior sequences in the domestic fowl: The significance of functional experience. *Dev. Psychobiol.* **2000**, *37*, 5–12. [CrossRef]
17. Blokhuis, H.J. The effect of a sudden change in floor type on pecking behaviour in chicks. *Appl. Anim. Behav. Sci.* **1989**, *22*, 65–73. [CrossRef]
18. Tahamtani, F.M.; Pedersen, I.J.; Toinon, C.; Riber, A.B. Effects of environmental complexity on fearfulness and learning ability in fast growing broiler chickens. *Appl. Anim. Behav. Sci.* **2018**, *207*, 49–56. [CrossRef]
19. Baxter, M.; Bailie, C.L.; O'Connell, N.E. Play behaviour, fear responses and activity levels in commercial broiler chickens provided with preferred environmental enrichments. *Animal* **2019**, *13*, 171–179. [CrossRef]
20. De Jonge, J.; van Trijp, H.C.M. Meeting heterogeneity in consumer demand for animal welfare: A reflection on existing knowledge and implications for the meat sector. *J. Agric. Environ. Ethics* **2013**, *26*, 629–661. [CrossRef]
21. Almuhanna, E.A.; Ahmed, A.S.; Al-Yousif, Y.M. Effect of air contaminants on poultry immunological and production performance. *Int. J. Poult. Sci.* **2011**, *10*, 461–470. [CrossRef]

22. Valentine, H. A study of the effect of different ventilation rates on the ammonia concentrations in the atmosphere of broiler houses. *Br. Poult. Sci.* **1964**, *5*, 149–159. [CrossRef]
23. Ihrig, A.; Hoffmann, J.; Triebig, G. Examination of the influence of personal traits and habituation on the reporting of complaints at experimental exposure to ammonia. *Int. Arch. Occup. Environ. Health* **2006**, *79*, 332–338. [CrossRef] [PubMed]
24. Wu, Y.N.; Yan, F.F.; Hu, J.Y.; Chen, H.; Tucker, C.M.; Green, A.R.; Cheng, H.W. The effect of chronic ammonia exposure on acute-phase proteins, immunoglobulin, and cytokines in laying hens. *Poult. Sci.* **2017**, *96*, 1524–1530. [CrossRef] [PubMed]
25. Miles, D.M.; Branton, S.L.; Lott, B.D. Atmospheric ammonia is detrimental to the performance of modern commercial broilers. *Poult. Sci.* **2004**, *83*, 1650–1654. [CrossRef] [PubMed]
26. Wang, Y.M.; Meng, Q.P.; Guo, Y.Z.; Wang, Y.Z.; Wang, Z.; Yao, Z.L.; Shan, T.Z. Effect of atmospheric ammonia on growth performance and immunological response of broiler chickens. *J. Anim. Vet. Adv.* **2010**, *22*, 2802–2806. [CrossRef]
27. Xin, H.; Gates, R.S.; Green, A.R.; Mitloehner, F.M.; Moore, P.A.; Wathes, C.M. Environmental impacts and sustainability of egg production systems. *Poult. Sci.* **2011**, *90*, 263–277. [CrossRef]
28. Li, H.; Lin, C.; Collier, S.; Brown, W.; White-Hansen, S. Assessment of frequent litter amendment application on ammonia emission from broilers operations. *J. Air Waste Manag. Assoc.* **2013**, *63*, 442–452. [CrossRef] [PubMed]
29. Van Breemen, N.; van Dijk, H.F.G. Ecosystem effects of atmospheric deposition of nitrogen in The Netherlands. *Environ. Pollut.* **1988**, *54*, 249–274. [CrossRef]
30. Boggia, A.; Paolotti, L.; Antegiovanni, P.; Fagioli, F.F.; Rocchi, L. Managing ammonia emissions using no-litter flooring system for broilers: Environmental and economic analysis. *Environ. Sci. Policy* **2019**, *101*, 331–340. [CrossRef]
31. Almeida, E.A.; Arantes de Souza, L.F.; Sant'Anna, A.C.; Bahiense, R.N.; Macari, M.; Furlan, R.L. Poultry rearing on perforated plastic floors and the effect on air quality, growth performance, and carcass injuries—Experiment 1: Thermal comfort. *Poult. Sci.* **2017**, *96*, 3155–3162. [CrossRef]
32. Li, H.; Wen, X.; Alphin, R.; Zhu, Z.; Zhou, Z. Effects of two different broiler flooring systems on production performances, welfare, and environment under commercial production conditions. *Poult. Sci.* **2017**, *96*, 1108–1119. [CrossRef]
33. Schmithausen, A.J.; Trimborn, M.; Büscher, W. Methodological comparison between a novel automatic sampling system for gas chromatography versus photoacoustic spectroscopy for measuring greenhouse gas emissions under field conditions. *Sensors* **2016**, *16*, 1638. [CrossRef]
34. Schmithausen, A.J.; Schiefler, I.; Trimborn, M.; Gerlach, K.; Südekum, K.-H.; Pries, M.; Büscher, W. Quantification of methane and ammonia emissions in a naturally ventilated barn by using defined criteria to calculate emission rates. *Animals* **2018**, *8*, 75. [CrossRef]
35. Welfare Quality. *Welfare Quality®Assessment Protocol for Poultry (Broilers, Laying Hens)*; Welfare Quality®Consortium: Lelystad, The Netheralands, 2009.
36. Akpobome, G.O.; Fanguy, R.C. Evaluation of cage floor systems for production of commercial broilers. *Poult. Sci.* **1992**, *71*, 274–280. [CrossRef]
37. Cengiz, Ö.; Hess, J.B.; Bilgili, S.F. Effect of protein source on the development of footpad dermatitis in broiler chickens reared on different flooring types. *Arch. Geflügelk.* **2013**, *77*, 166–170.
38. Çavuşoğlu, E.; Petek, M.; Abdourhamane, İ.M.; Akkoc, A.; Topal, E. Effects of different floor housing systems on the welfare of fast-growing broilers with an extended fattening period. *Arch. Anim. Breed.* **2018**, *61*, 9–16. [CrossRef]
39. Çavuşoğlu, E.; Petek, M. Effects of different floor materials on the welfare and behaviour of slow- and fast-growing broilers. *Arch. Anim. Breed.* **2019**, *62*, 335–344. [CrossRef]
40. Groot Koerkamp, P.W.G. Review on emissions of ammonia from housing systems for laying hens in relation to sources, processes, building design and manure handling. *J. Agric. Eng. Res.* **1994**, *59*, 73–87. [CrossRef]
41. Meda, B.; Hassouna, M.; Aubert, C.; Robin, P.; Dourmad, J.Y. Influence of rearing conditions and manure management practices on ammonia and greenhouse gas emissions from poultry houses. *Worlds Poult. Sci. J.* **2011**, *67*, 441–456. [CrossRef]
42. Ivoš, J.; Asaj, A.; Marjanović, L.; Madžirov, Ž. A contribution to the hygiene of deep litter in the chicken house. *Poult. Sci.* **1966**, *45*, 676–683. [CrossRef]
43. Miles, D.M.; Brooks, J.P.; McLaughlin, M.R.; Rowe, D.E. Broiler litter ammonia emissions near sidewalls, feeders, and waterers. *Poult. Sci.* **2013**, *92*, 1693–1698. [CrossRef] [PubMed]
44. Reece, F.N.; Bates, B.J.; Lott, B.D. Ammonia control in broiler houses. *Poult. Sci.* **1979**, *58*, 754–755. [CrossRef]
45. Xue, S.K.; Chen, S.; Hermanson, R.E. Measuring ammonia and hydrogen sulfide emitted from manure storage facilities. *Trans ASAE* **1998**, *41*, 1125–1130. [CrossRef]
46. Li, H.; Xin, H.; Burns, R.T.; Roberts, S.A.; Li, S.; Kliebenstein, J.; Bregendahl, K. Reducing ammonia emissions from laying-hen houses through dietary manipulation. *J. Air Waste Manag. Assoc.* **2012**, *62*, 160–169. [CrossRef] [PubMed]
47. Hartung, J.; Phillips, V.R. Control of gaseous emissions from livestock buildings and manure stores. *J. Agric. Eng. Res.* **1994**, *57*, 173–189. [CrossRef]
48. Kirchmann, H.; Lundvall, A. Treatment of solid animal manures: Identification of low NH_3 emission practices. *Nutr. Cycl. Agroecosyst.* **1998**, *51*, 65–71. [CrossRef]
49. Kirchmann, H.; Witter, E. Ammonia volatilization during aerobic and anaerobic manure decomposition. *Plant Soil* **1989**, *115*, 35–41. [CrossRef]

50. Mahimairaja, S.; Bolan, N.S.; Hedley, M.J.; Macgregor, A.N. Losses and transformation of nitrogen during composting of poultry manure with different amendments: An incubation experiment. *Bioresour. Technol.* **1994**, *47*, 265–273. [CrossRef]
51. Amon, B.; Amon, T.; Boxberger, J.; Alt, C. Emissions of NH3, N2O and CH4 from dairy cows housed in a farmyard manure tying stall (housing, manure storage, manure spreading). *Nutr. Cycl. Agroecosyst.* **2001**, *60*, 103–113. [CrossRef]
52. Acharya, C.N. Studies on the anaerobic decomposition of plant materials. *Biochem. J.* **1935**, *29*, 1116–1120. [CrossRef]
53. Sommer, S.G.; Husted, S. The chemical buffer system in raw and digested animal slurry. *J. Agric. Sci.* **1995**, *124*, 45–53. [CrossRef]
54. Christensen, M.L.; Sommer, S.G. Manure characterization and inorganic chemistry. In *Animal Manure Recycling: Treatment and Management*, 1st ed.; Sommer, S.G., Christensen, M.L., Schmidt, T., Jensen, L., Eds.; John Wiley & Sons, Ltd.: Hoboken, NJ, USA, 2013; pp. 41–65.
55. Sistani, K.R.; Brink, G.E.; McGowen, S.L.; Rowe, D.E.; Oldham, J.L. Characterization of broiler cake and broiler litter, the by-products of two management practices. *Bioresour. Technol.* **2003**, *90*, 27–32. [CrossRef]
56. Chadwick, D.R. Emissions of ammonia, nitrous oxide and methane from cattle manure heaps: Effect of compaction and covering. *Atmos. Environ.* **2005**, *39*, 787–799. [CrossRef]
57. Parkinson, R. Effect of turning regime and seasonal weather conditions on nitrogen and phosphorus losses during aerobic composting of cattle manure. *Bioresour. Technol.* **2004**, *91*, 171–178. [CrossRef]
58. Siegford, J.M.; Powers, W.; Grimes-Casey, H.G. Environmental aspects of ethical animal production. *Poult. Sci.* **2008**, *87*, 380–386. [CrossRef]
59. Leinonen, I.; Williams, A.G.; Wiseman, J.; Guy, J.; Kyriazakis, I. Predicting the environmental impacts of chicken systems in the United Kingdom through a life cycle assessment: Broiler production systems. *Poult. Sci.* **2012**, *91*, 8–25. [CrossRef] [PubMed]
60. Schwean-Lardner, K.; Herwig, E. Poultry Welfare: Future Directions and Challenges? *Meat Muscle Biol.* **2020**, *4*, 1–5. [CrossRef]
61. Nicks, B. Ammonia reduction in pigs. In *Livestock Production and Society*; Geers, R., Madec, F., Eds.; Wageningen Academic Publishers: Wageningen, The Netherlands, 2006; pp. 179–190. ISBN 9789076998893.
62. Lagadec, S.; Landrain, B.; Landrain, P.; Robin, P.; Hassouna, M. Ammonia and greenhouse gas emissions in pig fattening on slatted floor with excrementdischarge by flat scraping. In *Emissions of Gas and Dust from Livestock*; IFIP—Institut du Porc: Paris, France, 2013.
63. Nicholson, F.A.; Chambers, B.J.; Walker, A.W. Ammonia emissions from broiler litter and laying hen manure management systems. *Biosyst. Eng.* **2004**, *89*, 175–185. [CrossRef]
64. Fournel, S.; Pelletier, F.; Godbout, S.; Lagacé, R.; Feddes, J. Odour emissions, hedonic tones and ammonia emissions from three cage layer housing systems. *Biosyst. Eng.* **2012**, *112*, 181–191. [CrossRef]
65. Fangueiro, D.; Hjorth, M.; Gioelli, F. Acidification of animal slurry—A review. *J. Environ. Manag.* **2015**, *149*, 46–56. [CrossRef] [PubMed]
66. Misselbrook, T.; Hunt, J.; Perazzolo, F.; Provolo, G. Greenhouse gas and ammonia emissions from slurry storage: Impacts of temperature and potential mitigation through covering (pig slurry) or acidification (cattle slurry). *J. Environ. Qual.* **2016**, *45*, 1520–1530. [CrossRef] [PubMed]
67. Nahm, K.H. Environmental effects of chemical additives used in poultry litter and swine manure. *Crit. Rev. Environ. Sci. Technol.* **2005**, *35*, 487–513. [CrossRef]
68. Cockerill, S.A.; Gerber, P.F.; Walkden-Brown, S.W.; Dunlop, M.W. Suitability of litter amendments for the Australian chicken meat industry. *Anim. Prod. Sci.* **2020**, *60*, 1469. [CrossRef]

MDPI
St. Alban-Anlage 66
4052 Basel
Switzerland
Tel. +41 61 683 77 34
Fax +41 61 302 89 18
www.mdpi.com

Animals Editorial Office
E-mail: animals@mdpi.com
www.mdpi.com/journal/animals

www.ingramcontent.com/pod-product-compliance
Lightning Source LLC
LaVergne TN
LVHW071023170726
843515LV00006B/1189
* 9 7 8 3 0 3 6 5 4 5 8 6 8 *